Torsten Fließbach

Statistical Physics

Publishing: BoD · Books on Demand GmbH, Überseering 33, 22297 Hamburg,
bod@bod.de
Printing: Libri Plureos GmbH, Friedensallee 273, 22763 Hamburg

Originally published in the German language:
Statistische Physik by Torsten Fließbach
Copyright © Springer-Verlag GmbH Germany,
a part of Springer Nature, 2018. All rights reserved.

ISBN: 978-3-8192-4475-9

Preface

The present book *Statistical Physics* is the English version of the 6th edition of my German textbook *Statistische Physik* [4]. This book is the fourth one of a series of textbooks [1,2,3,4] which all developed from repeatedly held lectures. These textbooks are complemented by *Workbook for Theoretical Physics: Repetitorium and Exercises* [5] where the solutions of the numerous exercises are presented in detail.

The presentation is on an average university level in theoretical physics. The approach is rather intuitive instead of deductive. Formal derivations and proofs are carried out without emphasis on mathematical rigor. The material is divided into chapters which roughly correspond to a one to two hour lecture. Each chapter starts with a short abstract and is devoted to a specific subject. As far as possible, each chapter is kept self-contained.

Many textbooks on this subject start with thermodynamics and turn to the statistical approach only later. We start with the statistical basics, first by some mathematical probability preludes (including the central limit theorem, Part I), and then with the probability aspects of many-body systems (microstates, macrostates and statistical ensembles). Basic concepts like entropy and temperature are thereby introduced on a microscopic basis. On this way, the laws of thermodynamics are introduced (Part II). Only then the thermodynamics in a narrower sense is investigated (Part III). This procedure is comparable to that presented by Reif [6].

In Part IV we consider various statistic ensembles corresponding to different conditions (like closed system or given temperature). This is followed by numerous applications (like spin systems, ideal gas, van der Waal gas, ideal Bose and Fermi gases, Part V). Part VI presents an elementary introduction into phase transitions (van der Waals gas, liquid helium). Part VII finally considers non-equilibrium systems and uses the kinetic gas model for calculating some basic transport coefficients (friction, diffusion).

All over the world, one finds different conventions for the typesetting of formulas or mathematical expressions. This book follows the European style that numbers are written as upright (roman) letters, for example Euler's number e $= 2.71...$, the imaginary unit i $= \sqrt{-1}$ or the number $\pi = 3.14...$ In contrast to this, physical quantities (like the elementary charge e or the time t) are written in italic (slanted fonts). Physical quantities consist of a number times a (physical) unit (like $t = 3\,\mathrm{s}$, units like s for second are written in roman). If the physical quantity is a vector, then it is written in bold italic. In the US, one normally finds e $= 2.71...$ (roman) but $\pi = 3.14...$ (italic), and physical quantities are written in italic except for capital

Greek letters (roman) or vectors (bold roman). — The derivative of a function $f(t)$ may be found as df/dt or as df/dt. The d in dt is written in roman in order to point out to the mathematical limit (d stands for differential). Alternatively (from a more physical point of view), one may consider df/dt just as a fraction with a sufficiently small interval dt; then dt is a physical quantity and is written in italic. This book uses the form df/dt. — For my figures I use a Latex package named pictex. The modest graphical capacities of this package are sufficient for my purposes. Its full integration into Latex has the advantage that the symbols in the figures are identical to those in the text and in displayed equations.

I want to thank Dr. Andrea Wolf (former acquisitions editor of World Scientific) who initiated the translations of my German textbooks. Last but not least, many readers of various German editions helped improving my book.

May 2025   Torsten Fließbach

Preface of the 6th German edition

The present book is part of a lecture elaboration [1, 2, 3, 4] of the cycle Theoretical Physics I to IV. It presents the material of my lecture on Statistical Physics [4]. This lecture is often held for physics students in their 6th semester.

The presentation is at the average level of a course lecture in theoretical physics. The approach is rather intuitive instead of deductive; formal derivations and proofs are carried out without particular mathematical meticulousness.

Closely based on the text, but also partly for its continuation and supplementation, more than 100 exercises are presented. These exercises only fulfill their purpose if they are only if they are worked out by the student as independently as possible. This should be done before reading the sample solutions, which can be found in the *Workbook for Theoretical Physics* [5]. In addition to the solutions the workbook contains a compact repetitorium of the material covered in the text books [1, 2, 3, 4].

The scope of this book goes somewhat beyond the material that is usually covered during a semester of physics studies at German universities. The material is divided into chapters, which on average correspond to a one or two hour lecture. Of course, different chapters build on each other. However, we have tried to organize the individual chapters in such a way that they are as self-contained as possible. This facilitates the selection of chapters for a specific course (for example in a Bachelor's degree program) in which the material is to be more limited. On the other hand, students can more easily read the chapters that meet their specific interest.

There are many good presentations of statistical physics that are suitable for in-depth study. I will only list a few books that I have preferred to consult myself and which are occasionally cited in the text. At first, i would like to mention the *Statistical Physics and Theory of Heat* by Reif [6], which has significantly influenced my introduction into the subject. In addition, the books by Brenig [7], Becker [8] and Landau-Lifschitz [9] are particularly familiar to me.

Compared to the fifth edition of this book, numerous minor corrections have been made in this sixth edition. My thanks got to João da Providência Jr. (translator of the Portuguese edition) and to other readers of earlier editions for valuable comments. Error messages, comments and other suggestions are always welcome, for example via the contact link on my homepage `www2.uni-siegen.de/~flieba/`. On this homepage you will also find correction lists.

August 2018 Torsten Fließbach

Literature references

[1] T. Fliessbach, *Mechanics for Physicists*, subtitle: An Introduction, including Special Relativity, 1st edition, World Scientific Publishing Company, Singapore 2024

T. Fließbach, *Mechanik*, 8th edition, Springer Spektrum, Heidelberg 2020 (German original)

[2] T. Fliessbach, *Electrodynamics for Physicists*, subtitle: An Introduction, emphasizing Special Relativity, 1st edition, World Scientific Publishing Company, Singapore 2025

T. Fließbach, *Elektrodynamik*, 7th edition, Springer Spektrum, Heidelberg 2022 (German original)

[3] T. Fliessbach, *Quantum Mechanics*, subtitle: An Introduction for Physicists, 1st edition, World Scientific Publishing Company, Singapore 2025

T. Fließbach, *Quantenmechanik*, 6th edition, Springer Spektrum, Heidelberg 2018 (German original)

[4] T. Fließbach, *Statistical Physics*, 1st edition, BoD – Books on Demand, Hamburg 2025 (this book)

T. Fließbach, *Statistische Physik*, 6th edition, Springer Spektrum, Heidelberg 2018 (German original)

[5] T. Fließbach and H. Walliser, *Workbook for Theoretical Physics*, subtitle: Repetitorium and Exercises, 1st edition, BoD – Books on Demand, Hamburg 2025

T. Fließbach and H. Walliser, *Arbeitsbuch zur Theoretischen Physik – Repetitorium und Übungsbuch*, 4th edition, Springer Spektrum, Heidelberg 2020 (German original)

[6] F. Reif, *Fundamentals of Statistical and Thermal Physics*, various editions, e.g. Waveland 2009

[7] W. Brenig, *Statistische Theorie der Wärme*, 4th edition, Springer Verlag, Berlin 1996

[8] R. Becker, *Theorie der Wärme*, Springer Verlag, Berlin 1966

[9] L. D. Landau, E. M. Lifschitz, *Statistical Physics*, (vol. 5 of *Course of theoretical physics*, part I), Pergamon 1980

Table of contents

Introduction

In *statistical physics* we deal with systems consisting of very many particles. Examples are the atoms of a gas or a liquid, the phonons of a solid or the photons in a plasma. The laws governing the motion of individual particles are given by mechanics or quantum mechanics. Due to the large number of particles (for example $N = 6 \cdot 10^{23}$ for one mole of a gas), the equations of motion, however, cannot be evaluated. The result of such an evaluation, i.e. for example the trajectories of $6 \cdot 10^{23}$ particles, would also be of no interest. The treatment of these systems is therefore *statistical*, i.e. on the basis of assumptions about the probability of different trajectories or states.

In contrast to the dynamics of individual particles, the behavior of *macroscopic* quantities, such as the energy, pressure or temperature of the considered system is of interest. With the help of a few assumptions, in particular the three laws of thermodynamics, the relations between these quantities are investigated. This discipline is called thermodynamics (TD), or also macroscopic or phenomenological TD. The thermodynamics was developed in the 19th century without reference to the microscopic basis.

Since the beginning of the 20th century, the statement "heat is the disordered motion of atoms" has been an established fact. Important contributions to this were Boltzmann's gas theory (1898) and the treatment of Brownian motion by Einstein (1905). As a hypothesis, this statement is much older (Bernoulli, 1738). Based on this insight, the *Statistical physics* derives the macroscopic quantities and their relationships on the basis of the *microscopic* structure. This field is also referred to as *statistical mechanics*, whereby the term mechanics includes quantum mechanics.

Statistical physics is based on microscopic quantities and the basic laws of mechanics and quantum mechanics. It also uses plausible, but unproven hypotheses. These include in particular the assumption that "all accessible states are equally probable". Using statistical methods, macroscopic variables are then defined and relations between them are derived.

Macroscopic quantities that are defined microscopically in this way include the pressure and the temperature. One can establish general relations for these macroscopic quantities that apply to a large class of systems and processes, in particular the laws of thermodynamics. On the other hand, relations may be derived for specific systems, such as the equation of state (like $PV = Nk_{\mathrm{B}}T$) and the heat capacity (like $C_V = 3Nk_{\mathrm{B}}/2$) for an ideal gas.

The emphasis of this book lies on the one hand on the microscopic structure of statistical physics, and on the other hand on its application to special systems.

These systems are usually model systems that can be solved exactly or approximately. With the help of such models, many properties of real systems can be understood and treated quantitatively. Examples of such properties are the specific heat of solids, the radiation spectrum of hot bodies (such as the Sun) and the behavior during phase transitions (such as the boiling of a liquid).

Most of the results of statistical physics that we will discuss are related to equilibrium states. For the treatment of non-equilibrium states, we introduce the master equation and the Boltzmann equation and give an elementary derivation of the most familiar transport equations (e.g. for heat conduction or diffusion).

The necessary elementary knowledge of mathematical statistics are compiled in Part I. This is followed by the fundamentals of statistical physics (Part II). Part III presents phenomenological thermodynamics in the narrower sense. Part IV completes the tools of statistical physics, which are then used in Part V to treat numerous specific systems statistically. Parts VI and VII provide an introduction to phase transitions and to non-equilibrium processes.

I Mathematical statistics

1 Probability

The concepts of probability, mean value and mean deviation are introduced and discussed. These concepts are illustrated by the standard example of "dice rolling".

We first specify the concept of *probability*. To do this, we consider the statement: "The probability p of rolling a 1 with a dice is $p = 1/6$". This means that on average $1/6$ of the throws leads to a 1. In a real experiment, with N attempts (throws), N_i times the event i (i.e. the result 1, 2, 3, 4, 5 or 6) will occur. Then N_i/N is the relative *frequency* of the event i and

$$\boxed{p_i = \lim_{N \to \infty} \frac{N_i}{N} \qquad \text{probability}} \tag{1.1}$$

its probability. Here, $N \to \infty$ means "sufficiently often"; this condition will be quantified later. Since one of the events i must occur (this is how dice rolling is defined), the following applies

$$\sum_{i=1}^{6} N_i = N \tag{1.2}$$

From this follows

$$\sum_{i=1}^{6} p_i = 1 \tag{1.3}$$

The determination of the probabilities p_i by N throws can be realized in two ways, which we call *time average* and *ensemble average*. Here, time average means that the same dice is thrown N times under the same conditions. For the ensemble average, we take N dice of the same kind and carry out one roll for each.

The statement $p_i = 1/6$ for the dice is used in two different meanings. It means either a *physical* or a *hypothetical* property of the system. We discuss these two possibilities.

Regarding the first meaning: For a real cube or dice, (1.1) will in generally lead to $p_i \neq 1/6$. In a trick dice the center of mass might be shifted in the direction of 1; this would lead to $p_6 > 1/6$. Every real dice will have some deviations from the symmetry of an ideal cube (slightly shifted center of mass, differently rounded edges), which lead to small deviations from $p_i = 1/6$. The experimental determination of the probabilities p_i according to (1.1) is therefore the *measurement of a physical property* of the system. Since N is practically finite, this measurement (like any other) is subject to some inaccuracy. This inaccuracy will be quantified in the next chapter.

The second meaning: For a symmetrically constructed cubic dice we can make the *assumption* that all events i are equally probable, i.e. $p_i = 1/6$. This assumption is reasonable and plausible, at least before (a priori) measurements are available. Statistical physics is based on just such an *a priori hypothesis*. This hypothesis states that all accessible states of a closed system are equally probable. In contrast to the dice, however, the basic assumption of statistical physics cannot be directly tested experimentally. Only conclusions from this assumption can be verified or falsified.

Instead of the dice, let us consider a physical a gas of atoms. The kinetic energy ε of an individual atom is measured with the accuracy $\Delta\varepsilon$. We can then assign each atom a discrete value $\varepsilon_i = i \cdot \Delta\varepsilon$ where i denotes the possible values $0, 1, 2, \ldots$. Through collisions with other atoms often change their energy; in air, the average collision time of the atoms is $\tau \approx 2 \cdot 10^{-10}\,\mathrm{s}$ (Chapter 43). We now ask with what probabilities p_i the energies ε_i occur in the gas.

We can determine the ensemble average by measuring the energies of N atoms (at a certain point in time); in doing so we will find N_i times the energy ε_i. A time average obtained if we measure the energy of a specific atom at N different times; the number of measurements that yield ε_i is again denoted by N_i. In both cases, the measurement determines the probabilities

$$p_i = p(\varepsilon_i) = \frac{N_i}{N} \tag{1.4}$$

Let the number N be so large (about $N = 10^{24}$) that we can omit to the limit $N \to \infty$. The intervals $\Delta\varepsilon$ must be chosen so that also $N_i \gg 1$.

In general, the probabilities of the ensemble and time averages are the same. For the time average we require that the time intervals between measurements are large compared to the collision time τ. Only then the energies of the one selected atom are representative of all the atoms. Later we will find that the probabilities (1.4) are given by $p_i = \exp(-\beta\varepsilon_i)/\sum_j \exp(-\beta\varepsilon_j)$, where β is a constant.

Addition and multiplication of probabilities

In the following, we consider N systems of the same kind, they shall be in various discrete *state* i. The probability of finding a particular system in state i, is defined as in (1.1); here we replace $N \to \infty$ by the condition "N sufficiently large". In Equations (1.2) and (1.3), the sum runs over all possible system states.

In the above examples, the system consisted of a dice or a gas atom. In the case of the dice, the state is a certain dice eye, for the gas atom a certain energy interval. Instead of a state, we also speak of an *event i* that results from rolling the dice or from the energy measurement.

We examine the meaning of addition and multiplication for probabilities. From the definition of p_i in (1.1) follows immediately

$$p_i + p_j = p_{(i \text{ or } j)} \qquad \begin{array}{l}(\text{events } i \text{ and } j \\ \quad \text{exclude each other})\end{array} \qquad (1.5)$$

This holds for two discrete, mutually exclusive states or events i and j. For this, we count $N_{(i \text{ or } j)} = N_i + N_j$ occurrences. For an ideal dice, for example, $p_{(1 \text{ or } 2)} = p_1 + p_2 = 1/6 + 1/6 = 1/3$.

So far we have considered elementary states such as the number of dice eyes. For this, the specification "exclude each other" for $i \neq j$ is fulfilled from the outset. Examples of more general states a and b are: a = "even number of points" and b = "number of points less than 4". These states a and b are not mutually exclusive. Therefore, $p_{(a \text{ or } b)}$ is not equal to $p_a + p_b$.

The probability of finding a first system in state i and a second one in state j is

$$p_{ij} = p_i\, p_j \qquad \begin{array}{l}(\text{events } i \text{ and } j \text{ are} \\ \quad \text{independent of each other})\end{array} \qquad (1.6)$$

For the dice, this is the probability to roll first the number i and afterwards (second rolling) j, or the probability that of two distinguishable dice the first shows i and the second j. This assumes that the first dice result has no influence on the second, or that the two dice do not influence each other.

To justify (1.6), we assume that there are N first and M second systems, of which N_i are in state i and M_j in j. We assume the same probabilities, i.e. $p_i = \lim (N_i/N) = \lim (M_i/M)$. There are NM system pairs, of which $N_i\, M_j$ are in the state ij. From this follows the probability of finding the combined state ij:

$$p_{ij} = \lim_{M,N \to \infty} \frac{N_i\, M_j}{NM} = \lim_{N \to \infty} \frac{N_i}{N} \lim_{M \to \infty} \frac{M_j}{M} = p_i\, p_j \qquad (1.7)$$

Mean value and mean deviation

For N systems with the discrete states i, the sum runs

$$\sum_i p_i = 1 \qquad (1.8)$$

runs over all possible states. We now consider an arbitrary system variable x, which has the value x_i in state i. Examples are the number of eyes of a dice ($x_i = i$) or the energy of an atom ($x_i = \varepsilon_i$). For this we define the *mean value* (or average value) by

$$\boxed{\,\overline{x} = \sum_i p_i\, x_i \qquad \text{mean value}\,} \qquad (1.9)$$

Obviously, $\bar{x}$ gives the average value of x for all N systems (in the limit $N \to \infty$). The definition (1.9) can be transferred to a function $f(x)$:

$$\overline{f(x)} = \sum_i p_i f_i = \sum_i p_i f(x_i) \tag{1.10}$$

The actual values x_i deviate from the mean value $\bar{x}$. However, the mean (or average) value of this deviation disappears:

$$\overline{x - \bar{x}} = \sum_i p_i (x_i - \bar{x}) = \sum_i p_i x_i - \bar{x} \sum_i p_i = \bar{x} - \bar{x} = 0 \tag{1.11}$$

This is because positive and negative deviations cancel each other out. A sensibel measure of the deviation from the mean, is the *mean square deviation* Δx, which is defined by

$$\boxed{\Delta x = \sqrt{\overline{(x - \bar{x})^2}} \qquad \text{mean square deviation}} \tag{1.12}$$

Other designations for this are mean (or average) squared displacement or mean square fluctuation. We can express Δx by $\overline{x^2}$ and $\bar{x}^2$:

$$(\Delta x)^2 = \sum_i p_i (x_i - \bar{x})^2 = \sum_i p_i x_i^2 - 2\bar{x} \sum_i p_i x_i + \bar{x}^2 \sum_i p_i = \overline{x^2} - \bar{x}^2 \tag{1.13}$$

From $(\Delta x)^2 \geq 0$ follows

$$\overline{x^2} \geq \bar{x}^2 \tag{1.14}$$

As a simple example, let us consider a fair dice where $x = $ number of eyes, i.e. $x_i = i$. The mean value is

$$\bar{x} = \sum_{i=1}^{6} p_i \, i = \frac{1}{6} \left(1 + 2 + 3 + 4 + 5 + 6\right) = \frac{7}{2} \tag{1.15}$$

Off

$$\overline{x^2} = \sum_{i=1}^{6} p_i \, i^2 = \frac{1}{6} \left(1 + 4 + 9 + 16 + 25 + 36\right) = \frac{91}{6} \tag{1.16}$$

follows

$$(\Delta x)^2 = \overline{x^2} - \bar{x}^2 = \frac{35}{12} \tag{1.17}$$

i.e. the mean fluctuation $\Delta x \approx 1.7$.

The term *mean value* is often used for the mean value of a measurement. For N events (such as N times throwing the dice), this mean value is $\sum_i n_i x_i / N$. From this, the quantity (1.9) is obtained for $N \to \infty$, which is also called *expectation value*.

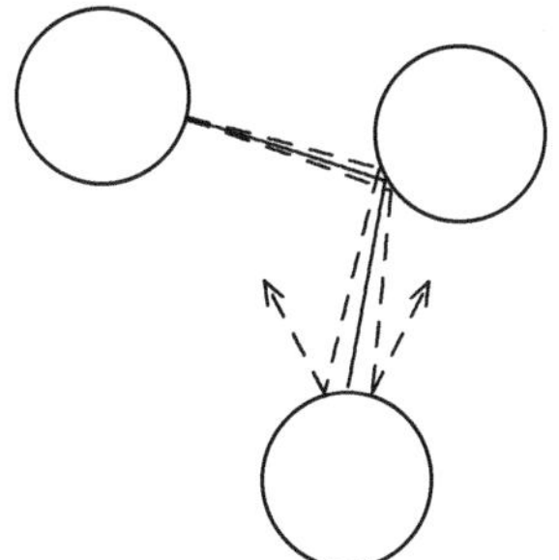

Figure 1.1 The circles stand for billiard balls, lottery balls or the atoms of a gas. The next two collisions of the upper left sphere are sketched schematically. This illustrates that small differences in the initial conditions can very quickly become large and then lead to completely different motion sequences. For a few collisions, the course is still predictable (billiards), but not for many collisions (lottery, gas).

In this book, we are dealing with very large N values (like $N = 10^{24}$). Then this distinction between mean and expectation value is then neither essential and nor common. In quantum mechanics (Chapter 5 in [3]), on the other hand, both terms are introduced separately in order to then establish their equality for $N \to \infty$. Also in general statistical, in particular also in non-physical investigations, a distinction between these two terms is made.

On the physics of dice rolling

We have implicitly assumed that the result of a dice rolling is random, i.e. unpredictable. We discuss this point with regard to the deterministic equations that govern the motion of the dice.

We imagine a dice machine in which several dice initially rest at certain positions, and in which predetermined time-dependent forces act on the dice. After some time, the forces cease and the dice come to rest. This description applies in particular to a common lottery machine. The motion of the lottery balls (or dice) is determined by equations of motion of classical mechanics, which include the given time-dependent forces (for this purpose we have eliminated the person rolling the dice). For given initial conditions, these equations uniquely determine the state of the system at any later point in time. Why then is the lottery result unpredictable?

A characteristic feature of such systems is that very small changes in an initial state can quickly become large. For example, if the impact parameter changes slightly when two spheres collide changes, the scattering angle also changes slightly. On the distance to the next collision, this small change can already result in a significantly larger change in the impact parameter for the next collision (Figure 1.1). In short, after a few collisions, two neighboring states may evolve into completely different states. Now the initial conditions for the lottery apparatus cannot be set arbitrarily precisely. Indeed, there are many initial conditions which lie very close together. Then the equations of motion may quickly lead to quite different trajectories.

Due to the chaotic behavior (neighboring states evolve quickly to quite different states), other small disturbances or uncertainties (apart from those in the initial conditions) are sufficient to make the result unpredictable. The disturbances can be uncertainties in the external forces (e.g. due to voltage fluctuations in the power grid), coughing of a person nearby to the lottery machine, flapping of a butterfly wing (butterfly effect), thermal fluctuations, or quantum mechanical uncertainty.

It should be emphasized, that for the random behavior of the lottery machine, neither the finite temperature nor the quantum mechanical uncertainty is necessary. One can –in principle– throw dice also at temperature $T = 0$ or in a (fictitious) world with $\hbar = 0$.

The chaotic behavior of the lottery balls can be compared to that of atoms of a classical gas. Here too, an ordered initial state (comparable to the initial position of the lottery balls) would very quickly become chaotic. External perturbations are not necessary for a gas; the kinetic energy of the atoms takes care of this.

Exercises

1.1 Undiscovered printing errors

Two proofreaders read a book. Proofreader A finds 200 printing errors, proofreader B only 150. Only 100 of the printing errors found match. Estimate how many printing errors remained undiscovered.

1.2 Common birthday

What is the probability P_N that out of N students, at least two have birthday on the same day? What results for $N = 10$? For what minimum number $N_{\min}$ does the probability that at least two students have their birthday on the same day exceeds the value $1/2$?

2 Law of large numbers

A mechanism that leads to random distributions is the one-dimensional random walk. For this, the probability distribution and its properties are easy to derive. This leads to the law of large numbers (this chapter) and to the normal distribution (Chapters 3 and 4).

Random walk

In the one-dimensional case, a random walk consists of N steps that are executed on a straight line. We use the x-axis for this straight line. The random walk starts at $x = 0$. One step consists of a jump by $\Delta x = +1$ or -1. The jump by $+1$ takes place with the probability p, that by -1 with q, and

$$p + q = 1 \tag{2.1}$$

The individual steps are independent of each other. The number of positive jumps is n_+, the number of negative jumps is n_-. Together they result in the total number of steps:

$$n_+ + n_- = N \tag{2.2}$$

After N steps we calculate the following quantities:

1. The probability $P_N(m)$ for arriving at $x = m = n_+ - n_-$ This is equal to the probability $W_N(n)$ that $n = n_+$ positive steps have been performed:

$$P_N(m) = W_N(n) \qquad (m = n_+ - n_-, \quad n = n_+) \tag{2.3}$$

2. The mean values $\overline{m}$ and $\overline{n}$.

3. The means square deviations Δm and Δn.

4. The continuous distribution for large N (Chapter 3).

We give some application examples for the random walk:

- In a crystal, one unpaired electron sits at each lattice point. It can adjust its spin and thus its magnetic moment $\boldsymbol{\mu}$ independently of the other electrons (Figure 2.1). In an external magnetic field $\boldsymbol{B} = B\,\boldsymbol{e}_z$, the z-component of the quantum mechanical spin can assume the values $\pm\hbar/2$; this implies $\boldsymbol{\mu} \cdot \boldsymbol{e}_z \approx \mp\mu_B$.

Figure 2.1   In an external magnetic field B the magnetic moment μ_i of a particle is parallel to B with the probability p, and antiparallel with the probability $q = 1 - p$. With what probability $W_N(n)$ are then n of N spins aligned with the magnetic field?

As will be derived later, the two possible spin settings occur with the probabilities

$$p = C \, \exp(+\mu_\mathrm{B} B / k_\mathrm{B} T) \,, \qquad q = C \, \exp(-\mu_\mathrm{B} B / k_\mathrm{B} T) \tag{2.4}$$

Here T denotes the temperature, k_B the Boltzmann constant and μ_B the Bohr magneton. The constant C is defined by $p + q = 1$.

The probability that the number of parallel magnetic moments exceeds number of antiparallel ones by m is given by $P_N(m)$. The mean magnetic moment of all N electrons is equal to $\overline{m} \, \mu_\mathrm{B} = N(p - q) \mu_\mathrm{B}$. The deviations from this mean value are of the size $\Delta m \, \mu_\mathrm{B}$.

- The N gas atoms in the volume V move statistically, the all positions of a specific atom are equally likely. We imagine the volume ro be divided into two partial volumes $p V$ and $q V$ (Figure 2.2). Then $W_N(n)$ gives the probability that there are exactly n atoms in the left subvolume. On average, these are $\overline{n} = Np$ atoms.

  In the real system, gas atoms continuously change from one to the other subvolume. The actual number n of atoms in the left-hand area is therefore time-dependent and fluctuates around the mean value $\overline{n}$. The mean square fluctuation (deviation) is given by Δn.

- The one-dimensional random walk is equivalent to all problems of the type: With probability p a certain property can be found for one of N objects. Then $\overline{n}$ is the mean number of objects with this property and Δn is the corresponding mean deviation.

  A standard example is dice rolling: One may ask how often the 1 occurs in N throws with one dice (or in one throw of N dice) occurs. For an ideal (or fair) dice, $p = 1/6$. For $N = 1000$ throws, one expects $\overline{n} = Np \approx 167$ times the result 1. In an actual experiment, one will get deviating results, for example $n_1 = n = 159$ or $n_1 = 180$. The likely values for the deviation $|n_1 - \overline{n}|$ are of the size Δn.

A problem related to the random walk is the random motion of an atom or molecule in a gas. Between two collisions, the particle travels the path $\boldsymbol{\ell}$ where the directions of $\boldsymbol{\ell}$ are randomly distributed, and the lengths $\ell = |\boldsymbol{\ell}|$ have a distribution around a

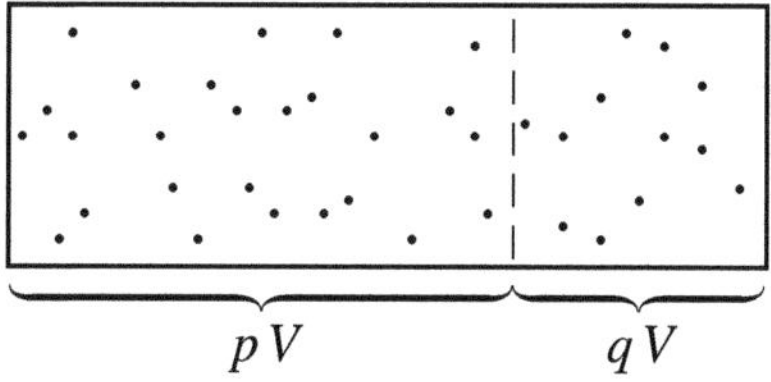

Figure 2.2   The gas-filled volume V is thought to be divided into two parts pV and qV. A selected atom of the ideal gas is located with probability p in the left-hand volume, and with probability q in the right-hand one. What is the probability $W_N(n)$ for finding exactly n atoms (out of N atoms) in the left-hand partial volume?

mean value $\overline{\ell}$. Here $\overline{\ell}$ is the mean free path length; in air under normal conditions, $\overline{\ell} \sim 10^{-7}$ m (Chapter 43). Where is the particle after N collisions? The answer is the probability distribution of a three-dimensional random walk with a variable step length.

The *Brownian motion* of small particles on the surface of a liquid is a random walk with variable step length and direction in two dimensions.

Probability distribution

Since the individual steps of a random walk are independent of each other, the corresponding probabilities must be multiplied. The probability for the path $(+1, -1, -1, +1, +1)$ is therefore

$$p\,q\,q\,p\,p = p^3 q^2$$

There are several other ways that also lead to $m = 1$ or $n = n_+ = 3$ for example $(-1, -1, +1, +1, +1)$. The various paths for $m = 1$ are mutually exclusive; therefore, the corresponding probabilities add up. There are $5! = 1 \cdot 2 \cdot 3 \cdot 4 \cdot 5$ possibilities to string 5 things together. However, if 3 of these ways are identical, then the $3!$ permutations (of these 3 things) do not lead to a different result. In our case there are therefore $5!/(3!\,2!) = 10$ different ways to $m = 1$, each of which has the probability $p^3 q^2$. The probability of getting to $m = 1$ in 5 steps on any path is therefore

$$P_5(1) = W_5(3) = \frac{5!}{3!\,2!}\, p^3 q^2 \tag{2.5}$$

This is also the probability that exactly 3 of the 5 steps go in a positive direction.

We repeat these considerations for a random walk consisting of $N = n_+ + n_-$ steps. Each particular path with n_+ positive and n_- negative steps has the probability $p^{n_+} q^{n_-}$. For this, there are

$$\frac{N!}{n_+!\,n_-!} = \frac{N!}{n!\,(N-n)!} = \binom{N}{n} \tag{2.6}$$

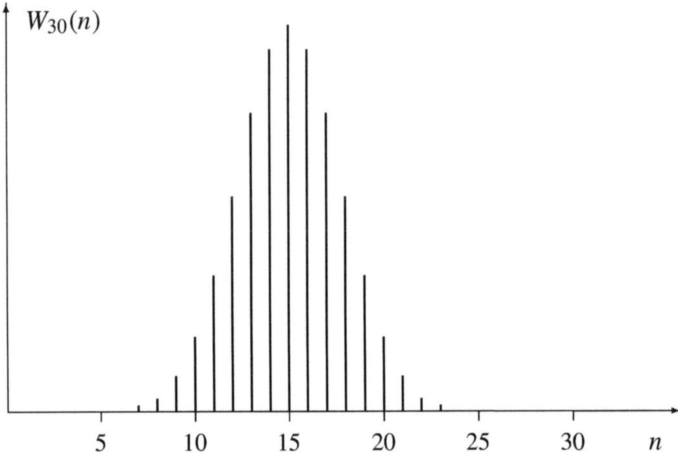

Figure 2.3 A one-dimensional random walk consisting of 30 steps contains just n positive steps with the probability $W_{30}(n)$. For $p = q = 1/2$, the distribution $W_{30}(n)$ is sketched.

different ways. This binomial coefficient "N over n" is equal to the number of ways to put $n_+ = n$ identical objects and $n_- = N - n$ identical (other) objects in a certain order. Thus we obtain

$$P_N(m) = W_N(n) = \binom{N}{n} p^n q^{N-n} \tag{2.7}$$

If we insert

$$n = \frac{N + m}{2} \tag{2.8}$$

the result can be written as a function of m. Figure 2.3 sketches the distribution $W_N(n)$ for another example.

In (2.7), n can take the values of 0, 1, 2,..., N, and m the values $-N, -N + 2,..., N$. For all other values, $W_N(n)$ and $P_N(m)$ vanish, for example $P_5(0) = 0$. The sum of all probabilities $W_N(n)$ is 1:

$$\sum_{n=0}^{N} W_N(n) = \sum_{n=0}^{N} \binom{N}{n} p^n q^{N-n} = (p + q)^N = 1 \tag{2.9}$$

With regard to this formula, we can also derive the $W_N(n)$ in a somewhat different way. In

$$1 = \underbrace{(p + q)(p + q) \cdot \ldots \cdot (p + q)}_{N \text{ factors}} \tag{2.10}$$

each factor $(p + q)$ stands for one of the N steps of the random walk, which is executed with the probability p to the right or with q to the left. The multiplication of the product (without swapping the order of the factors p and q) results in 2^N

summands, each representing a specific path. Now one may arrange the factors and combine all terms with $p^n q^{N-n}$. From the elementary algebra it is known that the term $p^n q^{N-n}$ occurs exactly $N!/(n!\,(N-n)!)$ times, i.e.

$$(p+q)^N = \sum_{n=0}^{N} \binom{N}{n} p^n q^{N-n} = \sum_{n=0}^{N} W_N(n) \tag{2.11}$$

The coefficient of $p^n q^{N-n}$ yields $W_N(n)$. Because of this relationship, the distribution $W_N(n)$ is also called *binomial distribution*.

Mean value and mean deviation

We calculate the mean value and the mean square deviation for a random walk of N steps. The probabilities p_i of Chapter 1 become $p_n = W_N(n)$ here. The possible states of the system are the positions at the end of the random walk; they are given by the values $n = 0, 1, 2, \ldots, N$. We calculate the mean value for the number $n = n_+$ of positive steps,

$$\bar{n} = \sum_{n=0}^{N} n\, W_N(n) = \sum_{n=0}^{N} n \binom{N}{n} p^n q^{N-n} \tag{2.12}$$

This can be expressed in the form

$$\bar{n} = p\, \frac{\partial}{\partial p}\, f(p,q,N) \quad \text{with} \quad f(p,q,N) = \sum_{n=0}^{N} \binom{N}{n} p^n q^{N-n} \tag{2.13}$$

With $f(p,q,N) = (p+q)^N$ we get

$$\bar{n} = p\, \frac{\partial}{\partial p}\, (p+q)^N = pN(p+q)^{N-1} = Np \tag{2.14}$$

The result is obvious. Note that $p + q = 1$ can be used in the result, but not in $f(p,q,N) = (p+q)^N$. For the function $f(p,q,N)$ the functional dependence on the variables p and q is decisive. Analogously, we obtain

$$\overline{n_-} = N - \overline{n_+} = N - \bar{n} = Nq \tag{2.15}$$

and

$$\bar{x} = \bar{m} = \overline{n_+} - \overline{n_-} = N(p-q) \tag{2.16}$$

We now calculate $\overline{n^2}$:

$$\overline{n^2} = \sum_{n=0}^{N} n^2 \binom{N}{n} p^n q^{N-n} = p\, \frac{\partial}{\partial p}\, p\, \frac{\partial}{\partial p}\, \sum_{n=0}^{N} \binom{N}{n} p^n q^{N-n}$$

$$= p\, \frac{\partial}{\partial p}\, p\, \frac{\partial}{\partial p}\, (p+q)^N = p\, \frac{\partial}{\partial p}\, Np\, (p+q)^{N-1}$$

$$= pN + p^2 N(N-1) = Npq + \bar{n}^2 \tag{2.17}$$

From this we obtain

$$\Delta n \overset{(1.13)}{=} \sqrt{\overline{n^2} - \overline{n}^2} = \sqrt{Npq} \tag{2.18}$$

Analogously,

$$\Delta m = 2\sqrt{Npq} \tag{2.19}$$

For the relative width of the distribution $W_N(n)$ follows

$$\boxed{\frac{\Delta n}{\overline{n}} = \sqrt{\frac{q}{p}}\frac{1}{\sqrt{N}} \overset{N\to\infty}{\longrightarrow} 0 \qquad \text{law of large numbers}} \tag{2.20}$$

For large N, the distribution $W_N(n)$ is sharply localized at $\overline{n}$. This result is referred to as the *law of large numbers*. Chapter 4 will show that this statement applies under rather general conditions. We illustrate the meaning of (2.20) by two examples:

- Let there be $N = 10^{24}$ atoms in the gas box of Figure 2.2; let the two sub-regions be of equal size, $p = q = 1/2$. On average, there are $\overline{n} = Np = N/2$ in the left subvolume. The actual number n of atoms in the left region deviates by values of the size $\Delta n = (Npq)^{1/2} = 0.5 \cdot 10^{12}$ from $\overline{n}$. This is an extremely small relative deviation, namely $\Delta n/\overline{n} = 10^{-12}$.

  If there were larger deviations from the mean value (such as 10%) one could build a perpetuum mobile of the second kind: At the time of such a deviation, one may insert a wall from the side (i.e. without work) between the sub-volumes. The higher particle density in one partial volume means a higher pressure. Due to this increased pressure, the gas can perform work; thereby the gas cools down. The missing heat, which is equal to the work performed, can now be absorbed from the environment. This means that by cooling the environment, work (such as electrical energy) can be generated.

  The impossibility of such a perpetuum mobile of the 2nd kind follows from the law of large numbers. The fluctuations that occur according to this law are so small that they cannot be utilized to perform work. The generalization of this statement leads to the second law of thermodynamics.

- For a dice, we use (2.20) to calculate the error of the measurement of p_1 for N throws. If we use $p = p_1$ and $q = 1 - p = p_2 + p_3 + p_4 + p_5 + p_6$, we can apply the random walk formulas directly. A concrete experiment with $N \gg 1$ throws may yield $n = n_1$ times to the result 1. The experimental value $p_{1,\exp}$ and the corresponding measurement uncertainty are

$$p_{1,\exp} = \frac{n_1 \pm \Delta n_1}{N} = \frac{n_1}{N} \pm \frac{\sqrt{Npq}}{N} \approx \frac{n_1}{N} \pm \frac{\sqrt{5}}{6}\frac{1}{\sqrt{N}} \tag{2.21}$$

A more detailed justification for the estimation of the experimental error by $\Delta n_1/\overline{n_1} = \Delta n/\overline{n}$ is given in the next chapter. In the small error term one may use the hypothetical values $p = 1/6$ and $q = 5/6$. An concrete experiment consisting of 10^3 dice rolling results in 175 times the 1, another experiment with 10^5 throws shall yield 17500 times the 1. With (2.21) we obtain

$$p_{1,\mathrm{exp}} = 0.175 \pm \begin{cases} 0.012 & (N = 10^3) \\ 0.001 & (N = 10^5) \end{cases} \tag{2.22}$$

The first experiment with 1000 throws is compatible with the assumption that the dice is "fair", i.e. a dice with $p = 1/6$. For the dice of the second experiment this is extremely unlikely. The statements "compatible" and "extremely unlikely" will be quantified in the next chapter.

Exercises

2.1 *Three correct numbers in the lottery*

Calculate the probability p_6 of getting six numbers right in the lottery (6 out of 49, $p = 6/49$ and $q = 43/49$). Calculate the probability p_3 of getting exactly three correct numbers in this lottery. Why is the probability p_3 is not equal to $W_6(3)$?

3 Normal distribution

A random walk of N steps leads to n positive steps with the probability $W_N(n)$ (previous chapter). Under certain conditions, the distribution $W_N(n)$ can be approximated by a Gaussian function. This Gaussian function is called normal distribution.

Functions of the form $f(n) = p^n$ or $f(n) = n! \approx (n/e)^n$ depend extremely sensitively on the n in the exponent. Therefore, an expansion into powers of $n - \bar{n}$ converges in a very small range only. However, the convergence range can be significantly increased if $\ln f(n)$ is expanded instead of $f(n)$; because $\ln f(n)$ varies much weaker with n. As an example consider the function $f(n) = p^n = \exp(n \ln p)$. A Taylor expansion around the value $\bar{n}$ yields:

$$f(n) = f(\bar{n}) + f'(\bar{n})(n - \bar{n}) + \ldots \approx p^{\bar{n}} + (\ln p)\, p^{\bar{n}}(n - \bar{n}) \tag{3.1}$$

As an approximation, we break of this series at the linear term. For $n = \bar{n} + 1$ we compare the approximation with the exact value:

$$f(\bar{n} + 1) = p^{\bar{n}+1} = p^{\bar{n}} \cdot \begin{cases} p & \text{(exact)} \\ 1 + \ln p & \text{(approximation (3.1)} \end{cases} \tag{3.2}$$

For $p = 1$ the approximation agrees with the exact result (because $f(n)$ does not depend on n). Otherwise (e.g. for small p), the approximation is useless. In contrast to this, the Taylor expansion of $\ln f(n) = n \ln p$ breaks off after the first term and yields the exact result:

$$\ln f(n) = n \ln p = \bar{n} \ln p + (n - \bar{n}) \ln p \tag{3.3}$$

This example shows: If a function $f(x)$ varies strongly with x (especially if it depends exponentially on x), it makes sense to expand the logarithm of the function instead of the function itself.

We want to expand the probability distribution $W_N(n)$, (2.7),

$$W_N(n) = \binom{N}{n} p^n q^{N-n} \tag{3.4}$$

into powers of $n - \bar{n}$. The function $W_N(n)$ contains terms such as p^n and $n!$, which depend strongly on n. Therefore, we consider the logarithm of this function:

$$\ln W_N(n) = \ln N! - \ln n! - \ln(N - n)! + n \ln p + (N - n) \ln q \tag{3.5}$$

For the Taylor expansion of this function, we need the derivatives at the point $\bar{n}$. The derivative of $\ln n!$ is approximated by the difference quotient with $dn = 1$:

$$\frac{d \ln n!}{dn} \approx \frac{\ln(n+1)! - \ln n!}{1} = \ln(n+1) \approx \ln n \qquad (n \gg 1) \qquad (3.6)$$

We calculate

$$\frac{d \ln W_N(n)}{dn} = -\ln n + \ln(N-n) + \ln p - \ln q \overset{n=\bar{n}}{=} 0 \qquad (3.7)$$

Here (3.6) was used for the derivatives of $\ln n!$ and $\ln (N-n)!$. Hereby, we assume $\bar{n} = Np \gg 1$ and $N - \bar{n} = Nq \gg 1$. These conditions can be summarized by

$$Npq \gg 1 \qquad \text{(prerequisite)} \qquad (3.8)$$

In the last step in (3.7), $\bar{n} = Np$ and $N - \bar{n} = Nq$ was used. The disappearance of the first derivative means that $\ln W_N(n)$ and thus also $W_N(n)$ are stationary $n = \bar{n}$. The second derivative results in

$$\frac{d^2 \ln W_N(n)}{dn^2} = -\frac{1}{n} - \frac{1}{N-n} \overset{n=\bar{n}}{=} -\frac{1}{Npq} = -\frac{1}{(\Delta n)^2} < 0 \qquad (3.9)$$

Since the second derivative is less than zero, $\ln W_N(n)$ and thus also $W_N(n)$ have a maximum at the mean value $\bar{n}$. For the later error estimation, we write on the third derivative, too:

$$\frac{d^3 \ln W_N(n)}{dn^3} = \frac{1}{n^2} - \frac{1}{(N-n)^2} \overset{n=\bar{n}}{=} \frac{q^2 - p^2}{N^2 p^2 q^2} \qquad (3.10)$$

The Taylor expansion of $\ln W_N(n)$ reads

$$\ln W_N(n) = \ln W_N(\bar{n}) + \left(\frac{d \ln W_N(n)}{dn} \right)_{\bar{n}} (n - \bar{n}) \qquad (3.11)$$

$$+ \frac{1}{2} \left(\frac{d^2 \ln W_N(n)}{dn^2} \right)_{\bar{n}} (n - \bar{n})^2 + \frac{1}{6} \left(\frac{d^3 \ln W_N(n)}{dn^3} \right)_{\bar{n}} (n - \bar{n})^3 + \dots$$

If we neglect the last term, we get from this

$$\ln W_N(n) \approx \ln W_N(\bar{n}) - \frac{1}{2} \frac{(n - \bar{n})^2}{Npq} = \ln W_N(\bar{n}) - \frac{(n - \bar{n})^2}{2 \Delta n^2} \qquad (3.12)$$

or

$$W_N(n) = W_N(\bar{n}) \, \exp\left(-\frac{(n - \bar{n})^2}{2 \Delta n^2} \right) \qquad (3.13)$$

Here and in the following, $\Delta n^2 = (\Delta n)^2$ (but not $\Delta(n^2)$).

We determine the prefactor $W_N(\overline{n})$. In the relevant range $n \sim \overline{n}$ is the relative change of $W_N(n)$ between n and $n \pm 1$ is small; this follows from the precondition (3.8). Therefore, we can replace the sum over n by an integral:

$$1 = \sum_{n=0}^{N} W_N(n) \approx \int_0^N dx\, W_N(x) \approx \int_{-\infty}^{+\infty} dx\, W_N(x)$$

$$= W_N(\overline{n}) \int_{-\infty}^{+\infty} dx\, \exp\left(-\frac{(x-\overline{n})^2}{2\,\Delta n^2}\right) = W_N(\overline{n})\,\sqrt{2\pi}\,\Delta n \tag{3.14}$$

At the integration limits 0 and N, the integrand is of the size $\exp(-\sqrt{N}\,)$; therefore the limits can be shifted to $\pm\infty$. With (3.14), (3.13) becomes

$$W_N(n) = \frac{1}{\sqrt{2\pi}\,\Delta n}\, \exp\left(-\frac{(n-\overline{n})^2}{2\,\Delta n^2}\right) \tag{3.15}$$

We now replace the step length 1 of the random walk on the x-axis by an arbitrary length ℓ. Then the following applies

$$x = n\ell, \qquad \overline{x} = \overline{n}\,\ell, \qquad \sigma \equiv \Delta x = \Delta n\,\ell \tag{3.16}$$

We define the *probability density* $P(x)$ on the the x-axis by

$$P(x)\,dx = \left\{ \begin{array}{l} \text{probability of finding the random} \\ \text{value between } x \text{ and } x + dx \end{array} \right. \tag{3.17}$$

For an interval dx that contains $dx/\ell \gg 1$ discrete n-values, the following applies

$$P(x)\,dx = \sum_{n' \text{ in } dx} W_N(n') \approx \frac{dx}{\ell}\, W_N(n) \tag{3.18}$$

Then (3.15) becomes the *normal distribution*

$$\boxed{\; P(x) = \frac{1}{\sqrt{2\pi}\,\sigma}\, \exp\left(-\frac{(x-\overline{x})^2}{2\sigma^2}\right) \qquad \text{normal distribution} \;} \tag{3.19}$$

The normal distribution is sketched in Figure 3.1. It can also be obtained under rather general conditions (Chapter 4). The quantity $\sigma = \Delta x$ is called *standard deviation*; it characterizes the width of the distribution. The step length ℓ introduced in (3.16) may have a physical dimension. Then x, $\overline{x}$ and σ have the same dimension.

In the random walk, x is restricted to the discrete values $x = n\ell$. We now disregard this restriction and consider $P(x)$ as a continuous function of the variable x. The summation over discrete values then becomes an integral. The normal distribution (3.19) is normalized:

$$\int_{-\infty}^{+\infty} dx\, P(x) = 1 \tag{3.20}$$

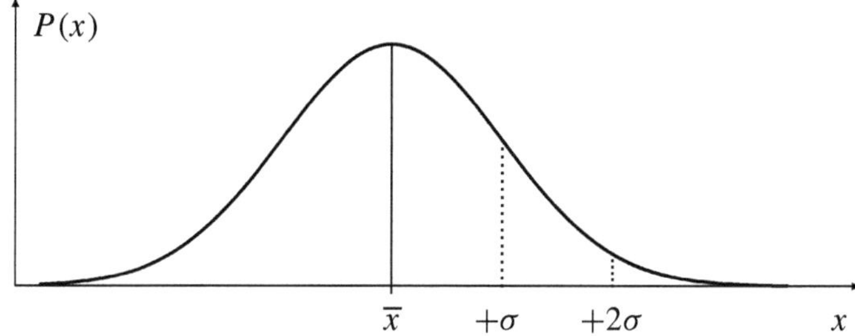

Figure 3.1   The normal distribution (3.19) is a Gaussian function. The drop from the central maximum at $\overline{x}$ may be characterized by $P(\overline{x} \pm \sigma) \approx 0.6\, P(\overline{x})$.

The probability to find an x-value somewhere is 1. We calculate the probabilities of finding an x-value in a range of one, two or three standard deviations around $\overline{x}$:

$$\int_{\overline{x}-\nu\sigma}^{\overline{x}+\nu\sigma} dx\, P(x) = \begin{cases} 0.683 & \nu = 1 \\ 0.954 & \nu = 2 \\ 0.997 & \nu = 3 \end{cases} \tag{3.21}$$

An event outside three standard deviations occurs with a probability of 0.3% and is therefore rather unlikely. Let us again look at the dice experiment from the previous chapter with the results (2.22): In (2.22), a standard deviation $\sigma = \Delta n$ was assumed as the experimental error. In the first experiment with $N = 10^3$ throws, the result was less than one standard deviation off the value $p_1 = 1/6$ for a fair dice. According to (3.21), deviations of one standard deviation or more occur with about 30% probability; the experimental result is therefore compatible with the assumption of an ideal dice. In the second experiment with $N = 10^5$ throws, the experimental result was almost 10 standard deviations off from the expected $p_1 = 1/6$. One can therefore be (practically) certain that the dice has been manipulated; it is not a fair dice.

In this chapter, we started from the probability distribution $W_N(n)$ for $n = n_+$ positive events. Therefore, $P(x)$ is the probability for positive values $x = n\ell$. If we admit the negative steps, then the random walk ends at $x = (n_+ - n_-)\ell$. In both cases, the result is given by (3.19) but the parameters must be chosen accordingly:

$$\overline{x} = Np\ell, \qquad \sigma = \sqrt{Npq}\,\ell \quad \text{for } x = n\ell = n_+\ell \tag{3.22}$$

$$\overline{x} = N(p-q)\ell, \qquad \sigma = 2\sqrt{Npq}\,\ell \quad \text{for } x = (n_+ - n_-)\ell \tag{3.23}$$

These mean values follow from (2.14) and (2.16), the deviations from (2.18) and (2.19).

Validity of the normal distribution

In the derivation, the higher terms of the Taylor expansion (3.12) were neglected. We require that the neglected 3rd order term is small compared to the 2nd order term:

$$\left| \frac{(\ln W)'''\,(n-\overline{n})^3}{(\ln W)''\,(n-\overline{n})^2} \right| \lesssim \frac{|n-\overline{n}|}{Npq} \ll 1 \tag{3.24}$$

The same condition is also obtained for the ratio of the $(i+1)$-th order term compared to the i-th order one. Provided that (3.24) holds, the termination of the Taylor expansion is justified, and the normal distribution is a valid approximation. Of interest are especially those n-values, which differ from the mean $\overline{n}$ by at least a few standard deviations $\sigma = \Delta n$,

$$|n-\overline{n}| = \nu\,\Delta n = \nu\,\sqrt{Npq}\,, \qquad \nu = \mathcal{O}(1) \tag{3.25}$$

For this, (3.24) is fulfilled if $Npq \gg 1$ applies. The normal distribution is therefore valid under the precondition (3.8). With $Npq \to \infty$ the errors tend towards zero.

The derivation is not valid for very large distances from the mean value, i.e. in particular for $n \approx 0$ or $n \approx N$. In this case, the relative error of the distribution (3.15) can become large. In these ranges, however, both the approximate value and the exact value are very small in absolute terms.

It should be emphasized that the condition $N \gg 1$ is not sufficient for the validity of the normal distribution; because for very small p the mean value $\overline{n} = Np$ can be of the size 1 or smaller. In this case, the result is a *Poisson distribution* instead of the normal distribution (Exercise 3.3).

Exercises

3.1　Approximate expression for factorials

The general definition of the *gamma function* reads

$$\Gamma(n+1) = n! = \int_0^\infty dx\; x^n\,\exp(-x) \tag{3.26}$$

where n is a real number. Convince yourself that $\Gamma(n+1)$ for positive integer n is equal to the factorial $n! = 1 \cdot 2 \cdot 3 \cdot \ldots \cdot n$.

The integrand is the function $f(x) = x^n\,\exp(-x)$. Expand $\ln f(x)$ into a Taylor series around the maximum of the function $f(x)$ (for $n \gg 1$). Insert this expansion into the integral and derive the *Stirling formula*:

$$n! \approx \sqrt{2\pi n}\; n^n\,\exp(-n) \qquad (n \gg 1) \tag{3.27}$$

Use this to determine the value of the random walk distribution $W_N(n) = \binom{N}{n} p^n q^{N-n}$ at $\overline{n} = Np$. Hereby $p + q = 1$ and $Npq \gg 1$.

3.2   *Estimation of a correlation*

In a group of N children whose parents are heavy smokers, 20% suffer from myopia. In contrast there are only 15% short-sighted children in the total population. Assuming there is no correlation between myopia and smoking: Then what is the probability w that 20% or more of N children are short-sighted?

The quantity $p_{\text{corr}} = 1 - w$ is a measure of the probability that a positive correlation exists. How can the data be assessed for $N = 100$? How large must the number N_0 of children be that there is a $p_{\text{corr}} = 99\%$ probability for a correlation between the myopia of the children and their parents' smoking?

3.3   *Poisson distribution*

The prerequisite for deriving the Gaussian distribution from (3.4) was the condition $Npq \gg 1$. For very small p (for example for $Np \sim 1$) this condition is violated. Show that the distribution (3.4) for $p \ll 1$ and $N \gg n$ can be approximated by the *Poisson distribution*

$$P(\lambda, n) = \frac{\lambda^n}{n!}\, \exp(-\lambda), \qquad (\lambda = Np) \tag{3.28}$$

Show that this distribution is normalized and determine $\bar{n}$ and Δn. Hint: With the help of $\ln(1 - p) \approx -p$ and $n \sim Np$ one gets $(1 - p)^{N-n} \approx \exp(-\lambda)$.

A book with 500 pages contains 500 randomly distributed printing errors. What are the probabilities that a randomly opened page does contain no error, or that it contains at least four errors?

3.4   *Random walk and diffusion equation*

After each time interval Δt, a step $s_i = \pm \ell$ of a random walk takes place with the probabilities $p = q = 1/2$. This results in a time-dependent probability distribution $P(x, t)$, where $x = \sum_i s_i$ is the distance from the starting point, and $N = t/\Delta t \gg 1$ is the number of steps. Show, that $P(x, t)$ fulfills the diffusion equation

$$\frac{\partial P(x, t)}{\partial t} = D\, \frac{\partial^2 P(x, t)}{\partial x^2} \tag{3.29}$$

and determine the diffusion constant D. Express the square of the mean distance from the starting point as a function of D and t. Show that the quantity

$$\Theta = \int_{-\infty}^{\infty} dx \left(\frac{\partial P}{\partial x}\right)^2 \geq 0$$

decreases monotonically with time, so that $d\Theta/dt \leq 0$.

4 Central limit theorem

If the variable $x = \sum s_i$ is a sum of many random variables s_i, then the probability distribution for x is a normal distribution. This is the statement of the central limit theorem, which will be derived in this chapter.

We consider a more general random walk, which consists of a sequence of N steps of length s_i. On the x-axis this random walk leads to the x-value

$$x = s_1 + s_2 + \ldots + s_N = \sum_{i=1}^{N} s_i \tag{4.1}$$

In the more specific random walk considered so far, we have only considered the two possible step

$$s_i = +1 \quad \text{or} \quad s_i = -1 \tag{4.2}$$

that take place with the probabilities p and q. As a generalization, we admit variable step lengths s_i occurring with the probabilities

$$w_i(s_i)\, ds_i = \begin{cases} \text{probability that the } i\text{th step has} \\ \text{a length between } s_i \text{ and } s_i + ds_i \end{cases} \tag{4.3}$$

The designation step length for the *random variable s_i* refers to an interpretation as a random walk. However, the s_i can be any variable, such as the kinetic energies of the atoms in a gas.

We assume that the probability densities $w_i(s)$ are normalized,

$$\int_{-\infty}^{\infty} ds\, w_i(s) = 1 \tag{4.4}$$

They shall have a finite mean value

$$\overline{s_i} = \int_{-\infty}^{\infty} ds\, s\, w_i(s) \tag{4.5}$$

and finite mean square deviations:

$$(\Delta s_i)^2 = \int_{-\infty}^{\infty} ds\, (s - \overline{s})^2\, w_i(s) = \overline{s_i^2} - \overline{s_i}^{\,2} \tag{4.6}$$

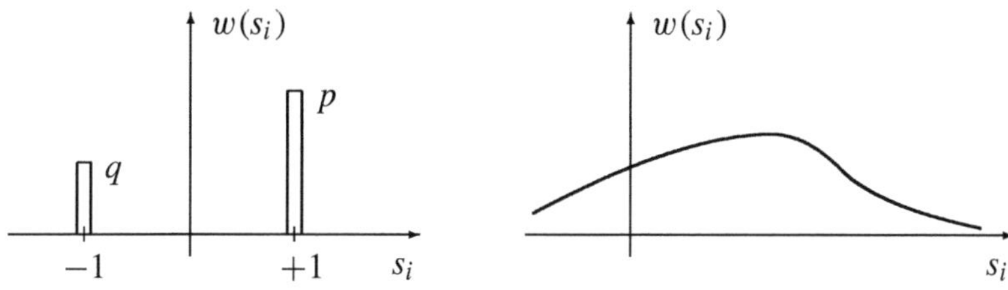

Figure 4.1   In the random walk, the ith step occurs with probability p to the right and with $q = 1 - p$ to the left. This corresponds to the probability distribution $w(s_i)$ sketched on the left. In this chapter we allow more general distributions $w(s_i)$ for the individual steps, such as the one shown on the right.

The last point excludes, for example, the distribution $w(s) \propto (a^2 + s^2)^{-1}$. A permissible distribution, on the other hand, is $w_i(s) \propto s^n \exp(-\beta_i s^2)$ with $n \geq 0$. The previously considered random walk (4.2) corresponds to the probability density

$$w_i(s) = w(s) = p\,\delta(s - 1) + q\,\delta(s + 1) \tag{4.7}$$

This special case and the generalization considered here are illustrated in Figure 4.1.

It is assumed that the random variables s_i are independent of each other. Therefore, the probabilities are to be multiplied:

$$w_1(s_1)\,ds_1 \cdot \ldots \cdot w_N(s_N)\,ds_N = \left\{ \begin{array}{l} \text{probability that the random variables} \\ \text{are between } s_i \text{ and } s_i + ds_i \end{array} \right. \tag{4.8}$$

This allows us to calculate the mean value of $x = \sum s_i$,

$$\bar{x} = \int_{-\infty}^{\infty} ds_1\, w_1(s_1) \ldots \int_{-\infty}^{\infty} ds_N\, w_N(s_N)\,(s_1 + s_2 \ldots + s_N) \tag{4.9}$$

$$= \int_{-\infty}^{\infty} ds_2\, w_2(s_2) \ldots \int_{-\infty}^{\infty} ds_N\, w_N(s_N)\,(\bar{s_1} + s_2 + \ldots + s_N) = \sum_{i=1}^{N} \bar{s_i}$$

The mean square deviation $(\Delta x)^2 = \overline{(x - \bar{x})^2}$ follows from

$$(\Delta x)^2 = \int_{-\infty}^{\infty} ds_1\, w_1(s_1) \ldots \int_{-\infty}^{\infty} ds_N\, w_N(s_N) \left(\sum_{i=1}^{N} (s_i - \bar{s_i}) \right)^2$$

$$= \int ds_1 \ldots ds_N \ldots \left(\sum_{i=1}^{N} (s_i - \bar{s_i}) \sum_{j=1}^{N} (s_j - \bar{s_j}) \right) \tag{4.10}$$

$$= \int ds_1 \ldots ds_N \ldots \left(\sum_{i=1}^{N} (s_i - \bar{s_i})^2 + \sum_{i,j,\, i \neq j}^{N} (s_i - \bar{s_i})(s_j - \bar{s_j}) \right)$$

$$= \int_{-\infty}^{\infty} ds_1\, w_1(s_1) \ldots \int_{-\infty}^{\infty} ds_N\, w_N(s_N) \sum_{i=1}^{N} (s_i - \bar{s_i})^2 = \sum_{i=1}^{N} (\Delta s_i)^2$$

From this we obtain the relative width of the distribution

$$\frac{\Delta x}{\overline{x}} = \frac{\sqrt{\sum (\Delta s_i)^2}}{\sum \overline{s_i}} = \mathcal{O}\left(\frac{1}{\sqrt{N}}\right) \qquad \text{(law of large numbers)} \qquad (4.11)$$

This estimate follows from $\sum_i \ldots = \mathcal{O}(N)$. Thus we have derived the law of large numbers under much more general conditions. For the case of equal distributions $w_i(s) = w(s)$ we get

$$\overline{s_i} = \overline{s} = \int_{-\infty}^{\infty} ds \, s \, w(s) \qquad (4.12)$$

$$(\Delta s_i)^2 = (\Delta s)^2 = \int_{-\infty}^{\infty} ds \, (s - \overline{s})^2 \, w(s) \qquad (4.13)$$

From (4.9) and (4.10) follow $\overline{x} = N\overline{s}$ and $\Delta x^2 = N \, \Delta s^2$. Thus (4.11) becomes

$$\boxed{\frac{\Delta x}{\overline{x}} = \frac{\Delta s}{\overline{s}} \frac{1}{\sqrt{N}}} \qquad \text{law of large numbers} \qquad (4.14)$$

We present two examples for this:

- We consider a gas of $N = 10^{24}$ atoms. The ith atom has with probability $w(\varepsilon_i) \propto \exp(-\varepsilon_i/k_{\mathrm{B}}T)$ the energy ε_i; here k_{B} is the Boltzmann constant and T is the temperature. Both the mean value $\overline{\varepsilon}$ as well as the fluctuation $\Delta \varepsilon$ of this distribution are of the size $\mathcal{O}(k_{\mathrm{B}}T)$. The energy of a single particle is therefore rather indeterminate,

$$\frac{\Delta \varepsilon}{\overline{\varepsilon}} = \mathcal{O}(1) \qquad (4.15)$$

  In contrast to this, the total energy $E = \sum \varepsilon_i$ is sharply defined:

$$\frac{\Delta E}{\overline{E}} = \frac{\Delta \varepsilon}{\overline{\varepsilon}} \frac{1}{\sqrt{N}} = \mathcal{O}\left(10^{-12}\right) \qquad (4.16)$$

  The law of large numbers implies that a given temperature (system in a heat bath) is practically equivalent to a given energy (closed system).

- We consider the oscillating element of a clock, for example the balance wheel or a quartz crystal. The ith oscillation period has with probability $w(t_i)$ the length t_i. The distribution $w(t_i)$ may have the mean value $\overline{t}$ and the relative width

$$\frac{\Delta t}{\overline{t}} = 10^{-2} \qquad (4.17)$$

  The natural frequency ω_0 of the oscillator determines the mean value $\overline{t} \approx 2\pi/\omega_0$. The frequency width and thus Δt depend on the quality of the oscillator. The clock performs a forced oscillation.

The clock adds up the individual periods and displays the total time $T = \sum t_i$. For N oscillation periods, the fluctuation of the time is equal to $\Delta T = \Delta t\, N^{1/2}$. The relative inaccuracy of the clock due to statistical fluctuations, however, decreases as the number of oscillations increases:

$$\frac{\Delta T}{T} = \frac{\Delta t}{t}\,\frac{1}{\sqrt{N}} \tag{4.18}$$

In a wristwatch, the quartz crystal oscillates about 10^4 times as fast as the conventional balance wheel; for a given period of time, N in (4.18) is therefore 10^4 times as large. The law of large numbers explains that a quartz watch runs much more accurate than a mechanical clock.

The quantity ΔT is the expected error due to *statistical* fluctuations of the duration of the individual oscillations. A concrete clock will generally have a larger error due to other effects. For example, the natural frequency of the oscillator (balance wheel, quartz) will be a function of the temperature. This leads to a systematic error in the time shown by the clock.

Derivation of the central limit theorem

We now turn to the actual central limit theorem. It states that the probability distribution for $x = \sum s_i$ (for sufficiently large N) is a normal distribution , i.e.

$$P(x) = \frac{1}{\sqrt{2\pi}\,\Delta x}\,\exp\left(-\frac{(x - \overline{x})^2}{2\,\Delta x^2}\right) \tag{4.19}$$

Here $\overline{x}$ and Δx are given by (4.9) and (4.10).

We first consider the case of equal distributions, $w_i(s) = w(s)$. Beyond (4.6) we assume that for $w(s)$ any moments $\overline{s^n}$ (with $n = 0, 1, 2,...$) are defined, i.e. that the quantities

$$\overline{s^n} = \int_{-\infty}^{\infty} ds\, s^n\, w(s) \tag{4.20}$$

are finite.

The probability of finding the random variables at the values $s_1,....,\,s_N$ is given by (4.8). From this results in the probability for an x-value if we integrate over all possible s_i-values that lead to that x:

$$P(x) = \int_{-\infty}^{\infty} ds_1\, w(s_1)\ldots \int_{-\infty}^{\infty} ds_N\, w(s_N)\, \delta\!\left(x - \sum_{i=1}^{N} s_i\right) \tag{4.21}$$

The δ function expresses the restriction to $\sum s_i = x$. When integrating over x, the δ function results in 1, leading to $\int dx\, P(x) = 1$.

To evaluate (4.21), we use the representation

$$\delta(y) = \frac{1}{2\pi} \int_{-\infty}^{\infty} dk \, \exp(-iky) \tag{4.22}$$

of the δ function. This gives us

$$
\begin{aligned}
P(x) &= \frac{1}{2\pi} \int_{-\infty}^{\infty} dk \int_{-\infty}^{\infty} ds_1 \, w(s_1) \ldots \int_{-\infty}^{\infty} ds_N \, w(s_N) \, e^{ik(s_1 + \ldots + s_N)} \, e^{-ikx} \\
&= \frac{1}{2\pi} \int_{-\infty}^{\infty} dk \, e^{-ikx} \int_{-\infty}^{\infty} ds_1 \, w(s_1) \, e^{iks_1} \int_{-\infty}^{\infty} \ldots \int_{-\infty}^{\infty} ds_N \, w(s_N) \, e^{iks_N} \\
&= \frac{1}{2\pi} \int_{-\infty}^{\infty} dk \, \left[W(k) \right]^N \exp(-ikx) \tag{4.23}
\end{aligned}
$$

In the last step, the Fourier transform $W(k)$ of the function $w(s)$ was introduced. We expand this $W(k)$ for small k:

$$W(k) = \int_{-\infty}^{\infty} ds \, w(s) \, \exp(iks) = 1 + ik\,\overline{s} - \frac{1}{2} k^2 \,\overline{s^2} \pm \ldots \tag{4.24}$$

and

$$\ln \left[W(k) \right]^N = N \ln \left(1 + ik\,\overline{s} - \frac{1}{2} k^2 \,\overline{s^2} \pm \ldots \right) \tag{4.25}$$

Here we use the expansion $\ln(1 + y) = y - y^2/2 + \ldots$,

$$\ln \left[W(k) \right]^N = N \left(ik\,\overline{s} - \frac{1}{2} k^2 \,\overline{s^2} - \frac{1}{2} (ik\,\overline{s})^2 \pm \ldots \right) = N \left(ik\,\overline{s} - \frac{1}{2} k^2 \,\Delta s^2 \pm \ldots \right) \tag{4.26}$$

We neglect the third and higher order terms. Then (4.26) yields

$$\left[W(k) \right]^N \approx \exp \left(iN\overline{s}\,k - \frac{N\,\Delta s^2}{2} k^2 \right) \tag{4.27}$$

We insert this, $\overline{x} = N\,\overline{s}$ and $\Delta x^2 = N\,\Delta s^2$ in (4.23) and carry out the integral:

$$P(x) \approx \frac{1}{2\pi} \int_{-\infty}^{\infty} dk \, \exp \left(-i(x - \overline{x})k - \frac{\Delta x^2}{2} k^2 \right) = \frac{1}{\sqrt{2\pi}\,\Delta x} \exp \left(-\frac{(x - \overline{x})^2}{2\,\Delta x^2} \right) \tag{4.28}$$

The result is the central limit theorem for the case of equal distributions ($w_i = w$).

Generalization

We generalize the derivation of (4.28) to the case of different probability distributions $w_i(s_i)$. First we obtain

$$P(x) = \frac{1}{2\pi} \int_{-\infty}^{\infty} dk \left(\prod_{j=1}^{N} W_j(k) \right) \exp(-ikx) \tag{4.29}$$

where $W_j(k)$ is defined as in (4.24). Thus (4.25) and (4.26) become

$$\ln \prod_{j=1}^{N} W_j(k) = \sum_{j=1}^{N} \ln\left(1 + i k \,\overline{s_j} - \frac{1}{2} k^2 \,\overline{s_j^2} \pm \ldots\right) \tag{4.30}$$

$$= i k \sum_{j=1}^{N} \overline{s_j} - \frac{k^2}{2} \sum_{j=1}^{N} (\Delta s_j)^2 \pm \ldots = i \overline{x} k - \frac{\Delta x^2}{2} k^2 \pm \ldots$$

We neglect the higher order terms and substitute (4.30) into (4.29). Performing the integral again results in (4.28). The sum of randomly distributed quantities thus yields a normal distribution under rather general conditions. This is the statement of the central limit theorem.

Validity range

The expansion (4.26) is possible if the $(n+1)$-th term is small compared to the n-th term:

$$\left| \frac{\overline{s^{n+1}} \, k^{n+1}}{\overline{s^n} \, k^n} \right| \ll 1 \qquad (n = 0, 1, 2, \ldots) \tag{4.31}$$

The range in which we want to use the result can be limited to a few standard deviations. The relevant k values are therefore of the size

$$k \sim \frac{1}{\Delta x} \sim \frac{1}{\sqrt{N}\,\Delta s} \tag{4.32}$$

For this, the left-hand side of (4.31) yields

$$\left| \frac{\overline{s^{n+1}} \, k^{n+1}}{\overline{s^n} \, k^n} \right| \sim \left| \frac{\overline{s^{n+1}}}{\overline{s^n} \, \Delta s} \right| \frac{1}{\sqrt{N}} \sim \frac{\mathcal{O}(1)}{\sqrt{N}} \tag{4.33}$$

Here we have assumed that the moments $\overline{s^n}$ scale with the width of the distribution; for common distributions this is the case. In the limiting case $N \to \infty$, the error of the Gaussian distribution becomes arbitrarily small; this statement applies to the relevant range of several standard deviations around the mean value.

We summarize the conditions under which we have derived the central limit theorem. First, we assumed that the individual random variables have a distribution with finite moments (4.20). Next we assumed a sufficiently large number of random variables, leading to $x = \sum s_i$ with $N \gg 1$. The resulting distribution (4.28) is quantitatively valid over several standard deviations $\sigma = \Delta x$ around the mean value. For large values of $|x - \overline{x}|$ the normal distribution (4.28) and the exact probability (4.21) are both very small in absolute terms. This means that in this the absolute (but not the relative) errors are small.

Exercises

4.1 Superposition of two Gaussian distributions

The independent random variables x and y satisfy Gaussian distributions with the mean values $\bar{x}$ and $\bar{y}$ and the mean deviations Δx and Δy. Calculate the probability distribution $P(z)$ for $z = x + y$. Determine the mean values $\bar{z}$ and $\overline{z^2}$ and the mean width Δz.

4.2 Sum of two random variables

The independent random variables x and y satisfy the probability distributions $P_1(x)$ and $P_2(y)$ with variable ranges from $-\infty$ to ∞. The distributions have the mean values $\bar{x}$ and $\bar{y}$ and the means deviation Δx and Δy. Otherwise the distributions $P_1(x)$ and $P_2(y)$ are arbitrary, they may also be non Gaussian. Determine the probability distribution $P(z)$ for $z = x + y$. Calculate the mean values $\bar{z}$ and $\overline{z^2}$ and the mean width Δz.

II Basics of statistical physics

5 Fundamental postulate

In Part II the basic concepts of statistical physics are introduced and discussed. The temperature and the entropy are defined microscopically. These quantities are then linked to measurable quantities (observables).

Statistical physics is based on an assumption about the probabilities of the microscopic states in the considered system. This assumption, the fundamental postulate, is formulated and explained in this chapter.

Microstate

We first introduce the concept *microstate*. A microstate is defined by a complete microscopic description of the system. Examples of such systems are a group of N dice, an ideal gas and a spin system.

For a system of N dice, a microstate r is defined by the numbers of eyes for each dice:

$$r = (n_1, n_2, \ldots, n_N) \qquad (n_i = 1, 2, ..., 6) \tag{5.1}$$

For N dice there are 6^N different microstates r.

The physical system under consideration shall depend on f degrees of freedom which are described by the generalized coordinates $q = (q_1, ..., q_f)$. The Hamilton function $H(q, p)$ of the system depends on these coordinates and the corresponding momentum $p = (p_1, ..., p_f)$. We consider closed systems, for which H does not depend on time. Let the Hamilton function be equal to the energy of the system. By the substitution $p \to p_{\mathrm{op}}$ the Hamilton function becomes the Hamilton operator $H(q, p_{\mathrm{op}})$. We consider both, the quantum mechanical as well as the classical description of microstates. We start with the quantum mechanical case.

As microstates we choose the eigenstates of the Hamilton operator. For a system with f degrees of freedom they depend on f quantum numbers n_k:

$$\text{microstate:} \quad r = (n_1, n_2, \ldots, n_f) \tag{5.2}$$

The eigenstates $|r\rangle$ are determined by $\hat{H}|r\rangle = E_r|r\rangle$. In the position representation this becomes the Schrödinger equation $H(q, p_{\mathrm{op}}) \psi_r(q, t) = E_r \psi_r(q, t)$, where $\psi_r(q, t)$ is of the form $\varphi_r(q)\exp(-iE_r t/\hbar)$. In the context of quantum mechanics, a wave function $\psi(q, t)$ represents a complete description of the system. The specification of $\psi(q, t_0)$ at a time t_0 determines the wave function $\psi(q, t)$ at any other time.

A closed system can be enclosed by a finte volume (without influencing the system itself). Then, at least in principle, all quantum numbers are discrete; this was assumed in (5.2). Each quantum number in itself has a countable number of values; this can be finite or infinite. For example, the m quantum number of the angular motion (wave function $Y_{lm}(\theta, \phi)$) for a given l only has $2l + 1$ values, whereas the l quantum number may adopt infinitely many values.

We consider two examples for (5.2), the ideal gas and a spin system. In the ideal gas N atoms move in a box with the volume V. The interaction between the atoms is neglected, so that each atom moves independently of the others within the box. We denote the momentum of the νth particle by $\boldsymbol{p}_\nu$, its Cartesian components with $p_{3\nu-2}, p_{3\nu-1}, p_{3\nu}$. In total, there are $3N$ Cartesian momentum components:

$$\boldsymbol{p}_1, \ldots, \boldsymbol{p}_N := p_1, \ldots, p_{3N} = \ldots, p_{3\nu+j-3}, \ldots \tag{5.3}$$

Here ν runs from 1 to N, and j from 1 to 3. The box shall have a cubic volume $V = L^3$. A particle can move freely inside the box, but its wave function must disappear at the boundaries. This restricts the momentum components to the values

$$p_k = \frac{\pi\hbar}{L}\, n_k \qquad (k = 3\nu + j - 3 = 1, 2, ..., 3N) \tag{5.4}$$

with $n_k = 1, 2, 3, ...$ This means that a microstate r is defined by

$$r = (n_1, n_2, \ldots, n_{3N}) \qquad (n_k = 1, 2, 3, ...) \tag{5.5}$$

The number of microstates r is infinite, because each quantum number n_k can adopt the values $1, 2, \ldots$. For given energy, the number of possible microstates is, however, finite.

A simple quantum mechanical example for (5.1) is a system of N independent particles with the spin $1/2$, where only the spin degrees of freedom are considered. (This can be applied to a a crystal with one unpaired electron per lattice site). Relative to the measurement direction (let it be the z-axis), the spin component $\hbar s_z$ is found, where s_z can have the values $+1/2$ or $-1/2$. A microstate of the system consisting of N spins is therefore defined by

$$r = (s_{z,1}, s_{z,2}, \ldots, s_{z,N}) \qquad (s_{z,\nu} = \pm 1/2) \tag{5.6}$$

There are 2^N such states.

If quantum mechanical effects do not play a role, we can treat the system classically. For this purpose, we define the classical microstates. In classical mechanics, the state of a system with the generalized coordinates $q_1, ..., q_f$ is defined by specifying $2f$ values for $q_1, ..., q_f$ and $\dot{q}_1, ..., \dot{q}_f$. This applies both for the initial state as

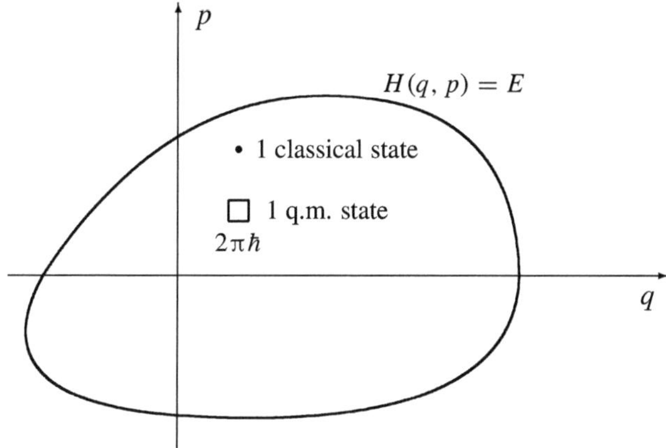

Figure 5.1   Sketch of the phase space for a one-dimensional system. A point in the phase space, i.e. the specification of q and p, defines a classical microstate. The condition $H(q, p) = E$ is a curve representing all microstates with the energy E. The points within of this closed curve represent all states with an energy below E. For each phase space area $2\pi\hbar$ there is one quantum mechanical (q.m.) state. The number of possible states with an energy less than E is therefore equal to the area enclosed by the curve $H(q, p) = E$ divided by $2\pi\hbar$.

well as for the state at any later time. Equivalent to the specification of $q_1, ..., q_f$ and $\dot{q}_1, ..., \dot{q}_f$ is that of $q_1, ..., q_f$ and $p_1, ..., p_f$, where p_j is the generalized momentum corresponding to q_j. We can therefore define the microstate of a mechanical system by

$$r = (q_1, ..., q_f, p_1, ..., p_f) \qquad \text{(classical microstate)} \qquad (5.7)$$

Specifically for the ideal gas (5.3) – (5.5), the classical microstate is given by

$$r = (\boldsymbol{r}_1, ..., \boldsymbol{r}_N, \boldsymbol{p}_1, ..., \boldsymbol{p}_N) \qquad\qquad (5.8)$$

We have combined the position and momentum components into vectors.

Phase space

We introduce an abstract $2f$-dimensional space, which is spanned by $2f$ Cartesian coordinate axes for the variables q_i and p_i. This space is called *phase space* (Figure 5.1). Each classical state r corresponds to a point in the phase space; conversely, each point in phase space corresponds to a state. For the three-dimensional motion of a single particle, the phase space has 6 dimensions, for N gas atoms it has $6N$ dimensions.

In contrast to (5.2), the right-hand side of (5.7) contains continuous variables. For a statistical treatment we have to count the states r in some way. An the exact specification of q_i and p_i is neither necessary nor feasible. According to the uncertainty principle, position and momentum cannot defined more precise than

$$\Delta p \; \Delta q \; \geq \; \frac{\hbar}{2} \tag{5.9}$$

Here we limit ourselves to $f = 1$. A quantum mechanical state implies a distribution of positions and momenta that is consistent with (5.9); the state therefore occupies an area of the size $\mathcal{O}(\hbar)$ in the phase space. The investigation of simple quantum mechanical systems (infinite box, one-dimensional oscillator) shows that for each phase space area $2\pi\hbar$ there is just one quantum mechanical state. In the $2f$-dimensional phase space, a state then corresponds to the $2f$-dimensional volume $(2\pi\hbar)^f$. We can therefore divide the phase space in cells of he size

$$\text{cell size} = (2\pi\hbar)^f \tag{5.10}$$

Each such cell each contains one state (Figure 5.1). We can number these cells; this makes the initially continuous states (5.7) *countable*.

Historically, the phase space was already used in this context before quantum mechanics. There was also the idea that the number of states is proportional to the considered phase space volume. The size of the phase space cell (here $2\pi\hbar$) remained open. However, in the classical limit the results are independent of this point.

Macrostate

In general, it is impossible to specify the actual microstate of a many-body system. For example, the microstate (5.8) of a classical gas changes continuously. Moreover, the positions and velocities of 10^{24} atoms are of no interest. Also between the quantum mechanical microstates r there are ongoing transitions $r \to r'$ between the numerous states of equal energy $E_r = E_{r'}$; arbitrarily small perturbations suffice for this. The transitions between microstates occur extremely quickly; this is obvious for the classical gas.

The individual microstates and their temporal sequence in the considered system are of no interest. Relevant is, however, which microstates occur at all, and with which statistical weight. If we restrict ourselves to this information, then we describe the state of the system by specifying the probabilities P_r for the microstates r. The so-defined state of the system is called *macrostate*:

$$\text{macrostate:} \;\; \{P_r\} = (P_1, P_2, P_3, \ldots) \tag{5.11}$$

The definition (1.1) of probability assumes a large number number M of systems, of which M_r are in the microstate r:

$$P_r = \lim_{M \to \infty} \frac{M_r}{M} \approx \frac{M_r}{M} \tag{5.12}$$

We replace the limit by the specification "M sufficiently large". The entirety of the M systems that define the P_r is called *statistical ensemble*. It is said that *the macrostate is represented by a statistical ensemble*. The statistical ensemble is a *conceptual* prerequisite for the definition of the P_r, i.e. for the statistical treatment.

A concrete macroscopic system (e.g. a classical gas) is at a specific time in a certain (undeterminable, not interesting) microstate. The concrete system goes through all possible microstates in some (non-interesting) way. In doing so, the system will be with probability P_r in the microstate r. In this picture the M systems we need for (5.12) are represented by the one system at M different points in time. Instead, we can also imagine a statistical ensemble of M identical systems that exist simultaneously. For these two possibilities the terms *time average* and *ensemble average* were introduced in Chapter 1.

In the following, we prefer to use *ensemble average*. The statistical ensemble is then a set of M systems of the same kind, of which M_r are each in state r. In the Chapters 1 – 4 we considered a dice or a gas particle as a system. There, the statistical ensemble was a large number N of identical dice or gas particles.

Let the system be a group of $N = 10$ dice. The statistical ensemble consists of a large number of groups of 10 dice. The number must be large enough to define the probability P_r of a certain microstate r. Since there are 6^N different states (5.1), the ensemble must comprise $M \gg 6^N$ systems. In contrast to this, a time average could be obtained by rolling M times a single group of N dice.

For the gas box, the statistical ensemble is a large number M of identical gas boxes (same volume, same amount of gas and energy). Since there are a large number of microstates, M is so large that this ensemble can only be an imaginary one. Its introduction is, however, conceptually necessary in order to define the probabilities P_r. A single system is in a certain microstate r_0 at a certain point in time; no quotients M_r/M can be formed from this. Alternatively as discussed above, one could consider the time average of a single system.

For the time average, the analogy between the dice group and the gas is incomplete. A group of dice must be thrown again and again, whereas the gas system moves itself from microstate to microstate.

We summarize: A microstate completely defines the microscopic state of the system. A macrostate, on the other hand only, defines the probabilities with which the microstates occur.

Formulation of the postulate

We formulate the basic postulate verbally and explain the occurring concepts. Then we determine the P_r following from the postulate in a form that is suitable for concrete calculations.

We consider closed many-body systems. Closed means that the system has no interaction with other systems; the system is isolated from its environment. Possible interactions will be considered only later (Chapter 7).

If a closed many-particle system is left to itself, the macroscopically measurable quantities tend towards constant values in time. This is an *empirical fact*. The macroscopic quantities are, for example, pressure, temperature, density or magnetization. The macrostate, in which the macroscopic variables have reached constant values is called *equilibrium state* or equilibrium for short. The equilibrium state is a special macrostate.

For the equilibrium state of the closed system, we postulate:

FUNDAMENTAL POSTULATE:

A closed system in equilibrium is equally probable
in each of its accessible microstates.

This postulate establishes the connection between the microscopic structure (the accessible microstates r) and the macroscopic quantities of the equilibrium state (represented by the probabilities P_r for the microstates r).

The fundamental postulate is an assumption on which statistical physics is built on. This assumption or hypothesis cannot be proven directly; the number of identical systems required to determine the P_r is far too large. From this hypothesis, however, empirically verifiable statements can be derived. The experimental confirmation of these statements then represents a verification of the fundamental postulate.

The question of whether the fundamental postulate can be derived from basic equations of physics is discussed in Chapter 41.

First examples

We start with some simple examples. For a group of N dice, each dice result represents an accessible state. The basic postulate is equivalent to the assumption that each of the 6^N outcomes (for distinguishable dice) is equally probable, i.e.

$$P_r = \frac{1}{6^N} \quad \text{for } r \text{ from (5.1)} \tag{5.13}$$

This is a plausible assumption for symmetrically constructed dice.

As a simple physical example, let us consider the spin setting of four electrons in the magnetic field. The energy E_r in the microstate (5.6) is $E_r = 2\mu_{\mathrm{B}} B \sum s_{z,v}$, where μ_{B} is the Bohr magneton and B is the magnetic field. The equilibrium state under consideration (macrostate with constant macroscopic quantities) shall have the $E = 2\mu_{\mathrm{B}} B$. Based on this information, the system must be in one of the four microstates

$$r = (\uparrow, \uparrow, \uparrow, \downarrow), \; (\uparrow, \uparrow, \downarrow, \uparrow), \; (\uparrow, \downarrow, \uparrow, \uparrow) \; \text{or} \; (\downarrow, \uparrow, \uparrow, \uparrow) \tag{5.14}$$

Here the spin settings in $r = (s_{z,1}, s_{z,2}, s_{z,3}, s_{z,4})$ are described by arrows. The basic postulate states that

$$P_r = \begin{cases} 1/4 & \text{for } r \text{ from (5.14)} \\ 0 & \text{otherwise} \end{cases} \tag{5.15}$$

In this case, all microstates are accessible that have the given energy. It follows from (5.15) that an arbitrary single spin is parallel to B with the probability $p = 3/4$.

Determination of the P_r

We now determine the P_r for the equilibrium state of in a form that is suitable for later applications.

For arbitrary microstates (5.2) or (5.7), the basic postulate can be expressed by

$$P_r = \begin{cases} \text{const.} & \text{all accessible states} \\ 0 & \text{all other states} \end{cases} \tag{5.16}$$

In the following, we specify the condition "accessible".

In addition to the coordinates and momentum, the Hamilton function of the system generally depends on a number of parameters x:

$$H = H(q, p; x) = H(q_1, ..., q_f, p_1, ..., p_f; x_1, x_2, ..., x_n) \tag{5.17}$$

This form applies accordingly to the Hamilton operator $H(q, p_{\text{op}}; x)$. We denote the quantities $x = x_1, ..., x_n$ as *external parameter*. For a gas, H depends on the volume V and the number of particles N, i.e. $x = (V, N)$. Which parameters besides V and N have to be taken into account depends on the system, the experimental conditions and also on the desired accuracy. For polarizable matter, for example, an external electric field is a potential external parameter. On the other hand, a weak electric field usually has little influence on the energy of the states and can then be disregarded. Other possible external parameters are a magnetic or gravitational field. For the equilibrium states considered here, we assume constant external parameters. Changes in the external parameters will be dealt with later.

The equilibrium state will generally depend on the external parameters x. In addition, it depends on all variables that limit the accessibility of microstates. These are in particular all conserved quantities of the system. It is known from mechanics and quantum mechanics that the energy E is a conserved quantity for the closed system. Thus, only microstates r with this energy are accessible. The total momentum and the angular momentum of the system are often not conserved because the external conditions (vessel or box) violate the corresponding symmetries.

The energy values of the microstates r

$$E_r = E_r(x) = E_r(x_1, \ldots, x_n) \tag{5.18}$$

follow from the Hamilton operator $\hat{H}$ or from the Hamilton function $H(q, p; x)$. In the quantum mechanical case, the microstates $r = (n_1, ..., n_f)$ are the eigenstates of the Hamilton operator, $\hat{H}(x)|r\rangle = E_r(x)|r\rangle$. The $E_r(x)$ are then the eigenvalues of the Hamilton operator. For the classical microstates $r = (q, p)$, the energy is given directly by the Hamilton function, $E_r = H(q, p; x)$.

For a closed system, the total energy E is a conserved quantity. *Accessible* in the sense of the fundamental postulate are therefore only the microstates r, whose energy values E_r coincide with the conserved energy E of the system. The energy E can only be determined with a finite accuracy δE; let $\delta E \ll E$ apply, for example $\delta E = 10^{-5} E$. We denote the number of states between $E - \delta E$ and E as *microcanonical partition function* $\Omega(E, x)$,

$$\Omega(E, x) = \sum_{r:\, E - \delta E \, \leq \, E_r(x) \, \leq \, E} 1 \tag{5.19}$$

With a suitable choice of δE, Ω only depends insignificantly on δE (Chapter 6). The partition function (5.19) is defined for both quantum mechanical and classical microstates; in the classical case, each phase space cell (5.10) is counted as one state.

The partition function Ω is equal to the number of accessible states of the closed system. According to the fundamental postulate, all $\Omega(E, x)$ states are are equally probable, i.e.

$$P_r(E, x) = \begin{cases} \dfrac{1}{\Omega(E, x)} & E - \delta E \leq E_r(x) \leq E \\ 0 & \text{otherwise} \end{cases} \tag{5.20}$$

The P_r, i.e. $\Omega(E, x)$, can be used to calculate all statistical mean values. The function $\Omega(E, x)$, (5.19), is determined by the $E_r(x)$, i.e. from the Hamilton operator or the Hamilton function. This makes it clear in principle clear how macroscopic quantities (mean values, calculated with P_r) can be derived from the microscopic structure (described by $H(x)$). This program is described below for some simple systems, first and foremost the ideal gas.

The statistical ensemble consists of a sufficiently large number M of systems, of which $M_r = M P_r$ are in state r. The statistical ensemble considered here ensemble is called *microcanonical ensemble*. Physically, this ensemble is defined by the condition that the system is closed. Another possible condition would be that the system is placed in a heat bath; then the temperature of the system is given and not the energy. The corresponding statistical ensembles will be introduced in Part IV.

Macrostate and equilibrium state

The physical system under consideration is characterized by a Hamilton operator $H(x)$, which depends on the external parameters x. The macrostates are described by probabilities $\{P_r\} = (P_1, P_2, \ldots)$ for the individual microstates r.

The equilibrium of a closed system is a special macrostate in which the P_r are given by (5.20),

$$\text{equilibrium state: } \{P_r\} \text{ with } P_r \overset{(5.20)}{=} P_r(E, x_1, \ldots, x_n) \tag{5.21}$$

The state of equilibrium is thus fixed by the variables E and x. This we can define it by

$$\text{equilibrium state:}\quad E, x_1, \ldots, x_n \tag{5.22}$$

All macroscopic quantities that are defined in the equilibrium state are called *state variables*. These include first of all E and $x = x_1, \ldots, x_n$, but also all quantities that are a functions of these quantities, $y_i = y_i(E, x)$ are state variables. Instead of $E, x_1, \ldots, x_n$ the state of equilibrium can also be determined by $n + 1$ other suitable state variables $y_1, \ldots, y_{n+1}$:

$$\text{equilibrium state:}\quad y_1, \ldots, y_{n+1} \tag{5.23}$$

Thus, the state of a gas can be determined by E, V and N (energy, volume, particle number), or alternatively by T, P and N. Here T is the temperature and P the pressure; these variables will be defined later.

In thermodynamics in the narrower sense, the microscopic basis of the macroscopic variables is ignored. Moreover, one often considers equilibrium states only. Then, the *thermodynamic states* are defined by (5.22) or (5.23).

The transition from the microscopic to the macroscopic description implies a drastic reduction in the number of variables considered. The macrostate of a gas, for example, is usually defined by three values for E, V and N. For the definition of a quantum mechanical microstate, on the other hand, $f + n$ numerical values are required (with $f \approx 2 \cdot 10^{24}$ degrees of freedom for one mole of a monatomic gas and $n = 2$ for V and N).

Concluding remark

In conclusion, we emphasize the plausibility of the fundamental basic postulate. The postulate (5.20) is of the same nature as the assumption $p_i = 1/6$ for a symmetrically built dice. Compare the following two statements:

- Each of the 6 possible dice states (eyes) occurs with the same probability, i.e. $p_i = 1/6$.

- Each of the $\Omega(E, x)$ accessible states occurs with the same probability, i.e. $P_r = 1/\Omega(E, x)$.

Both are plausible assumptions as long as none of the states is somehow distinguished. For the dice this hypothesis can be tested directly. For many-particle systems it can be tested only indirectly (by testing conclusions of this assumption).

Exercises

5.1 Phase space of the oscillator

The one-dimensional oscillator has the Hamilton function

$$H(q, p) = \frac{p^2}{2m} + \frac{m \omega^2 q^2}{2}$$

What is the shape of the curve $H(q, p) = E$ in phase space? Calculate the phase space volume $V_{\text{PS}}(E) = \int dq \int dp$ enclosed by this curve. The known energy eigenvalues $E_n = \hbar \omega (n + 1/2)$ determine the number N_E of states with $E_n \leq E$. Establish the relationship between this number N_E and the phase space volume $V_{\text{PS}}(E)$.

6 Partition function of the ideal gas

We calculate the partition function Ω for an ideal, monatomic gas. The external parameters are, as for many other systems, the volume V and the number of particles N.

In the Hamilton operator $H = H(q, p_{\mathrm{op}}; x)$ we suppress the arguments q and p_{op}. As external parameters we consider $x = (V, N)$, i.e. $H = H(V, N)$. The statistical treatment of equilibrium states can then be represented by the following scheme:

$$H(V, N) \xrightarrow{\;1.\;} E_r(V, N) \xrightarrow{\;2.\;} \Omega(E, V, N) \xrightarrow{\;3.\;} \left\{ \begin{array}{l} S = S(E, V, N) \\ E = E(T, V, N) \\ P = P(T, V, N) \end{array} \right. \qquad (6.1)$$

The individual steps are:

1. Determine the eigenvalues $E_r(V, N)$ of the Hamilton operator $H(V, N)$.

2. Calculate the partition function $\Omega(E, V, N)$ from the eigenvalues $E_r(V, N)$.

3. Determine the entropy $S = S(E, V, N) = k_B \ln \Omega(E, V, N)$ and all other macroscopic quantities and relationships. These relationships include in particular the caloric equation of state $E = E(T, V, N)$ and the thermal equation of state $P = P(T, V, N)$.

In the following, we carry out the steps 1 and 2 for the ideal gas. For these two steps, it is convenient to classify the states quantum mechanically, even if quantum mechanical effects do not play a role. We will carry out step 3 in Chapters 9 and 10; the new concepts used in this step are also defined there.

The Hamilton operator of the ideal monatomic gas of N particles is

$$H(V, N) = \sum_{\nu=1}^{N} \left(-\frac{\hbar^2}{2m} \Delta_\nu + U(\boldsymbol{r}_\nu) \right) = \sum_{\nu=1}^{N} h(\boldsymbol{r}_\nu, \boldsymbol{p}_{\mathrm{op},\nu}; V) \qquad (6.2)$$

Here Δ_ν is the Laplace operator, $\boldsymbol{r}_\nu$ is the position and $\boldsymbol{p}_{\mathrm{op},\nu}$ is the momentum operator for the ν-th particle. The potential $U(\boldsymbol{r}_\nu)$ is zero inside the volume V and infinite otherwise; it restricts the particles to the volume V. Compared to the Hamilton operator of a real gas, the following effects are neglected:

1. The interactions of the gas particles with each other are not taken into account. A gas without interaction is referred to as *ideal*. In practice, the interaction in a real gas can be neglected if the density is low and the temperature is high.

2. No rotations or vibrations of gas molecules are taken into account. The discussion therefore refers to a monatomic gas.

3. No internal degrees of freedom of the atoms are taken into account. For room temperature this is an excellent approximation. Typical excitation energies of the (electronic) degrees of freedom are of the order of electron volts; thermal excitation of such energies is only possible at very high temperatures (about 10000 °C).

As assumed in (6.2), H is a sum of N single particle operators $h(\nu)$. The solution of (6.2) is therefore reduces to the quantum mechanical one particle problem "particle in the infinitely high potential well". As in (5.3), we denote the momentum of the particles with $\boldsymbol{p}_\nu$ and momentum components with $p_k = p_{3\nu+j-3}$. For the sake of simplicity, we consider a cubic potential well with $V = L^3$. Then each of the $3N$ Cartesian momenta can take the discrete values (5.4):

$$p_k = p_{3\nu+j-3} = \frac{\pi\hbar}{L}\, n_k\,, \qquad \text{where} \quad \begin{cases} \nu = 1, 2, ..., N \\ j = 1, 2, 3 \\ k = 1, 2, ..., 3N \end{cases} \tag{6.3}$$

The microstates are given by (5.5),

$$r = (\ldots, n_k, \ldots) = (n_1, n_2, \ldots, n_{3N}) \qquad (n_k = 1, 2, ...) \tag{6.4}$$

They have the energy eigenvalues

$$E_r(V, N) = \sum_{\nu=1}^{N} \frac{p_\nu^2}{2m} = \sum_{k=1}^{3N} \frac{\pi^2\hbar^2}{2m L^2}\, n_k^2 \tag{6.5}$$

The wall potential $U(\boldsymbol{r}_\nu)$ in (6.2) yields no contribution to the energy. It merely ensures that the wave function is zero at the border of the volume. This leads to quantization (6.3) of the momenta.

With

$$\Phi(E, V, N) = \sum_{r:\, E_r(V, N) \le E} 1 \tag{6.6}$$

we first calculate the number $\Phi(E)$ of states whose energy is less than or equal to E. The microcanonical partition function (5.19) is then

$$\Omega(E) = \Phi(E) - \Phi(E - \delta E) \tag{6.7}$$

The sum over r is a multiple sum over $n_1, \ldots, n_k, \ldots$. We evaluate the individual sums under the assumption that they run over many units of n_k, i.e. for

$$\overline{p_k} = \frac{\pi\hbar}{L}\, \overline{n_k} \gg \frac{\hbar}{L} \qquad \text{(classical limit case)} \tag{6.8}$$

The momenta of the particles shall be much larger than the quantum mechanical minimum value $(\Delta p)_{\text{q.m.}} = \pi \hbar / L$, which results from the limitation to the box. Considering the indistinguishability of the particles, leads to the stronger condition $\overline{p} \gg N^{1/3} \hbar / L$ (section N-dependence below and Chapter 30). For ordinary gases, these conditions are almost always fulfilled. When the energy E decreases, the gas condenses long before these conditions are violated.

Because of (6.8) we can replace the sums in (6.6) by integrals. In doing so, we also allow negative n_k-values and correct this by 1/2 factors:

$$\underbrace{\sum_{r : E_r \leq E} 1 = \underbrace{\sum_{n_1 = 1, 2, \ldots} \cdots \sum_{n_{3N} = 1, 2, \ldots} 1}_{E_r \leq E} \approx \frac{1}{2^{3N}} \underbrace{\int dn_1 \ldots \int dn_{3N} \; 1}_{E_r \leq E}} \qquad (6.9)$$

The evaluation of this sum is illustrated in Figure 6.1. We replace the integration variables n_k by $p_k = (\pi \hbar / L) n_k$. This turns (6.6) into

$$\Phi(E, V, N) = \frac{V^N}{(2\pi\hbar)^{3N}} \underbrace{\int dp_1 \ldots \int dp_{3N} \; 1}_{\sum_k p_k^2 \leq 2mE} \qquad (6.10)$$

We have used $V = L^3$ and (6.5). The factor $V^N = L^{3N}$ can also be expressed by spatial integrals over the accessible space (within the box):

$$\Phi(E, V, N) = \frac{1}{(2\pi\hbar)^{3N}} \underbrace{\int dp_1 \ldots \int dp_{3N}}_{\sum_k p_k^2 \leq 2mE} \int_0^L dx_1 \ldots \int_0^L dx_{3N} \; 1$$

$$= \frac{\text{phase space volume}}{(2\pi\hbar)^{3N}} \qquad (6.11)$$

This shows that the quantum mechanical result (6.10) is consistent with the classical (right-hand side of (6.11)).

We now evaluate (6.10). The integration $\int dp_1 \ldots \int dp_{3N}$ gives the volume of a $3N$-dimensional sphere with the radius

$$R = \sqrt{2mE} \qquad (6.12)$$

This volume is proportional to R^{3N}. Except for one numerical prefactor, we obtain

$$\Phi(E, V) = c \, V^N \, E^{3N/2} \qquad (6.13)$$

This result displays the complete E and V dependency. To determine the N-dependence, we still have to calculate the prefactor of the $3N$-dimensional spherical volume and further effect (last section in this chapter). For this reason, N was not included here as an argument of Φ.

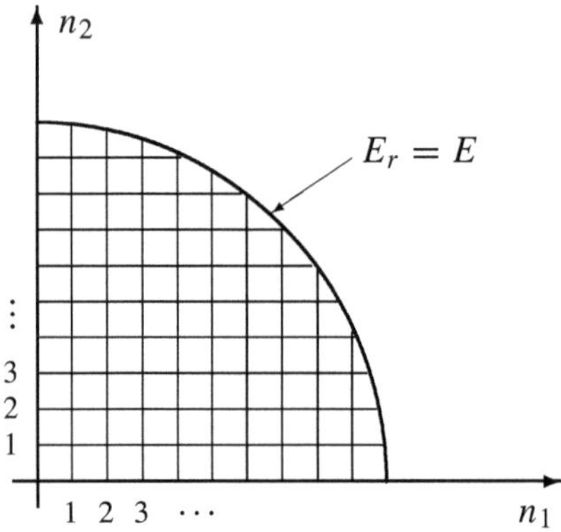

Figure 6.1 Each intersection point in n_1-n_2-...-space stands for a discrete quantum mechanical state. The energy of the state $r = (n_1, n_2,...)$ is given by $E_r = \text{const.} \cdot \sum n_i^2$. Geometrically, $E_r = E = \text{const.}$ represents a multidimensional sphere in the n_1-n_2-...-space, i.e. a circle in the section shown. The number of states with $E_r \leq E$ is equal to the volume of this sphere, since there is exactly one intersection point per unit volume.

For $3N/2 = 10^{24}$ and $\delta E / E = 10^{-5}$ we compare $\Phi(E)$ with $\Phi(E - \delta E)$:

$$\Phi(E) = \text{const.} \cdot E^{10^{24}} = \text{const.} \cdot \left(\frac{E}{E - \delta E} \right)^{10^{24}} (E - \delta E)^{10^{24}}$$

$$= \left(\frac{1}{1 - 10^{-5}} \right)^{10^{24}} \Phi(E - \delta E) \geq \left(\exp\left(10^{-5}\right) \right)^{10^{24}} \Phi(E - \delta E)$$

$$= \exp\left(10^{19}\right) \Phi(E - \delta E) \gg \Phi(E - \delta E) \tag{6.14}$$

Because of this, it is an excellent approximation to replace (6.7) by

$$\Omega(E, V) \approx \Phi(E, V) \tag{6.15}$$

This means that Ω practically does not depend on δE (at least not as long as a sensible value is chosen for δE).

From (6.15) and (6.13) we obtain the partition function Ω for the ideal gas. We consider the logarithm of Ω:

$$\ln \Omega(E, V) = \frac{3N}{2} \ln E + N \ln V + \ln c \tag{6.16}$$

In contrast to Ω itself, $\ln \Omega$ has a moderate dependence on the energy E and on the volume V; therefore, for example, Taylor expansions of $\ln \Omega$ are possible. Since Ω itself is defined as a number, all logarithms in (6.16) together yield the logarithm of a dimensionless number.

N-dependence

Equation (6.16) correctly reflects the E- and V-dependence. For the correct N-dependence, the following two points must be considered points must be taken into account:

1. The volume of a sphere with the radius $R = \sqrt{2mE}$ in $3N$ dimensions is (Exercise 6.3):

$$V_{3N}(R) = \frac{\pi^{3N/2}}{(3N/2)!}\, R^{3N} \approx a^{3N/2} \left(\frac{R^2}{N}\right)^{3N/2} \qquad (6.17)$$

The factorial was approximated by $n! \approx (n/e)^n$ (Exercise 3.1). This N-dependence can be taken into account by the substitutions $E \to E/N$ and $\ln c \to N \ln c'$ in (6.16).

2. Atoms of the same kind are *indistinguishable*. This is a quantum mechanical effect: When any two particle coordinates are interchanged, the wave function receives the factor $+1$ for bosons (particles with integer spin) and -1 for fermions (particles with half-integer spin). The permutation $\boldsymbol{p}_v \leftrightarrow \boldsymbol{p}_{v'}$ or

$$n_{3v+j-3} \longleftrightarrow n_{3v'+j-3} \qquad (v,\, v' \text{ arbitrary},\ j = 1,\, 2 \text{ and } 3) \qquad (6.18)$$

in $r = (n_1, ..., n_{3N})$ does therefore not result in a new state. For a state r of the form (6.4), such swapping results in $N!$ other states, all of which are counted separately (6.6). By the modification

$$\Omega = \frac{\Omega_{(6.16)}}{N!} \qquad (6.19)$$

we take into account that $N!$ states are the same. Because of $N! \approx (N/e)^N$, the additional factor can be taken into account by making the substitutions $V \to V/N$ and $\ln c \to N \ln c'$ in (6.16).

Both points together result in:

$$\boxed{\begin{array}{c} \text{Partition function of the ideal gas:} \\[4pt] \ln \Omega(E,\, V,\, N) = \frac{3N}{2}\, \ln\left(\frac{E}{N}\right) + N \ln\left(\frac{V}{N}\right) + N \ln c \end{array}} \qquad (6.20)$$

We renamed the new constant c' by c again. The now complete N dependence means that the number of states per atom is proportional to V/N and to $(E/N)^{3/2}$.

Generalization

We may express the energy dependence of the function $\Omega(E)$ by the number $f = 3N$ of degrees of freedom of the system,

$$\Omega(E) \propto \left(\frac{E}{N}\right)^{3N/2} \propto \left(\frac{E}{f}\right)^{\gamma f} \tag{6.21}$$

In this form, the energy dependence applies also to other systems with many degrees of freedom: If, for example, the energy is increased by a factor of 100, the energy available per particle or degree of freedom is 100 times larger. This increases the number of achievable states by a corresponding factor ($\sim (E/f)^{\gamma}$). The number of total states then receives this factor to the power of f.

A prerequisite for the form (6.21) is that with higher energy, more and more states are available. This does not apply to spin systems: For N electrons in the magnetic field B, the energy $E_r = 2\mu_B B \sum_{\nu} s_{z,\nu}$. Then the possible energy values are between $E_{\min} = -N\mu_B B$ and $E_{\max} = N\mu_B B$. For these systems, (6.21) only applies in the energy range $\mu_B B \ll E - E_{\min} \ll N\mu_B B$.

Exercises

6.1   *Exponential function with very large large exponent*

The function $f(E) = E^N$ with $N = \mathcal{O}(10^{24})$ shall be expanded around E_0 into a Taylor series. Which condition must be fulfilled so that the first order term in $(E - E_0)$ is small compared to the 0th order? What results if $\ln f(E)$ is expanded instead of $f(E)$?

6.2   *Partition function for gas mixture*

In a box with the volume V there are N_1 gas atoms of kind 1 and N_2 gas atoms of kind 2. Treat the system as a mixture of ideal monatomic gases and determine the partition function Ω.

6.3   Volume of the n-dimensional sphere

Calculate the volume

$$V_n(R) = C_n\, R^n$$

of an n-dimensional sphere with radius R. Use this to determine the partition function $\Omega(E, V, N)$ of an ideal gas.

Instructions: Start from the relationship

$$\int_{-\infty}^{\infty} dx_1 \ldots \int_{-\infty}^{\infty} dx_n\, \exp\left(-\sum_{i=1}^{n} x_i^2\right) = \int_{0}^{\infty} dR\, A_n(R)\, \exp\left(-R^2\right) \qquad (6.22)$$

Here, $R^2 = \sum_i x_i^2$, and $A_n(R) = dV_n/dR$ is the surface area of the sphere under consideration. The determination of both sides yield C_n.

6.4   Ideal spin system

In a crystal lattice, there is one unpaired electron at each lattice point. The spin s_ν (here without the factor $\hbar$) of the ν-th electron is linked to a magnetic moment $\mu_\nu = -2\mu_{\mathrm{B}} s_\nu$; where $\mu_{\mathrm{B}} = e\hbar/(2mc)$ is the Bohr magneton. In an magnetic field $\boldsymbol{B}$, an electron has the energy $\varepsilon = -\boldsymbol{\mu} \cdot \boldsymbol{B}$. Relative to the field $\boldsymbol{B} = B\, \boldsymbol{e}_z$, the spin can be parallel or antiparallel, $s_{z,\nu} = \pm 1/2$. The microstates $r = (s_{z,1}, s_{z,2}, \ldots, s_{z,N})$ have the energy

$$E_r(B) = 2\mu_{\mathrm{B}} B \sum_{\nu=1}^{N} s_{z,\nu} \qquad (6.23)$$

Calculate the partition function $\Omega(E, B)$.

Instructions: Which value E_n is obtained if exactly n magnetic moments are parallel to the magnetic field? Determine the number Ω_n of microstates with this energy E_n. If δE is chosen so that in the δE interval there is just one of the E_n values, then $\Omega(E, B) = \Omega_n$ with $E \approx E_n$. One may assume $n \gg 1$ and $N - n \gg 1$. Show

$$\ln \Omega(E, B) = -\frac{N}{2}\left(1 - \frac{E}{N\mu_{\mathrm{B}} B}\right) \ln\left(\frac{1}{2} - \frac{E}{2N\mu_{\mathrm{B}} B}\right) - \frac{N}{2}\left(1 + \frac{E}{N\mu_{\mathrm{B}} B}\right) \ln\left(\frac{1}{2} + \frac{E}{2N\mu_{\mathrm{B}} B}\right)$$

$$(6.24)$$

What changes if the interval δE comprises several E_n values?

7 First law of thermodynamics

The first law of thermodynamics (TD) describes the energy change of a macrostate. This energy change is divided into heat and work.

Mean value of the energy

A physical system is described by a Hamilton operator or a Hamilton function $H(x)$. This operator/function determines the microstates r and their energies $E_r(x)$. For a specific macrostate (5.22), the probabilities P_r and the external parameters x are given. This determines the mean value of the energy of the macrostate:

$$\overline{E_r} = \sum_r P_r\, E_r(x_1,...,x_n) = \sum_r P_r\, E_r(x) = E \qquad (7.1)$$

We discuss this in more detail:

- By (7.1) the mean value of the energy is given for an arbitrary macrostate, i.e. for any P_r. This macrostate can also be a non-equilibrium state.

- To apply (7.1) to the equilibrium state of a closed system, the $P_r(E, x)$ from (5.20) must be used. Then $\overline{E_r}$ lies between $E - \delta E$ and E. Indeed, $\overline{E_r}$ lies is very close to E, because there are much more microstates than at $E - \delta E$, (6.14).

- We also denote the mean value $\overline{E_r}$ for arbitrary macrostates (including non-equilibrium states) by E. Instead of "mean value of energy" we usually simply say "energy". This energy E is a macroscopic quantity, it is defined for the macrostate.

- In the mathematical introduction (Part I), we designated a mean value like (7.1) with $\overline{E}$. Now we omit the averaging bar for E. It is however retained for $\overline{E_r(x)}$. This is favorable for the later discussion of the x-dependence; it makes it clear that this dependence stems form the energies $E_r(x)$ of the microstates.

We consider a process that moves from a macrostate a to another macrostate b. For this we want to calculate the change ΔE of the mean value (7.1):

$$\Delta E = E_b - E_a \qquad \text{for the process} \quad a \to b \qquad (7.2)$$

By *process* we mean the transition from an initial state to a final state. The state a is defined according to (5.22) by $P_r(a)$ and x_a; this applies accordingly to b. For the process with $\Delta E \neq 0$, contact between the system under consideration and environment is necessary; for closed system the energy would be conserved.

According to (7.1), energy changes are possible if the external parameters $x = (x_1, \dots, x_n)$ are altered, or when the occupations in the statistical ensemble (i.e. the P_r) change. The functions $E_r(x)$ always remain the same, because they follow from the microscopic structure of the physical system described by $H(x)$.

Heat and work

The energy change $\Delta E = E_b - E_a$ can be divided into two contributions. This division is defined by the following *experimental* conditions:

1. Energy transfer with constant external parameters x.

2. Change in the external parameters under simultaneous thermal insulation of the system.

We explain the meaning of these conditions in some detail.

In the first case, the external parameters are fixed; for a gas these are, for example, the volume V and the particle number N. The energy transferred under these conditions ΔE is defined as the amount of *heat* supplied to the system

$$\Delta Q = \Delta E \qquad \text{(constant parameters } x) \tag{7.3}$$

The amount of heat (also called heat for short) ΔQ can be positive or negative; for $\Delta Q < 0$ the system releases energy. Such an energy transfer can occur through contact between the system and the warmer (or colder) environment (heat conduction), or also through targeted heating of the system (e.g. by a heat coil), or by the system emitting or absorbing heat radiation.

In the experimental investigation of heat transfer with constant external parameters, it is possible to determine the condition by which it can be prevented. We call this condition *thermal insulation* of the system; in concrete terms, it could be realized by thick polystyrene (styrofoam) walls.

The change in energy due to a change in the external parameters $x_1, \dots, x_n$ under simultaneous thermal insulation (i.e. $\Delta Q = 0$) is defined as the *work* performed on the system,

$$\Delta W = \Delta E \qquad \text{(thermal insulation)} \tag{7.4}$$

The standard example is the work done during the compression of a gas. In this case, the external parameter $x_1 = V$ is changed; for example, the piston of a gas cylinder is shifted. The work done ($\Delta W > 0$) increases the energy of the gas.

Heat and work input (output) can occur simultaneously. In general we obtain[1]

$$\boxed{\Delta E = \Delta W + \Delta Q \qquad \text{1st law of TD}} \tag{7.5}$$

We refer to this relationship as *first law of thermodynamics (TD)*. For a closed system, $\Delta Q = 0$ and $\Delta W = 0$ hold; the energy of the closed system is conserved. A process with $\Delta Q = 0$ is called *adiabatic*[2].

The introduction of the classification (heat or work) was based on the alternative "parameter constant" or "parameter not constant"; therefore, all possible energy transfers are taken into account. At the same time the experimental conditions ("parameter constant" or "adiabatic") exclude the respective other form of energy transfer. Therefore, both contributions must be added.

Microscopic discussion

The 1st law (7.5) is an energy balance equation for the macrostates of the considered system.

The 1st law applies to any macrostates a and b. For the sake of simplicity, we will limit the following discussion to the case that a and b are equilibrium states with the P_r from (5.20). The equilibrium states a and b are completely defined by E_a, x_a and E_b, x_b. In practice, such equilibrium states are obtained by isolating the physical system from the environment and leaving it to its own for some time.

The initial state a is represented by an ensemble of M identical systems, of which $M_r = MP_r$ are in the microstate r. The interaction with the environment may lead to transitions between microstates ($r \to r'$), and the $E_r(x)$ may change together with x. This usually turns the equilibrium ensemble into a non-equilibrium ensemble. The intermediate states are then macrostates whose P_r's deviate from (5.20).

Based on (7.1), the division (7.5) of the 1st law looks like this:

1. In the case of heat transfer, the parameters x and thus the energy eigenvalues $E_r(x)$ are kept constant. A change in energy $\Delta E \neq 0$ can then only result from the change of the P_r in (7.1). In the ensemble of M identical systems are $M_r = MP_r$ in the state r. A change in P_r means that transitions between the microstates of the ensemble take place. When heat is added, states of higher energy are more strongly occupied. This results in P_r which deviate from (5.20), i.e. non-equilibrium states. After the end of the heat transfer, the system relaxes again to an equilibrium state; this means another change of the P_r to $P_r(E_b, x_b)$.

---

[1]Occasionally, ΔW is also defined as the work done by the system (e.g. in [6]) yielding $E = -\Delta W + \Delta Q$. However, it seems more consistent, to use the same sign convention for ΔW and ΔQ.

[2]In other areas of physics the term "adiabatic" is also used for "slow with respect to internal processes of the system".

If the heat is supplied *quasi-statically* (very slowly), then at each stage of the process, the system can continuously relax to its equilibrium, i.e. to a macrostate with the $P_r(E, x)$ from (5.20).

2. If the external parameters are changed under thermal insulation, then both the E_r and the P_r are altered: The microscopic energy levels $E_r(x)$ shift due to their x-dependence. The time dependence of the external parameters can cause transitions between microstates. These processes will generally result in P_r's that deviate from (5.20). After the interaction with the environment has ended, the system relaxes back to an equilibrium state. This means another change of the P_r to $P_r(E_b, x_b)$.

   Particularly simple conditions arise when the change of the external parameters (under thermal insulation) take place *quasi-statically*. Then there are no transitions between different microstates, and at each stage of the process an equilibrium state is restored. In this quasi-static case, the energy change can be calculated from the x-dependence of the $E_r(x)$, Chapter 8.

We summarize: During the heat or work input (output), the system is generally in non-equilibrium states. For this, our discussion remains qualitative; no attempt is made to specify the P_r for these intermediate states. Especially for quasi-static processes, however, there are significant simplifications. Then the system then passes through macrostates with the $P_r(E, x)$ from (5.20), where E and x are respective values during the process.

Examples

To illustrate the previous discussion, let us consider two specific processes:

A) If a gas-filled vessel is heated at one end, time-dependent heat conduction processes take place. The system therefore passes through non-equilibrium states. After the heat supply, the now insulated system relaxes to its equilibrium. As this final state has the same external parameters as the initial state, the energy change ΔE defines the amount of the absorbed heat ΔQ.

B) If a piston is pulled out of a cylinder filled with gas, the external parameter "volume" changes. If the piston is pulled out slowly, the system passes through a sequence of equilibrium states. Alternatively, the volume can also be altered without doing any work (very fast extraction of the piston or lateral extraction of an intermediate wall between gas and vacuum). In both case, $\Delta W = \Delta E$ holds for the thermally insulated system.

Designation of the differential

Instead of arbitrary processes, we will often deal with infinitesimal changes. Then (7.5) becomes

$$dE = đW + đQ \qquad \text{1st law of TD} \qquad (7.6)$$

We denote the infinitesimal quantity with "d" for the energy, and "$\bar{d}$" for the amount of heat and the work. The reason for this is that E – but not Q and W – is a state variable. That means that dE constitutes a exact differential, whereas the quantities $\bar{d}W$ and $\bar{d}Q$ depend on the path (taken for the infinitesimal process). One also finds the alternative notation $dE = \delta W + \delta Q$.

State variables are, according to Chapter 5, all macroscopic variables that have fixed values in a given equilibrium state. State variables are those variables that are selected to determine the macrostate. The state variables can be the energy E and the external parameters $x_1,..., x_n$, or $n + 1$ other variables $y_1,..., y_{n+1}$, (5.22) or (5.23). If the energy eigenvalues are of the form $E_r = E_r(V, N)$, then E, V and N represent a possible set of state variables. Experimentally, however, it is often more convenient to define the equilibrium state by the temperature T, the pressure P and the particle number N.

The heat absorbed by the system and the work performed on the system are no state variables. To illustrate this, we consider a gas in an equilibrium state E_a, V_a and the following processes that lead to the same final state with $E_b = E_a + \Delta E$ and $V_b = V_a$ (volume unchanged):

1. The heat $\Delta Q = \Delta E$ is supplied.

2. Through oscillating piston motions with finite speed energy $\Delta W = \Delta E$ is transferred to the gas. The final position of the piston shall be the same as the start position. (This process is dealt with quantitatively in the last part of the next chapter, Figure 8.3).

In both cases, the process leads to the same final state (given by $x_b = x_a$ and $E_b = E_a + \Delta E$). The occurring quantities of heat and work are, however, different in the two cases. Therefore, ΔQ and ΔW are not defined by the states a and b; heat and work are no state variables.

Another example is a cycle process or thermodynamic cycle, which starts from an equilibrium state a an proceeds via other states back to a. For a gas, such a process is shown in Figure 19.3. Since the energy is a state variable, $\oint dE = 0$ holds. In contrast to this, $\oint \bar{d}Q$ and $\oint \bar{d}W$ are not equal to zero.

We summarize: The energy $E = E(y_1,..., y_{n+1})$ is a state variable. This means that the exact differential $dE = \sum(\partial E/\partial y_i)\, dy_i$ is defined. The variables W or Q are not state variables; the exact differentials dW and dQ are not defined. Infinitesimal changes of W or Q are therefore not denoted by dW and dQ, but by $\bar{d}W$ and $\bar{d}Q$.

8 Quasi-static process

A process is called quasi-static *(q.s.) if the system goes through a sequence of equilibrium states. We investigate how the energy (7.1) changes during a quasi-static process. This leads to a definition of the generalized forces. The expansion of an ideal gas and the corresponding generalized force (pressure) are discussed in detail.*

By $\tau_{\exp}$ we denote the time scale on which an external change is applied to the system (such as a change of external parameters or a heat input). The relaxation time τ_{relax}, on the other hand, is the time that the system needs to return to equilibrium after a sudden external perturbation. The condition "quasi-static" is fulfilled in the limit $\tau_{\exp}/\tau_{\text{relax}} \to \infty$, i.e. for an "infinitely slow" process. A process is approximately quasi-static if it is so slow that $\tau_{\exp} \gg \tau_{\text{relax}}$ holds. The condition $\tau_{\exp}/\tau_{\text{relax}} \to \infty$ is applied for the expansion of a gas and for other examples in Chapter 9.

The following applies to such a quasi-static process:

1. During a change in the external parameters and/or input of heat, the system undergoes a sequence of equilibrium states.

2. The change in the external parameters x_i does not cause any transitions between the microstates of the system. The quantum mechanical transition probability per time for $r \to r'$ is proportional to $\dot{x}_i^2$, i.e. arbitrarily small for a quasi-static change of x_i. The wave function $\psi_r(q; x)$ and the energy $E_r(x)$ adapt continuously to the change in x_i.

An example of a quasi-static process is the slow extraction of a piston from a thermally insulated cylinder filled with gas, Figure 8.1. The position of the piston is $x_1 = L$. The piston speed is $\dot{x}_1 = dL/dt$. By $\overline{v}$ we denote the mean velocity of the gas particles. The condition $\tau_{\exp}/\tau_{\text{relax}} \to \infty$ for "quasi-static" becomes $|\dot{x}_1|/\overline{v} \to 0$. The non quasi-static effects are proportional to $\dot{x}_1/\overline{v}$.

We consider a quasi-static process in which the parameter x_1 changes whereas the other parameters $x_2,..., x_n$ remain constant. We do not exclude the possibility that heat is also transferred during the process. The change in energy E follows from (7.1):

$$dE = d\sum_r P_r E_r(x_1, x_2,..., x_n) = \sum_r dP_r E_r + \sum_r P_r \frac{\partial E_r}{\partial x_1} dx_1$$

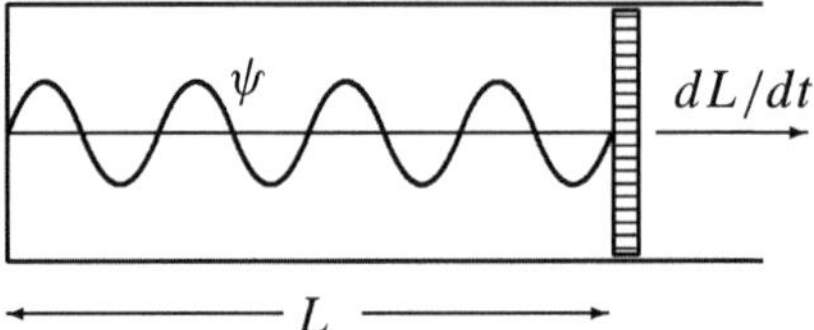

Figure 8.1 A quasi-static expansion: In an ideal gas, the motion of each individual atom is described by a wave function ψ. If the piston moves with a sufficiently small velocity dL/dt, the quantum numbers of this state ψ remain unchanged. From this follows the energy change of the gas during quasi-static expansion.

$$\overset{\text{(q.s.)}}{=} \ \ dQ_{\text{q.s.}} + \overline{\frac{\partial E_r}{\partial x_1}}\, dx_1 = dQ_{\text{q.s.}} + dW_{\text{q.s.}} \tag{8.1}$$

According to the first law of TD, $dE = dQ + dW$. Because of the condition "quasi-static" (q.s.), the quantities on the right-hand side are given a corresponding index.

The equations $dW_{\text{q.s.}} = \sum P_r\,(\partial E_r/\partial x_1)\,dx_1$ and $dQ_{\text{q.s.}} = \sum dP_r\,E_r$ in (8.1) are justified as follows: If in the thermally insulated system the parameter x_1 is changed quasi-statically, then the system remains in the same quantum state. Then dx_1 only leads to the change $dE_r = (\partial E_r/\partial x_1)\,dx_1$ of the energy values E_r, but not to a change in the probabilities $P_r = M_r/M$. A change in P_r may be caused by a heat transfer $dQ_{\text{q.s.}}$. Such a heat transfer does not contribute to the second term ($\propto dx_1$), because dQ was defined as the energy change under constant external parameters.

The mean value $\overline{\partial E_r/\partial x_1}$ is determined using the $P_r(E, x)$ of the equilibrium state. According to the precondition "quasi-static", the system passes through a sequence of such states. For a finite process (i.e. for the calculation of $\Delta W_{\text{q.s.}} = \int dW_{\text{q.s.}}$) the respective values of E and x for the sequence of equilibrium states are to be used in $P_r(E, x)$. We define

$$\boxed{X_i = -\,\overline{\frac{\partial E_r(x_1,..., x_n)}{\partial x_i}} \qquad \text{generalized force}} \tag{8.2}$$

as the *generalized force* X_i corresponding with the parameter x_i. The minus sign is a convention; with the specified sign, the generalized force associated with the volume $x_1 = V$ is equal to the pressure, $X_1 = P$. As a mean value, the generalized force could also be designated as $\overline{X_i}$. As in the case of energy ($E = \overline{E} = \overline{E_r}$) and other macroscopic quantities, however, we omit the bar.

The $E_r(x_1,..., x_n)$ considered here are the eigenvalues of quantum mechanical states r; for quasi-static process there are no transitions between these states. The derivation (8.1) and the definition (8.2) are therefore initially only applicable for quantum states r. Since quantum mechanics includes mechanics as a limiting case, this is only a practical but not a fundamental restriction. In Chapter 10, (8.2) is used to establish a relation between the generalized forces and the partition function $\Omega(E, x)$. In this relation, $\Omega(E, x)$ may be calculated classically.

From (8.1) and (8.2) it follows that $đW_{\text{q.s.}} = -X_1\, dx_1$. If several external parameters are changed $(\partial E_r/\partial x_1)\, dx_1$ in (8.1) is replaced by $\sum_i (\partial E_r/\partial x_i)\, dx_i$ and we obtain

$$đW_{\text{q.s.}} = -\sum_{i=1}^{n} X_i\, dx_i \tag{8.3}$$

This defines the generalized forces X_i as observables. The measurement of work and of the external parameters is assumed to be known.

The standard example of a generalized force is pressure; it is discussed in detail below. As a simple further example, we consider a system of N independent particles with spin 1/2. With the spin degree of freedom $s_z = \pm 1/2$ the magnetic moment $\pm\mu_0$ is linked. The system shall be located in an external magnetic field $x_1 = B$. For the energy E_r of the microstates defined in (5.6) we get

$$E_r = -2\mu_0 B \sum_{\nu=1}^{N} s_{z,\nu} \tag{8.4}$$

The external force

$$X_1 = -\overline{\frac{\partial E_r}{\partial B}} = 2\mu_0 \overline{\sum_{\nu} s_{z,\nu}} = 2N\mu_0\, \overline{s_z} = VM \tag{8.5}$$

is equal to the mean magnetic moment of all particles. Here M denotes the magnetization (magnetic moment per volume).

Pressure

We consider a gas in a cylinder with a mobile piston, Figure 8.1. The external parameter is the position L of the piston. The gas exerts a force F on the piston. For a displacement of the piston by dL the work

$$đW_{\text{q.s.}} = -F\, dL = -\frac{F}{A} A\, dL = -P\, dV \tag{8.6}$$

is performed. Alternative to L, we may use the volume V as the external parameter. The generalized force is then equal to the *pressure* $P = F/A = $ force/area.

By (8.6) the pressure is defined as a observable. For the microscopic definition of the pressure, we evaluate (8.3) and (8.2) for the external parameter $x_1 = V$:

$$đW_{\text{q.s.}} = -X_1\, dx_1 = \overline{\frac{\partial E_r(V)}{\partial V}}\, dV \tag{8.7}$$

The comparison with (8.6) shows

$$\boxed{P = -\frac{\overline{\partial E_r(V)}}{\partial V} \qquad \text{pressure}} \tag{8.8}$$

This defines the pressure microscopically. For a gas, the pressure is always positive. An increase in volume ($dV > 0$) means that the system is performing work ($dW_{\text{q.s.}} < 0$).

The relationship $dW_{\text{q.s.}} = -P\,dV$ is independent of the shape of the volume. As the pressure at equilibrium is the same everywhere, every small surface change results in a work contribution $-P\,\delta v$, where $\sum \delta v = dV$. The definition (8.8) of the pressure is therefore not limited to the special geometry of Figure 8.1.

The classical microscopic origin for the pressure is discussed below in detail (for an ideal gas). Thereby the quantum mechanical definition (8.8) is supplemented by the corresponding classical picture. We compare both derivations:

1. Ideal quantum mechanical gas: In the quasi-static process r and the quantum numbers $n_1,\ldots,n_{3N}$ are not changed. Pushing the piston moves the nodes the single-particle wave function together (Figure 8.1). This increases the momentum and energy of the particle; work is done on the system. The calculation is performed according to (8.8).

2. Ideal classical gas: The reflections of particles from the moving piston increases (decreases) the velocity of these particle (Figure 8.2). From this, $dW_{\text{q.s.}} = -P\,dV$ can be calculated, too.

Pressure of an ideal gas

Quantum mechanical treatment

We evaluate (8.8) for the ideal gas. For this purpose, we consider a cuboid volume with the side lengths L_1, L_2 and L_3. As variable external parameters we choose the side length $x_1 = L_1$. A microscopic state r for N gas particles is defined by (6.4),

$$r = (n_1, \ldots, n_{3N}) = (\ldots, n_{3v+j-3}, \ldots) \tag{8.9}$$

The particle index v runs from 1 to N, the index j from 1 to 3. The momentum $\boldsymbol{p}_v := (p_{3v-2}, p_{3v-1}, p_{3v})$ of the v-th particle has the components:

$$p_{3v+j-3} = \frac{\pi\hbar}{L_j}\, n_{3v+j-3} \tag{8.10}$$

The energy value of the state (8.9) is

$$E_r = \sum_{v=1}^{N} \frac{\boldsymbol{p}_v^2}{2m} = \sum_{v=1}^{N} \sum_{j=1}^{3} \frac{\hbar^2 \pi^2}{2m L_j^2} \left(n_{3v+j-3}\right)^2 \tag{8.11}$$

For quasi-static change of $x_1 = L_1$ we use (8.3),

$$\bar{dW}_{\text{q.s.}} = \overline{\frac{\partial E_r}{\partial L_1}}\, dL_1 = -\frac{2}{L_1} \sum_{v=1}^{N} \overline{\frac{\hbar^2 \pi^2}{2mL_1^2} \left(n_{3v-2}\right)^2}\, dL_1 \qquad (8.12)$$

In equilibrium, all possible states are equally probable. This implies that all momentum directions are equally probable for all particles. Therefore, the mean kinetic energy is the same for all three spatial directions

$$\sum_{v=1}^{N} \overline{\frac{\hbar^2 \pi^2}{2mL_1^2} \left(n_{3v-2}\right)^2} = \frac{1}{3} \sum_{j=1}^{3} \sum_{v=1}^{N} \overline{\frac{\hbar^2 \pi^2}{2mL_j^2} \left(n_{3v+j-3}\right)^2} = \frac{1}{3}\, \overline{E_r} = \frac{E}{3} \qquad (8.13)$$

This averaging presupposes that after each displacement of the piston an equilibrium state is restored. For this, scattering processes between the particles are required (Chapter 43) which are otherwise neglected in the ideal gas model.

With (8.13), (8.12) becomes

$$\bar{dW}_{\text{q.s.}} = -\frac{2}{L_1}\frac{E}{3}\, dL_1 = -\frac{2}{3}\frac{E}{V}\, dV \qquad (8.14)$$

During the volume change, the respective values of E/V must be used. The derivation of (8.14) does not exclude that heat is also added during the process. Since the averaging assumes equilibrium states, this heat must be supplied quasi-statically.

The pressure P was defined microscopically defined in (8.8). With (8.14) we get

$$P = \frac{2}{3}\frac{E}{V} \qquad \text{(ideal gas)} \qquad (8.15)$$

The pressure was determined here via the change in energy: For a quasi-static volume change, the quantum numbers remain the same. Then the momenta behave like $p_{3v-2} \propto 1/L_1$. This implies $dE = \bar{dW}_{\text{q.s.}} \propto -dV$; the proportionality coefficient is the pressure.

This example has shown, how the generalized force X_i can be calculated from the microscopic structure of the system, i.e. from the $E_r(x)$. Here the system was treated quantum mechanically.

Classical treatment

For deriving the pressure P in the framework of classical mechanics, we start from Figure 8.2. The reflection of an atom from the moving wall increases the kinetic energy when the volume is reduced and decreases it when the volume increases. We calculate the corresponding energy change.

In the laboratory system LS, the gas box is at rest, and the piston moves with the velocity $u\, \boldsymbol{e}_1$ and a picked-out particle with $\boldsymbol{v} = \sum v_i\, \boldsymbol{e}_i$. For a simple treatment of the reflection, we temporarily go into the rest system RS of the piston ($u' = 0$).

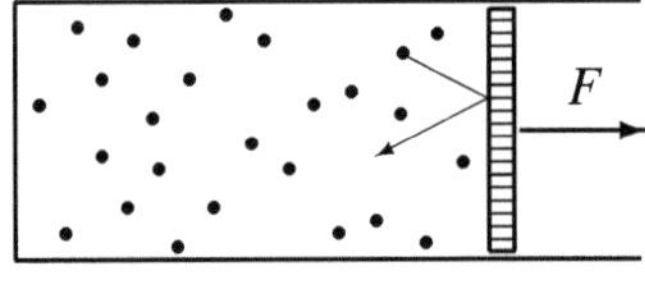
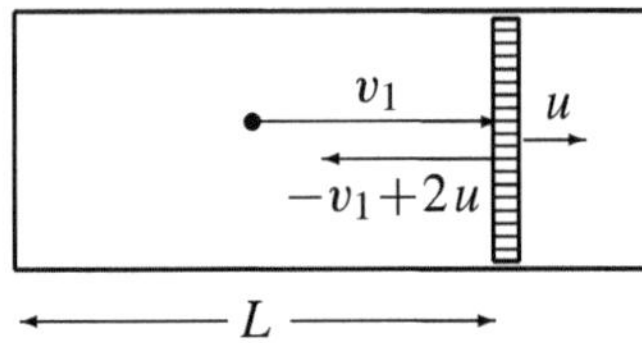

Figure 8.2 By a reflection at the piston, each particle transmits a force impulse. The time averaging over many such impulses results in a (practically constant) force F on the piston (left part). To determine this force, we consider the change in energy of the reflected particles when the piston moves with the speed $u = dL/dt$ (right part). For a selected particle, the right side shows how the velocity v_1 changes by a reflection.

In RS, the particle moves with the speed (v'_1, v'_2, v'_3) towards the piston ($v'_1 > 0$). After an elastic reflection, it then has the velocity $(-v'_1, v'_2, v'_3)$. We now go back from RS to LS; formally this is a Galilean transformation with the velocity $-u\,e_1$. From $v'_1 \to -v'_1$ in RS we get $v'_1 + u \to -v'_1 + u$ in LS. With $v_1 = v'_1 + u$ this yields

$$(v_1, v_2, v_3) \quad \overset{\text{reflection}}{\longrightarrow} \quad \big(-(v_1 - 2u), v_2, v_3\big) \tag{8.16}$$

For one reflection, the kinetic energy $\varepsilon = m\,v^2/2$ of the particle changes by

$$\Delta\varepsilon = -2m\,v_1 u + 2m u^2 \qquad (1 \text{ reflection}, v_1 > 0) \tag{8.17}$$

The time between two reflections of the particle is $\Delta t = (2L + \Delta L)/v_1 \approx 2L/v_1$. During this time the piston moves by $\Delta L = u\,\Delta t$. Thus

$$\frac{\Delta L}{L} = \frac{2u}{v_1} \qquad \text{(particle is reflected once)} \tag{8.18}$$

This shift is small, $|\Delta L|/L = 2|u|/v_1 \ll 1$. From (8.17), (8.18) and $\varepsilon_1 = m v_1^2/2$ follows

$$\Delta\varepsilon = -2\,\varepsilon_1 \frac{\Delta L}{L} + 2\,\varepsilon_1 \frac{u}{v_1} \frac{\Delta L}{L} \overset{\text{(q.s.)}}{=} -2\,\varepsilon_1 \frac{\Delta L}{L} \tag{8.19}$$

For last step the quasi-static limit case $u/v_1 \to 0$ was used. The term omitted here is dealt with in the next section; it leads to non quasi-static corrections.

From (8.18) for one reflection, we get a total of N reflection for N particles. We average over all N particles or reflections. In the quasi-static limit the system passes through a sequence of equilibrium states; therefore $\overline{\varepsilon_1} = \overline{\varepsilon}/3$. We now average (8.19), divide by $\overline{\varepsilon}$ and use $E = N\,\overline{\varepsilon}$ and $dE = N\,\overline{\Delta\varepsilon}$. This results in

$$\frac{dE}{E} = -\frac{2}{3}\frac{dL}{L} = -\frac{2}{3}\frac{dV}{V} \tag{8.20}$$

At this point we have replaced ΔE and ΔL with dE and dL; for $N \sim 10^{24}$, even for very small dL there are still many reflections. The change in energy due to the quasi-static change of an external parameter is equal to the work supplied:

$$dE = đW_{\text{q.s.}} = -\frac{2}{3}\frac{E}{V}\,dV \tag{8.21}$$

This agrees with the quantum mechanical result (8.14).

In a real gas, there are continuous collisions between the gas particles. These collisions are implicitly assumed in our derivation: If no collisions took place in the gas, then the piston motion would only transfer the kinetic energy in e_x direction; this would lead to a non-equilibrium state. In the averaging over (8.19) however, we have assumed an equilibrium state (for example by using $\overline{\varepsilon_1} = \overline{\varepsilon}/3$ has been used). Such an equilibrium is achieved by the collisions between the particles.

For (8.18), we have pretended that the picked-out particle traverses the distance $2L$ at a constant speed. In the real gas, however, collisions occur after very short time intervals; in air under normal conditions, a mean free path is of the length $\lambda \approx 10^{-7}$ m. However, this does not change the *number of reflections* on the piston: The number of reflections in the next time interval depends only on the density and velocity distribution in the immediate vicinity of the piston; these quantities are independent of λ.

Non quasi-static effects

We calculate the non quasi-static effects for a volume change that occurs at finite speed (Figure 8.3). This is a particularly transparent example for the condition "quasi-static". The piston enclosing the gas enclosing the gas shall oscillate:

$$L = L_0 + A\,\sin(\omega t) \qquad (A \ll L_0)$$
$$u = dL/dt = A\,\omega\,\cos(\omega t) \qquad (A\,\omega \ll \overline{v}) \tag{8.22}$$

As in (8.16), we consider the reflection of one particle. We determine its energy change (8.19):

$$\Delta\varepsilon = -2\,\varepsilon_1\,\frac{\Delta L}{L} + 2\,\varepsilon_1\,\frac{u}{v_1}\,\frac{\Delta L}{L} \tag{8.23}$$

In the quasi-static limit case, the second term is omitted. We now calculate the non quasi-static corrections due to this term. To do this, we average over the N particles of the gas or over N reflections at the piston. We assume a slow piston motion, $|u| \ll \overline{v}$. Then the occurring states are not far from equilibrium. This is sufficient for the averaging procedure.

For $v_1 > 0$ we obtain $\overline{\varepsilon_1/v_1} = m\,\overline{v_1}/2 \approx \overline{\varepsilon_1}/\overline{v} = \overline{\varepsilon}/(3\,\overline{v})$; a closer examination would result in an additional factor $\mathcal{O}(1)$, which we ignore here. We now proceed as in the step from (8.19) to (8.20) and obtain from (8.23)

$$\frac{dE}{E} = \underbrace{-\frac{2}{3}\frac{dL}{L}}_{\text{reversible}} + \underbrace{\frac{2}{3}\frac{u}{\overline{v}}\frac{dL}{L}}_{\text{irreversible}} \tag{8.24}$$

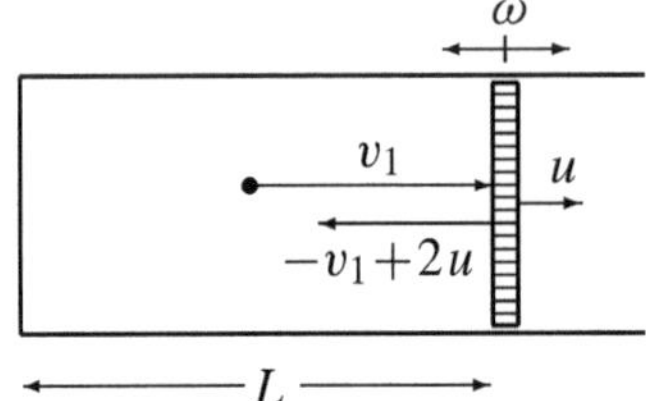

Figure 8.3 A classical ideal gas is enclosed by a piston that oscillates back and forth, $L = L_0 + A \sin(\omega t)$. For finite piston speed, the energy of the gas gradually increases. The system is thermally insulated.

We integrate this from $t = 0$ to $t = 2\pi/\omega$, i.e. over one period of the piston oscillation:

$$\ln\left(\frac{E(2\pi/\omega)}{E(0)}\right) = -\frac{2}{3}\left(\ln\frac{L(2\pi/\omega)}{L(0)}\right) + \frac{2}{3\overline{v}}\int_0^{2\pi/\omega} dt\,\frac{u(t)^2}{L(t)} \tag{8.25}$$

The terms dE/E and dL/L were integrated directly; in the last term $dL = u\,dt$ was used.

The first term on the right side of (8.25) vanishes because of $L(2\pi/\omega) = L(0)$. The term $(-2/3)\,dL/L$ in (8.24) increases the energy of the gas in the compression phase; in the expansion phase, however the energy decreases again by the same amount. This part of the process is *reversible*.

The second term on the right-hand side of (8.25) leads to an energy change $\Delta E^{(1)}_{\text{irrev}} = E(2\pi/\omega) - E(0)$; the upper index of $\Delta E^{(1)}_{\text{irrev}}$ shows the number of oscillation periods. Since this energy change is small, we may use $\ln[1 + \Delta E^{(1)}_{\text{irrev}}/E(0)] \approx \Delta E^{(1)}_{\text{irrev}}/E$. This gives us

$$\frac{\Delta E^{(1)}_{\text{irrev}}}{E} \approx \frac{2}{3\overline{v}}\int_0^{2\pi/\omega} dt\,\frac{A^2\,\omega^2\cos^2\omega t}{L_0} = \frac{2\pi}{3}\,\frac{A\,\omega}{\overline{v}}\,\frac{A}{L_0} \tag{8.26}$$

In the integral, we have set $L(t) \approx L_0$, i.e. terms of relative size $A/L_0 \ll 1$ neglected. For the remaining integral $\int dt\,\cos^2\omega t = \pi/\omega$ holds.

We evaluate (8.26) under the following assumptions: Let the gas be air at room temperature; for this $\overline{v} \approx 400\,\text{m/s}$. The piston oscillates with the speed $A\,\omega = 20\,\text{cm/s}$ and the amplitude $A = L_0/10$. Then the energy E of the gas increases by 1% after 100 piston oscillations:

$$\frac{\Delta E^{(100)}_{\text{irrev}}}{E} \approx 0.01 \tag{8.27}$$

The energy E consists of the (disordered) kinetic energy of the gas molecules. Therefore, in this process, work (piston motion) is converted into heat. Such a process is *irreversible*; the heat cannot be converted back into work (at least not completely). The increase in energy comes from the term labeled *irreversible* in (8.24); accordingly, the index "irrev" was used in (8.26) and (8.27). Additional irreversible effects may result from the friction of the piston on the vessel wall.

Formally, the irreversibility is caused by the averaging procedure in the step from (8.23) to (8.24). The averaging implies that the energy increase due to the second term in (8.23) is continuously distributed to all degrees of freedom. This happens in the real gas by collisions.

The reversible/irreversible classification made here is generalized in Chapter 12 for arbitrary processes. The concepts of reversible (irreversible) and quasi-static (non quasi-static) processes is discussed in more detail there.

The energy $\Delta E_{\text{irrev}}^{(1)}$ is transferred to the gas during an oscillation period $T = 2\pi/\omega$. From this follows

$$\frac{dE_{\text{irrev}}}{dt} = \Delta E_{\text{irrev}}^{(1)} \frac{\omega}{2\pi} \propto A^2 \omega^2 \propto \left(\frac{dL}{dt}\right)^2 \tag{8.28}$$

In a quantum mechanical perturbation theory, the time dependence of a parameter $x(t)$ in $H(x(t))$ leads to transition probabilities per time, which are proportional to $(dx/dt)^2$. These transitions result in an irreversible energy transfer per time, which is proportional to $(dx/dt)^2$. In this respect, the result (8.28) is representative for a large class of processes.

To summarize, we state:

- Exact reversibility only applies for the limit $dx/dt \to 0$.

This is also the condition required for a quasi-static process. In addition to "very slow", we require a sequence of equilibrium states.

Work for arbitrary expansion

We discuss the possible values of the work $\text{d}W$ for an arbitrary expansion process. Without irreversible parts we obtain $\text{d}W = \text{d}W_{\text{q.s.}}$. Irreversible energy transfers due to the piston motion are according to (8.24) of the form $dE_{\text{irrev}}/E \sim u\,dL/(\overline{v}L) \geq 0$. This means that they are non-negative (the pre-factor, $\overline{v}$ and L are positive, and $u = dL/dt$ and dL have the same sign). Therefore we obtain in general

$$\boxed{\text{d}W \geq \text{d}W_{\text{q.s.}} \qquad \text{arbitrary process}} \tag{8.29}$$

For such a process, several external parameters could be altered, or additional heat could be transferred.

We consider (8.29) for an expansion process from V_a to $V_b > V_a$, which is realized with the piston velocity ($u = \Delta L/\tau_{\text{exp}}$). For $u \ll \overline{v}$ the process is (approximately) quasi-static. For $|u|/\overline{v} \to \infty$, the piston moves so quickly that no particles are reflected at all when the piston is pulled out. Then the work is zero, because without reflections no force acts on the piston. (In practice, this free expansion is realized with a throttle valve, Figure 16.1). We supplement these two limiting cases with a third scenario:

$$\Delta W = \begin{cases} \Delta W_{\text{q.s.}} < 0 & u \ll \overline{v} \\ 0 & u \gg \overline{v} \\ n \cdot \Delta E_{\text{irrev}}^{(1)} & n \text{ oscillations} \end{cases} \tag{8.30}$$

The last case refers to (8.26). It means that the work has no upper limited; there is only a lower limit, (8.29).

Exercises

8.1   Ideal gas in a sphere

An ideal gas enclosed in a sphere (radius R) shall be treated quantum mechanically. The radial part of the Schrödinger equation is solved by the spherical Bessel functions whose zeros can be assumed to be known. For this special geometry, determine the pressure $P = -\overline{\partial E_r(V)/\partial V}$ as a function of E and V.

8.2   Heating in winter

When asked "Why do we heat in winter", the layman answers: "To increase the temperature of the indoor air". With regard to the first law, a physicist could perhaps answer: "We add the amount of heat ΔQ to increase the internal energy E of the air in the room". Calculate the energy change of the room air for the temperature increase ΔT, where the air is treated as an ideal gas with $c_V = $ const.. Note that the pressure $P = P_0$ is kept constant at ambient pressure due to the cracks in doors and windows.

9 Entropy and temperature

The entropy and temperature are defined microscopically. The central result is Boltzmann's formula $S = k_B \ln \Omega$ which connects the entropy S with the number Ω of the accessible microstates.

Historically, temperature and entropy were introduced as macroscopic observables (Chapter 14). Their microscopic significance ("heat is the disordered motion of atoms") was recognized only later.

We consider two systems A and B that are in thermal contact (Figure 9.1). Both subsystems shall have many degrees of freedom. Examples are a copper rod A in a volume of water B, or two volumes with the gases A and B. We neglect the heat capacity of the enclosing boxes. The external parameters of A and B shall be fixed.

Let the entire system be closed, so that the energy E is conserved:

$$E = E_A + E_B = \text{const.} \tag{9.1}$$

We investigate the question of how the energy is divided into E_A and E_B in equilibrium. Since all external parameters are constant, the energy exchange between A and B is a pure heat exchange.

In equilibrium, all $\Omega_0(E)$ microstates of the overall system are equally probable, $P_r = 1/\Omega_0(E)$. We determine the probability $W(E_A)$ that system A has the energy E_A. This is the sum of the probabilities P_r for all microstates r, for which the energy is divided according to $E = E_A + E_B$:

$$W(E_A) = \sum_{r\,:\,E_A, E_B} P_r = \sum_{E_{r,A}=E_A} \sum_{E_{r,B}=E_B} \frac{1}{\Omega_0(E)} = \frac{\Omega_A(E_A)\,\Omega_B(E - E_A)}{\Omega_0(E)} \tag{9.2}$$

Here, Ω_A and Ω_B are the partition functions of the subsystems. The constant external parameters were not written on. By $E_{r,A} = E_A$ the restriction of the sum to $E_A - \delta E \leq E_{r,A} \leq E_A$ is meant, where the index of $E_{r,A}$ denotes a microstate of system A. This sum results in $\Omega_A(E_A)$; the sum over the microstates of B with $E_{r,B} = E_B$ then results in $\Omega_B(E_B)$. The product $\Omega_A \Omega_B$ is the number of states of the total system with a given distribution $E_A + E_B$. The probability $W(E_A)$ is the number of positive events (microstates with a given energy distribution) divided by the total number of events (Ω_0 microstates).

To discuss the function $W(E_A)$ we use the energy dependence (6.21):

$$\Omega(E) = c\,E^{\gamma f} \tag{9.3}$$

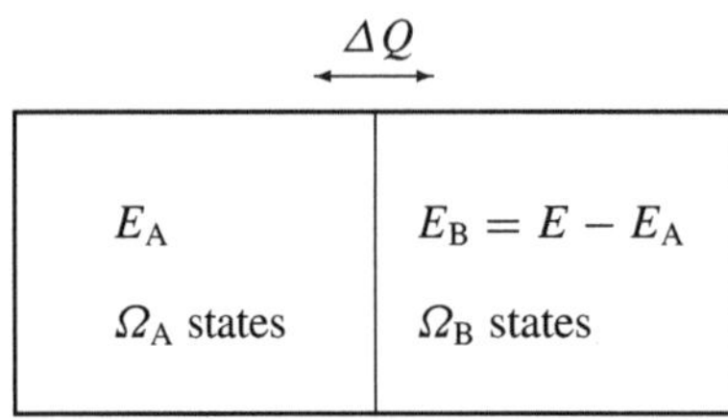

Figure 9.1   A closed system consists of two macroscopic subsystems that can exchange heat. How is the total energy $E = E_A + E_B$ distributed into A an B in thermal equilibrium?

Here f is the number of degrees of freedom, and $\gamma = \mathcal{O}(1)$ is a number. Since $\Omega(E)$ depends strongly on E we consider the logarithm of $W(E_A)$, i.e.

$$
\begin{aligned}
\ln W(E_A) &= \ln \Omega_A(E_A) + \ln \Omega_B(E - E_A) - \ln \Omega_0(E) \\
&= \gamma f_A \ln E_A + \gamma f_B \ln(E - E_A) + \text{const.} \tag{9.4}
\end{aligned}
$$

Since we are only considering the dependency on E_A, $\Omega_0(E)$ is a constant. The function $\ln W(E_A)$ is sketched in Figure 9.2. The range to be examined is $0 \leq E_A \leq E$; at the edge of this range, $\ln W(E_A)$ goes towards $-\infty$. In between, $\ln W(E_A)$ is continuously differentiable. Since

$$
\frac{d \ln W(E_A)}{d E_A} = \frac{\gamma f_A}{E_A} - \frac{\gamma f_B}{E - E_A} = 0 \tag{9.5}
$$

has only one solution, $\ln W(E_A)$ must have a maximum here. The position $\overline{E_A}$ of the maximum follows from (9.5)

$$
\frac{\overline{E_A}}{f_A} = \frac{E - \overline{E_A}}{f_B} = \frac{\overline{E_B}}{f_B} \tag{9.6}
$$

We expand the function $\ln W(E_A)$ into a Taylor series:

$$
\ln W(E_A) = \ln W(\overline{E_A}) - \frac{(E_A - \overline{E_A})^2}{2\,\Delta E_A^2} \pm \ldots \tag{9.7}
$$

The error due to the truncation of the higher terms is negligible for macroscopic systems; the reasoning for this is the same as for the derivation of the normal distribution for $W_N(n)$ in Chapter 3. The quantity ΔE_A is given by the second derivative of $\ln W(E_A)$:

$$
\frac{1}{\Delta E_A^2} = -\left(\frac{d^2 \ln W}{d E_A^2} \right)_{\overline{E_A}} \overset{(9.4)}{=} \frac{\gamma f_A}{E_A^2} + \frac{\gamma f_B}{E_B^2} \tag{9.8}
$$

From (9.7) we obtain

$$
W(E_A) = W(\overline{E_A}) \, \exp\left(-\frac{\left(E_A - \overline{E_A}\right)^2}{2\,\Delta E_A^2} \right) \tag{9.9}
$$

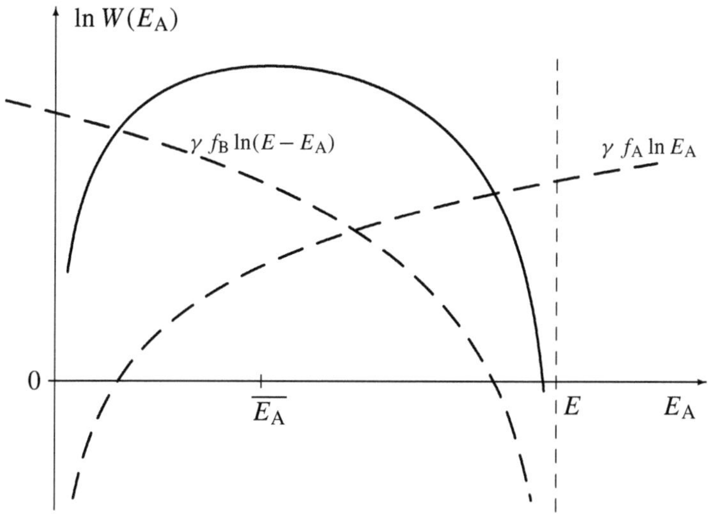

Figure 9.2 The function $\ln W(E_A)$ from (9.4) has exactly one maximum in the range $0 < E_A < E$. In the neighborhood of the maximum the function can be approximated by a Taylor series.

From this one can see that the maximum position $\overline{E_A}$ is also the mean value, and that the ΔE_A is the width of the distribution. With the estimate

$$\Delta E_A = \left(\frac{\gamma f_A}{\overline{E_A}^2} + \frac{\gamma f_B}{\overline{E_B}^2} \right)^{-1/2} < \frac{\overline{E_A}}{\sqrt{\gamma f_A}} \tag{9.10}$$

we see that for a many-body system (e.g. with $f_A = \mathcal{O}(10^{24})$), the relative width of the distribution is extraordinarily sharp:

$$\frac{\Delta E_A}{\overline{E_A}} < \frac{1}{\sqrt{\gamma f_A}} \sim 10^{-12} \tag{9.11}$$

The distribution (9.9) is sketched in Figure 9.3.

The following argument is based on the sharpness of the distribution, i.e. the smallness of $\Delta E_A / \overline{E_A}$. Therefore, we exclude systems with $f_A < 100$ (for example an atomic nucleus); on the other hand, a tiny piece of matter with 10^{-10} grams is readily acceptable ($f_A = \mathcal{O}(10^{12})$).

Figure 9.3 and equation (9.11) show that (almost) all of the $\Omega_A \Omega_B = \Omega_0 W(E_A)$ microstates are located at $E_A = \overline{E_A}$. Therefore, the *equilibrium* is given by

$$\ln(\Omega_A \Omega_B) = \ln \Omega_A(E_A) + \ln \Omega_B(E - E_A) = \text{maximum} \tag{9.12}$$

The maximum lies at $E_A = \overline{E_A}$. According to (9.6) we obtain

$$\frac{\overline{E_A}}{f_A} = \frac{\overline{E_B}}{f_B} \qquad \begin{array}{l}\text{(equilibrium condition}\\ \text{under heat exchange)}\end{array} \tag{9.13}$$

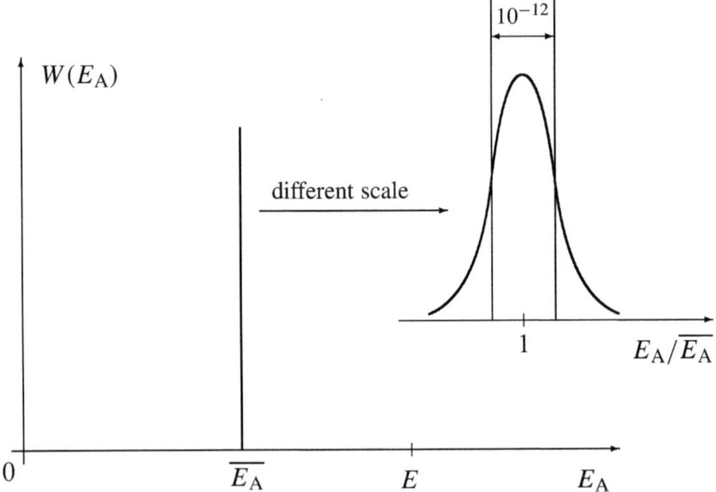

Figure 9.3   The Gaussian function $W(E_{\rm A})$ is so narrow that it appears as a line. The enlarged section (upper right, different scale) illustrates that the width is even much smaller than the line width shown.

This means: Admitting heat exchange, the energy is distributed such that the number of possible states is maximized. With this distribution, the energy per degree of freedom is the same in both subsystems.

We now express the central results of this chapter, (9.12) and (9.13), by the entropy and the temperature. To do this, we define the *entropy S* of an equilibrium system by

$$\boxed{S = S(E, x) = k_{\rm B} \ln \Omega(E, x) \qquad \text{definition of entropy}} \tag{9.14}$$

and the *temperature T* by

$$\boxed{\frac{1}{T} = \frac{1}{T(E, x)} = \frac{\partial S(E, x)}{\partial E} \qquad \text{definition of temperature}} \tag{9.15}$$

Here $\Omega(E, x)$ is the partition function of the considered system. In many applications, the external parameters $x = (V, N)$ are the volume and the number of particles; in general, they are the parameters on which the Hamilton operator $H(x)$ depends. The *Boltzmann constant* $k_{\rm B}$ in (9.14) is an initially arbitrary constant.

With this definition, (9.12) becomes

$$S(E_{\rm A}) = S_{\rm A}(E_{\rm A}, x) + S_{\rm B}(E - E_{\rm A}, x') = \text{maximum} \tag{9.16}$$

Here $S(E_{\rm A}) = k_{\rm B} \ln(\Omega_{\rm A}\Omega_{\rm B})$ denotes the sum of the entropies as a function of $E_{\rm A}$. The external parameters (x for A and x' for B, constant in the present discussion) were not written on in $S(E_A)$. From $\partial S(E_A)/\partial E_A = 0$ then follows $\partial S_A/\partial E_A - \partial S_B/\partial E_B = 0$, and thus

$$T_A = T_B \qquad (9.17)$$

Equations (9.16) and (9.17) are the equilibrium conditions for the case that the two systems can exchange heat. If we bring two systems with different temperatures into contact, then the temperature is equalized by heat exchange. This means that the temperature difference can be seen as the driving force for heat exchange.

For the entropy and temperature introduced in this way, we state:

1. The definitions of entropy and temperature refer to the equilibrium state of a many-particle system. Only for this macrostate the probabilities $P_r = 1/\Omega$ are given by the partition function Ω. However, these definitions can be applied to local equilibria, too (point 4).

2. Entropy and temperature are observables, they can be measured macroscopically (Chapter 14).

3. By $S(E, x) = k_B \ln \Omega(E, x)$ the connection between the microscopic structure (Ω) and the macroscopic observables S and T is established. This central equation for statistical physics was established in 1877 by Boltzmann.

4. Entropy is additive. For two subsystems $\Omega = \Omega_A \Omega_B$ holds, i.e.

$$S = k_B \ln \Omega = k_B \ln(\Omega_A \Omega_B) = k_B \left(\ln \Omega_A + \ln \Omega_B \right) = S_A + S_B \qquad (9.18)$$

In Figure 9.1, there might be initially separate equilibria in the volumes A and B. Then the temperatures (T_A and T_B) and the entropies (S_A and S_B) are defined for each subsystem. In this way, the temperature and entropy definition can be applied to *local equilibria*.

For $T_A \neq T_B$ the overall system is in a non equilibrium state; no temperature T is defined for of the combined system. The entropy S of the overall system can be expressed as a sum, $S = S_A + S_B$.

If an energy exchange is allowed for the two systems with initially separate equilibria (with different temperatures), then the overall system adjusts itself to a global equilibrium with $T = T_A = T_B$ and $S = S_A + S_B = $ maximum.

5. *Entropy is a measure of the disorder of the system.*

   Perfect order is given if there is only one possible microstate of the system, $\Omega = 1$ and $S = 0$. The more microstates Ω are accessible, the more disordered the equilibrium state is; because in the equilibrium state, all Ω states equally probable.

6. *The temperature corresponds the energy per degree of freedom.*

With (9.3) we obtain from (9.15):

$$\frac{1}{\beta} = k_{\rm B}T = \frac{E}{\gamma f} \tag{9.19}$$

Here β is the commonly used abbreviation for $1/(k_{\rm B}T)$. Except for the constant $k_{\rm B}$ and the numerical factor γ, the temperature is equal to the energy per degree of freedom. It would therefore be natural and reasonable to set $k_{\rm B} = 1$ and to measure the temperature in joules. Any other definition for T implies a some constant $k_{\rm B}$; its dimension is then $[k_{\rm B}] = [\text{energy}]/[T]$.

7. In (9.19), E is the energy available for excitations; it has the minimum value zero. Therefore

$$T \geq 0 \tag{9.20}$$

We refer to the temperature $T = 0$ as (absolute) zero.

For an ideal monatomic gas, $\Omega \propto E^{3N/2}$, i.e. $\gamma f = 3N/2$. From (9.19) then follows $E/N = 3k_{\rm B}T/2$. On the other hand, E/N is equal to the mean kinetic energy $\overline{p^2}/2m$ per particle. From $\overline{p^2} \geq 0$ follows $T \geq 0$.

The statement "$S = $ maximum at equilibrium" implies that the entropy is smaller for non-equilibrium states. Let us look again at Figure 9.1. The energy exchange between A and B shall take place so slowly that each of the two subsystems passes through a sequence of equilibrium states. In this case, the entropies $S_{\rm A}$ and $S_{\rm B}$ of the subsystems are defined during the process, i.e. for different values of $E_{\rm A}$. We can then calculate the entropy of the overall system $S(E_{\rm A}) = S_{\rm A}+S_{\rm B}$ at any time although the system as a whole is not in equilibrium. Then the total entropy S has initially not its maximum value, but it approaches it by heat exchange.

Prerequisites for the definition of entropy and temperature

The definitions (9.14), (9.15) of S and T assume a many-particle system in (at least local) equilibrium.

Only the equilibrium state is defined via $P_r = 1/\Omega$ where Ω is the number of accessible microstates. An arbitrary macrostate, on the other hand, can have arbitrary P_r's. The quantities defined via $\Omega(E, x)$, i.e. the entropy S and the temperature T have a physical meaning for equilibrium states. Therefore, these variables can be connected to observables for equilibrium states (Chapter 14).

The definitions of S and T require a system with a very large number of particles. Only then is the width (9.11) is so small, so that (almost) all microstates lie at the maximum of the distribution $W(E_{\rm A})$. Only in this case, the equilibrium is defined by the position of the maximum itself, i.e. by $S = $ maximum.

We discuss the equilibrium condition using an example. A copper rod is heated at one end. Then the temperature in the rod is initially position and time dependent; the system is not in equilibrium. Since we have defined the temperature for equilibrium states, the question arises what we mean by temperature in this case. For this purpose, one may consider small regions (for example, subvolumes with $V = 1\,\text{mm}^3$ for the copper rod) in which a *local* equilibrium against heat exchange is established relatively quickly. In such a subvolume there are still very many particles, so that the temperature definition can be applied. The temperature defined in this way is then position and time dependent. The variation with time shall be so slow that local subvolumes pass through a sequence of equilibrium states.

A position and time dependent entropy can be defined in the same way. This is comparable to the specification of S_A and S_B in (9.18) for $E_A \neq \overline{E_A}$. For a specific division of the energy $E = E_A + E_B$, the subsystems can each be in equilibrium separately, but not the overall system. These two subsystems can be seen as a schematic simplification of the heated copper rod; instead of different small regions with a local equilibrium, only two parts are considered. If the copper rod is isolated, a global equilibrium establishes itself after some time. This adjustment takes place through heat exchange. At equilibrium, the entropy of the rod is then maximum and the temperature is the same everywhere.

Another example is the atmosphere. Obviously the temperature here depends on position and time. First of all, the temperature can again be defined for a small volume (let us say for $1\,\text{m}^3$) in which there is equilibrium under heat exchange. In this sense, the specification of a position and time dependent temperature $T = T(\mathbf{r}, t)$ in meteorology is to be understood. Because of $T \neq$ const., the atmosphere as a whole is not in equilibrium with respect to heat exchange. This is not surprising, as it is not a closed system (Sun!). The copper rod would also not be in equilibrium if it is continuously heated at one end. In the atmosphere, the pressure $P = P(\mathbf{r}, t)$ is also position and time dependent on position and time; the system is not in equilibrium with respect to volume exchange (Chapter 10).

Relaxation times

In a quasi-static process, the system runs through a sequence of equilibrium states. The process could be triggered by a heat supply or by the change of external parameters (Chapter 8). Only if these changes are very slow, the temperature and the generalized forces (especially the pressure) of the system are defined at all times. We discuss the condition "quasi-static" using a few examples. This discussion is qualitative and uses terms (heat conduction) that are introduced later.

We denote the time scale on which the external intervention occurs as by τ_{exp}; this could be, for example, the time during which the volume of a gas changes by $\Delta V = V/10$ or during which the heat $\Delta Q = E/10$ is supplied. The time after which a system returns to equilibrium (after a sudden or temporary disturbance) is referred to as *relaxation time* τ_{relax}. The condition "quasi-static" then means

$$\boxed{\quad \frac{\tau_{\text{exp}}}{\tau_{\text{relax}}} \to \infty \qquad \begin{array}{l}\text{condition for} \\ \text{quasi-static}\end{array}\quad} \tag{9.21}$$

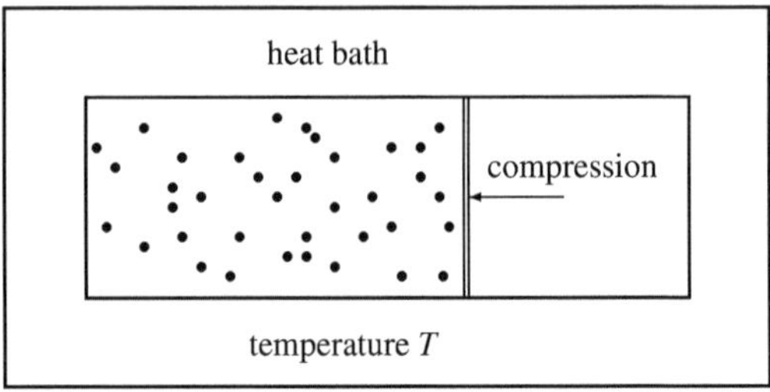

Figure 9.4   A gas that has reached the temperature T in contact with the heat bath, is compressed within a time span of one second. The condition "quasi-static" is fulfilled for the gas itself; the gas passes through a sequence of equilibrium states. Since the gas heats up during compression, it is no longer in equilibrium with the heat bath; the temperature equalization with the heat bath takes much longer than one second. For the overall system (gas and heat bath), the process is therefore not quasi-static.

As an example, we consider a gas whose volume is reduced by 10% during the time span

$$\tau_{\exp} \sim 1\,\mathrm{s} \tag{9.22}$$

First, the gas is compressed at the position of the piston; this compression propagates as a density wave with the speed of sound $c_s \approx 300\,\mathrm{m/s}$. For a system with an expansion of $L = 30\,\mathrm{cm}$, this takes about

$$\tau_{\mathrm{relax}} = \mathcal{O}(L/c_s) \approx 10^{-3}\,\mathrm{s} \tag{9.23}$$

Since the sound waves are attenuated, after a few time spans L/c_s a homogeneous macroscopic density is reached; therefore L/c_s is a rough estimate for the relaxation time of the gas. As shown in (8.22) – (8.27), there are non quasi-static (irreversible) effects of the magnitude $\tau_{\mathrm{relax}}/\tau_{\exp} \approx 10^{-3}$. (The damping of the sound wave implies a conversion of the work performed on the piston into heat). The process under consideration is approximately quasi-static because the irreversible effects are small.

We now consider the same experiment, but place the gas vessel in a heat bath (Figure 9.4). The system is initially in equilibrium; the gas has the temperature T of the of the heat bath. If we now compress the gas within $\tau_{\exp} = 1\,\mathrm{s}$, it comes to a higher temperature T'; it is no longer in equilibrium with the heat bath. The temperature equalization might take place in the time span

$$\tau'_{\mathrm{relax}} \approx 10^2\,\mathrm{s} \tag{9.24}$$

With poor thermal contact (for example polystyrene walls), $\tau'_{\mathrm{relax}} \gg 10^2\,\mathrm{s}$ is also possible. In any case, $\tau'_{\mathrm{relax}} \gg \tau_{\exp}$ applies. The system "gas and heat bath" is therefore not in equilibrium during compression, the process is not quasi-static with respect to the total system. Because of

$$\tau_{\exp} \ll \tau'_{\mathrm{relax}} \tag{9.25}$$

however, we can consider the experiment as if there were a thermal insulation between the gas and the heat bath. Then the process of compression can be treated approximately as quasi-static and adiabatic. If the subsequent temperature equalization between the gas and the heat bath is slow, then the gas passes again goes through a sequence of equilibrium states.

We can therefore treat processes with $\tau_{\mathrm{exp}} \gg \tau_{\mathrm{relax}}$ as well as $\tau_{\mathrm{exp}} \ll \tau_{\mathrm{relax}}$ with the help of equilibrium states. Processes with $\tau_{\mathrm{exp}} \approx \tau_{\mathrm{relax}}$, on the other hand, are comparatively complicated. In such cases, we usually limit ourselves to statements about the initial and the final state.

Infinitely large relaxation times

Up to now, we have assumed that a closed system reaches the equilibrium state after a sufficiently long time. There are, however, special systems with an infinitely long relaxation time. We give two examples.

There are systems that have a practically stable state at a local maximum entropy. An example of this is a diamond. Converting the diamond into graphite would increase the entropy. However, this transformation is hindered and does not occur by itself; the associated relaxation time is infinite. Within the diamond (as a closed system), however, an equilibrium is reached after a relatively short time; this refers to all degrees of freedom of the system (e.g. for the lattice vibrations).

Another case are systems with hysteresis effects, in which the actual state depends on the previous history. The best known example is a ferromagnet. The magnetization (generalized force) can vary for different, practically stable values for given parameters (magnetic field). In this case, too, the relaxation time to (absolute) equilibrium is infinite. The change in the external magnetic field B leads to irreversible effects that do not disappear even in the limit $\dot{B} \to 0$.

With regard to these special cases, the experimental condition condition "infinitely slow" does not necessarily imply "sequence of equilibrium". The term "quasi-static" defined in Chapter 8 is therefore a stronger condition than "infinitely slow". Only processes that are quasi-static in this sense (sequence of equilibrium states) are also reversible (Chapter 12).

Exercises

9.1   Entropy change during mixing

Two ideal monatomic gases both have the temperature T, the particle number N and the volume V. The two volumes are adjacent to each other. The wall between them is now pulled out to the side so that the gases can mix. How large is the entropy change in this process if the gas atoms are (i) of the same kind or (ii) different kind (e.g. helium and argon gas)? The partition function $\Omega(E, V, N)$ of an ideal gas is assumed to be known.

10 Generalized forces

The last chapter derived the statement "$S = maximum$" for the equilibrium under heat exchange. Here we show that "$S = maximum$" holds generally for the equilibrium of a closed system.

Temperature is the driving force for heat exchange. Similarly, the generalized force X_i (such as the pressure) is the driving force for an x_i exchange (such as a volume exchange). The temperature is given by the derivative of the entropy $S(E, x)$ with respect to the energy E. Analogously, the generalized force X_i is the derivative of the entropy with respect to x_i.

General probability distribution

We generalize (9.9) to the probability distribution $W(\xi)$ for an arbitrary extensive macroscopic quantity ξ. By extensive (or additive) we understand that ξ is proportional to the number of particles N. The equilibrium is determined by the maximum of $W(\xi)$ or of the entropy. We refer to this equilibrium as *statistical* because it is based on the assumption of equal probabilities for all accessible microstates, i.e. on the fundamental postulate.

The macroscopic state of a closed system shall depend (in addition to the energy E) on the macroscopic quantity ξ. All other variables shall be constant and are suppressed in the notation. We start from the entropy

$$S = S(E, \xi) \tag{10.1}$$

and investigate which values ξ assumes in statistical equilibrium. In equilibrium, all $\Omega(E, \xi)$ microstates are equally probable. The probability $W(\xi)$ for a certain ξ value is therefore proportional to the number $\Omega(E, \xi)$ of microstates with this value:

$$W(\xi) = C\,\Omega(E, \xi) = C\,\exp\left(\frac{S(E, \xi)}{k_\mathrm{B}}\right) \tag{10.2}$$

Here we used $S = k_\mathrm{B} \ln \Omega$ where C is a constant. We expand the logarithm of $W(\xi)$ around an initially arbitrary point $\bar{\xi}$:

$$\ln W(\xi) = \ln W(\bar{\xi}) + \frac{1}{k_\mathrm{B}}\left(\frac{\partial S}{\partial \xi}\right)_{\bar{\xi}}(\xi - \bar{\xi}) + \frac{1}{2k_\mathrm{B}}\left(\frac{\partial^2 S}{\partial \xi^2}\right)_{\bar{\xi}}(\xi - \bar{\xi})^2 + \ldots \tag{10.3}$$

We now choose $\bar{\xi}$ as the value of the maximum of $S(E, \xi)$:

$$\left(\frac{\partial S(E, \xi)}{\partial \xi}\right)_{\xi = \bar{\xi}} = 0 \tag{10.4}$$

The result will show that $\bar{\xi}$ is equal to the *mean value*. The second derivative defines the *fluctuation* around the mean value (or the deviations from $\bar{\xi}$):

$$\Delta \xi = \sqrt{-\frac{k_B}{(\partial^2 S/\partial \xi^2)_{\bar{\xi}}}} \qquad \text{(mean deviation)} \tag{10.5}$$

From (10.3)–(10.5) we obtain the desired probability distribution

$$W(\xi) = \frac{1}{\sqrt{2\pi}\,\Delta \xi}\, \exp\left(-\frac{(\xi - \bar{\xi})^2}{2\,\Delta \xi^2}\right) \tag{10.6}$$

The prefactor follows from the normalization $\int d\xi\, W(\xi) = 1$.

Since the macroscopic quantity ξ is extensive, $\xi = \mathcal{O}(N)$. From (10.5) and $S = \mathcal{O}(N)$ then follows $\Delta \xi = \mathcal{O}(N^{1/2})$ and thus

$$\frac{\Delta \xi}{\bar{\xi}} = \mathcal{O}\left(1/\sqrt{N}\right) \tag{10.7}$$

This is also the size of the ratio of the $(n+1)$-th to n-th term in the Taylor expansion (10.3). Therefore the termination at the quadratic term is justified.

The smallness of the relative fluctuation (10.7) implies that almost all microstates are at the maximum (mathematically, this is another outcome of the law of large numbers). Therefore, the maximum determines the equilibrium state

$$\boxed{\; S(E, \xi) = \text{maximum} \qquad \begin{array}{l}\text{equilibrium state}\\ \text{in closed system}\end{array} \;} \tag{10.8}$$

This refers to the maximum of $S(E, \xi)$ with respect to ξ; the energy E is constant in the closed system.

Discussion

If ξ is an exchange quantity between two subsystems the condition (10.4) states the equality of the corresponding forces. This means in particular: Same temperature for heat exchange (Chapter 9), the same pressure for volume exchange (below) and the same chemical potential for particle exchange (Chapter 20).

In statistical equilibrium, the possible values of ξ are close to $\bar{\xi}$. However, the ξ values may slightly deviate from $\bar{\xi}$; they may fluctuate around the mean value $\bar{\xi}$. The size of these *fluctuations* is defined by the mean square deviation $\Delta \xi$.

The probability $W(\xi)$ refers to a statistical ensemble of many identical systems (Chapter 5). It can be also understand as a time average: We consider the gas shown in Figure 2.2. In this system, individual gas particles continuously change between the subvolumes. The left subvolume sometimes contains more, sometimes fewer particles. The actual number $\xi = N_{\text{left}}$ of particles therefore fluctuates over time around the mean value $\overline{N}_{\text{left}}$.

For macroscopic quantities, the relative fluctuations are small, (10.7). This does not apply to microscopic quantities. For example, the relative fluctuation of the energy of a single gas particle is of the size 1. The particle repeatedly collides with other particles and changes its energy thereby significantly. A probability distribution can also be given for this single particle (Chapter 22); but it is not a Gaussian distribution like (10.6).

Because of the smallness of the relative fluctuations (10.7), they may be neglected in most applications. In particular, the state of equilibrium is completely determined by specifying the mean values of a suitable set of macroscopic variables.

The fluctuations $\xi = \overline{\xi} + \delta\xi$ imply fluctuations of the entropy:

$$S(E, \overline{\xi} + \delta\xi) \approx S(E, \overline{\xi}) + \frac{1}{2}\left(\frac{\partial^2 S}{\partial \xi^2}\right)_{\overline{\xi}} (\delta\xi)^2 = S_{\text{max}} - k_{\text{B}} \left(\frac{\delta\xi}{\Delta\xi}\right)^2 \qquad (10.9)$$

According to (10.6), deviations from the mean value are of the size $\delta\xi = \mathcal{O}(\Delta\xi)$. These deviations of the entropy from its maximum and equilibrium value are of the size k_{B}. Due to $S_{\text{max}} = \mathcal{O}(Nk_{\text{B}})$, these deviations are very small, $\delta S / S_{\text{max}} = \mathcal{O}(1/N)$.

Definition of the generalized forces

We consider the dependence of the entropy on the external parameters x_i and determine the partial derivatives $\partial S/\partial x_i = k_{\text{B}}\, \partial \ln \Omega / \partial x_i$. This leads to a general definition of the generalized forces.

The microcanonical partition function was defined in (5.19),

$$\Omega(E, x) = \Omega(E, x_1, x_2, ..., x_n) = \sum_{r:\, E-\delta E\, \leq\, E_r(x)\, \leq\, E} 1 \qquad (10.10)$$

For the parameter x_1, the partial derivative is given by

$$\frac{\partial \ln \Omega(E, x)}{\partial x_1} = \frac{\ln \Omega(E, x_1 + dx_1, x_2, ..., x_n) - \ln \Omega(E, x_1, x_2, ..., x_n)}{dx_1} \qquad (10.11)$$

We first calculate

$$\Omega(E, x_1 + dx_1, ..., x_n) = \sum_{r:\, E-\delta E\, \leq\, E_r(x_1+dx_1,...,x_n)\, \leq\, E} 1$$

$$= \sum_{r:\, E-\delta E \,\leq\, E_r(x)+dE_r \,\leq\, E} 1 \;=\; \sum_{r:\, E-\overline{dE_r}-\delta E \,\leq\, E_r(x) \,\leq\, E-\overline{dE_r}} 1$$

$$= \Omega(E - \overline{dE_r}, x_1,..., x_n) \tag{10.12}$$

The displacement $dE_r = (\partial E_r/\partial x_1)\,dx_1$ was replaced by its mean value $\overline{dE_r}$. This is possible because the sum runs over the very many microstates in an interval of size δE around E. This corresponds to an average with the $P_r(E, x)$ from (5.20). The mean energy shift $\overline{dE_r}$ can be expressed by the generalized force X_1, (8.2):

$$\overline{dE_r} = \overline{\frac{\partial E_r(x)}{\partial x_1}}\, dx_1 = -X_1\, dx_1 \tag{10.13}$$

We set $dx_1 = -\overline{dE_r}/X_1$ and insert (10.12) into (10.11):

$$\frac{\partial \ln \Omega(E, x)}{\partial x_1} = -\frac{\ln \Omega(E - \overline{dE_r}, x) - \ln \Omega(E, x)}{\overline{dE_r}/X_1}$$

$$= \frac{\partial \ln \Omega(E, x)}{\partial E}\, X_1 = \beta\, X_1 \tag{10.14}$$

Here $\beta = \partial \ln \Omega/\partial E = 1/k_\mathrm{B} T$. We now replace x_1 with x_i and obtain

$$\boxed{X_i = k_\mathrm{B} T\, \frac{\partial \ln \Omega(E, x)}{\partial x_i} \qquad \text{generalized force}} \tag{10.15}$$

In the following, we consider this as the basic definition of the generalized forces. The previous definition (8.2) referred to the energies $E_r(x)$ of the microstates. The definition (10.15) is more general and also easier to evaluate. It is of the same kind as the temperature definition (9.15).

From the microscopic structure (i.e. from $\Omega(E, x)$) we can now determine the following macroscopic quantities: The entropy $S(E, x) = k_\mathrm{B} \ln \Omega$, the temperature $1/T(E, x) = \partial S/\partial E$ and the generalized forces $X_i(E, x) = T\, \partial S/\partial x_i$. All these macroscopic quantities are functions of the energy E and the external parameters x. We will often consider $x = V$ as the only external parameter. In this case, the relevant macroscopic quantities are

$$S(E, V) = k_\mathrm{B} \ln \Omega(E, V)\,, \qquad \frac{1}{T} = \frac{\partial S(E, V)}{\partial E}\,, \qquad \frac{P}{T} = \frac{\partial S(E, V)}{\partial V} \tag{10.16}$$

Other examples of generalized forces are the magnetization M for the magnetic field $x = B$, see (8.5), and the chemical potential μ for the number of particles $x = N$ (Chapter 20, 21).

Volume exchange

If the external parameter x_i is an exchange parameter between two subsystems, then the equilibrium condition (10.8) becomes the equality of the generalized forces of the two systems. One of the most important external parameters is the volume. We consider the volume exchange between two subsystems.

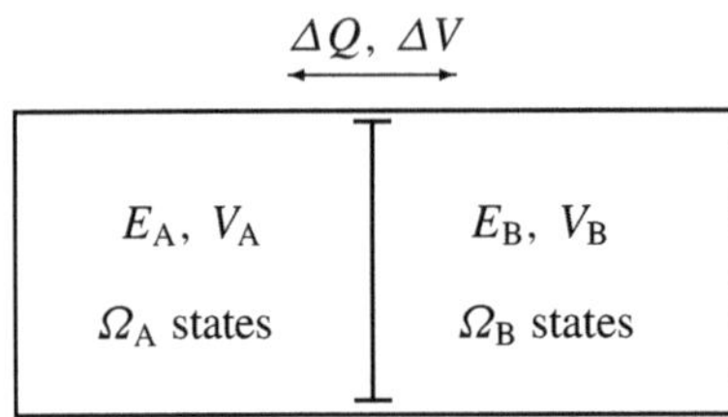

Figure 10.1 A closed system consists of two macroscopic subsystems that can exchange heat and volume. How are the energy and the volume distributed in thermal equilibrium?

Volume and heat exchange

We derive the equilibrium condition for two systems that can exchange heat *and* volume, Figure 10.1. The closed overall system has the energy

$$E = E_A + E_B = \text{const.} \tag{10.17}$$

and the volume

$$V = V_A + V_B = \text{const.} \tag{10.18}$$

The entropy of the closed system is maximum at equilibrium:

$$S(E_A, V_A) = S_A(E_A, V_A) + S_B(E - E_A, V - V_A) = \text{maximum} \tag{10.19}$$

In this case, the variable ξ consists of two variables, $\xi = (E_A, V_A)$. Other external parameters are constant and are not displayed. The necessary conditions for the existence of a maximum are:

$$\frac{\partial S}{\partial E_A} = 0 \quad \longleftrightarrow \quad \frac{1}{T_A} - \frac{1}{T_B} = 0 \tag{10.20}$$

and

$$\frac{\partial S}{\partial V_A} = 0 \quad \longleftrightarrow \quad \frac{P_A}{T_A} - \frac{P_B}{T_B} = 0 \tag{10.21}$$

That there is actually a maximum at these points follows from the form of the E and V dependence of $\Omega(E, V)$; in Chapter 9 this was shown for the energy dependence. For the equilibrium under heat *and* volume exchange we obtain thus

$$P_A = P_B, \quad T_A = T_B \qquad \text{(heat and volume exchange)} \tag{10.22}$$

Volume exchange via an adiabatic wall

A movable wall as shown in Figure 10.1 shall not allow any heat exchange; for example, it could contain a layer of polystyrene. The subsystems can therefore only exchange volumes. In equilibrium, the resulting forces on the movable wall must disappear, i.e.

$$P_A = P_B \qquad \text{(only volume exchange)} \tag{10.23}$$

Condition (10.21), on the other hand, does not apply, because it does not take into account the energy changes that are inevitably associated with volume exchange.

We consider a possible scenario for a finite pressure difference. The movable and heat-impermeable wall is released in an initial state with $P_A \neq P_B$. The wall (such as a piston) then swings back and forth. Due to the friction (of the piston), part of the energy $E_A + E_B$ is converted into heat. (The irreversible energy transfer considered in (8.26) may be left away in the present discussion). It depends on the special structure of the system how this heat is distributed to the gases and the vessel. After more or less many oscillations the piston comes to rest; the closed system has reached equilibrium.

How is this process be understood in the context of the general condition $S =$ maximum? We construct a closed system in which, in addition to the gases A and B, we provide a device C that can absorb or release energy via the piston motion. Then we get

$$E_A + E_B + E_C = \text{const.} \quad \text{and} \quad V = V_A + V_B = \text{const.} \tag{10.24}$$

The equilibrium condition for the closed system is

$$S = S_A(E_A, V_A) + S_B(E_B, V_B) + S_C(E_C) = \text{maximum} \tag{10.25}$$

Let the device C be coupled to the piston in such a way that the piston motion is quasi-static. Under a quasi-static and adiabatic volume change, the entropies of the gases entropies of the gases do not change[1], i.e. $S_A =$ const. and $S_B =$ const. Thus becomes (10.25) becomes $S_C(E_C) =$ maximum or $E_C =$ maximum. From $E_C =$ maximum and $E_A + E_B + E_C =$ const. it follows that $E_A + E_B =$ minimal, thus

$$E_{A+B}(V_A) = E_A(S_A, V_A) + E_B(S_B, V - V_A) = \text{minimum} \tag{10.26}$$

The minimum is to be determined with respect to the dependence on V_A (where S_A, S_B, V are constant); this results in (10.23). In this case under, the extremal condition $S =$ maximum for the closed system leads to $E_A + E_B =$ minimal for the subsystem (from A and B).

General case

The discussion in the last section can be applied to the exchange of an arbitrary external parameter x_i between two systems A and B for which $x_i = x_{i,A} + x_{i,B} =$ const. holds. Analogous to (10.23), we get the equilibrium condition $X_{i,A} = X_{i,B}$.

The equilibrium conditions associated with the arguments of $\Omega(E, x)$ are thus:

$$X_{i,A} = X_{i,B} \qquad \text{(equilibrium for } x_i\text{-exchange)} \tag{10.27}$$

$$T_A = T_B \qquad \text{(equilibrium for heat exchange)} \tag{10.28}$$

---

[1]For $S(E, V)$, we get $dS = (\partial S/\partial E)\,dE + (\partial S/\partial V)\,dV = 0$ because $dE = đW_{\text{q.s.}} = -P\,dV$. The quasi-static and adiabatic expansion will be discussed in more detail in the next chapter.

In Chapter 5, we stated as a fact of experience that a closed system approaches the equilibrium (all macroscopic quantities are constant) by itself. The equilibrium with respect to heat exchange means the same temperature, that for x_i exchange means equal forces X_i.

The approach to equilibrium implies that unequal temperatures are equalized by heat exchange, and unequal generalized forces by exchange of the associated external parameter. A temperature difference is the driving force for the heat exchange, just as a pressure difference is the driving force for volume exchange.

Exercises

10.1   Pressure contributions in a gas mixture

In a volume V there is a mixture of ideal gases (each with each N_i particles of the i-th kind, $i = 1,..., m$). Determine the partition function $\Omega(E, V, N_1,..., N_m)$ and calculate the pressure from this. How do the individual components contribute to the pressure?

11 Second and third law of thermodynamics

We formulate and discuss the second and third law of thermodynamics. The second law gives a lower limit for an entropy change. The third law states that the entropy goes to zero for $T \to 0$.

For the sake of completeness, we will start with the 1st law of thermodynamics (TD) from Chapter 7:

$$\boxed{dE = \boldsymbol{\dj}Q + \boldsymbol{\dj}W \qquad \text{1st law of TD}} \qquad (11.1)$$

This means that energy changes dE of a system are divided into the absorbed heat $\boldsymbol{\dj}Q$ and into the work performed on the system $\boldsymbol{\dj}W$. In practice, this division is made by the experimental conditions "no change of the external external parameters" or "thermal insulation". The 1st law applies to arbitrary processes, in particular also to non quasi-static processes. In general, the considered system is not closed; otherwise we had $\boldsymbol{\dj}Q = \boldsymbol{\dj}W = 0$.

Second law of TD

Closed system

As an experimental fact, we have stated that a closed system strives for equilibrium by itself. Phenomenologically, the equilibrium is defined by the condition that the macroscopic variables have reached constant values. Theoretically, the equilibrium is described by an ensemble in which all accessible microstates are equally probable. Practically, all microstates are located at the point where the entropy has a maximum. Thus, on the approach to equilibrium, the entropy S moves towards its maximum value. Therefore, the entropy change in a closed system obeys the following 2nd law of thermodynamics,

$$\boxed{\Delta S \geq 0 \qquad \begin{array}{l} \text{2nd law of TD} \\ \text{(closed system)} \end{array}} \qquad (11.2)$$

This statement refers to the entropy change $\Delta S = S_b - S_a$ in a process $a \to b$. In the Chapters 9 (heat exchange), 10 (ξ exchange) and 12 (particle exchange) the statistical basis of the statement $\Delta S > 0$ is examined in detail for the respective process.

The entropy definition of Chapter 9 admitted local equilibria, i.e. separate equilibria for the subsystems A and B in Figure 9.1; the entropy of the overall system is then $S = S_A + S_B$. For the statement $\Delta S = S_b - S_a \geq 0$, the initial and final states, a and b, are limited to macrostates with local equilibria.

In Chapter 41, the statement $dS/dt \geq 0$ will be derived from the master equation; for this purpose, the entropy S is defined for an arbitrary macrostate. Chapter 41 might be read in advance.

When the entropy in the closed system reaches its maximum value, it fluctuates around this value according to (10.9). This means that there are violations of condition (11.2) on the scale k_B. Compared with $S_{\max} = \mathcal{O}(Nk_B)$, however, these deviations are negligibly small. In particular, they do not allow the construction of a perpetuum mobile of the second kind (Chapter 19).

Open system

The entropy $S(E, x) = k_B \ln \Omega(E, x)$ of an equilibrium state depends on the energy E and the external parameters $x = (x_1,..., x_n)$. We calculate the exact differential of the entropy:

$$dS = \frac{\partial S(E, x)}{\partial E}\, dE + \sum_{i=1}^{n} \frac{\partial S(E, x)}{\partial x_i}\, dx_i = \frac{dE}{T} + \sum_{i=1}^{n} \frac{X_i}{T}\, dx_i \qquad (11.3)$$

Here we insert the 1st law of TD and (8.3), $\sum X_i\, dx_i = -đW_{\text{q.s.}}$,

$$dS = \frac{1}{T}\left(đQ + đW - đW_{\text{q.s.}} \right) \qquad (11.4)$$

For a quasi-static process, this yields

$$\boxed{dS = \frac{đQ_{\text{q.s.}}}{T}} \qquad (11.5)$$

According to (8.29)

$$đW \geq đW_{\text{q.s.}} \qquad (11.6)$$

holds. Thus (11.4) becomes

$$\boxed{dS \geq \frac{đQ}{T} \qquad \begin{array}{l}\text{2nd law of TD} \\ \text{(open system)}\end{array}} \qquad (11.7)$$

This statement includes (11.5) as a special case. If the equals sign in (11.7) does not apply, this is due to non quasi-static work; this follows from (11.4) with (11.6). As an example, we have introduced the non quasi-static piston oscillation in Figure 8.3, for which $dS > đQ/T = 0$. This process was identified in Chapter 8 as irreversible; we will return to this designation in the next chapter.

The formulation (11.7) is restricted by the condition that the temperature T of the system must be defined. Since the temperature change $dT = T_b - T_a$ is infinitesimal it does not matter whether T_a or T_b is used in the denominator in (11.7).

When we introduced the entropy and temperature in Chapter 9 we considered a system with local equilibria (in the subsystems A and B respectively). For the application of (11.7) it is sufficient that the temperature is defined locally (e.g. as a function of position, see section "Prerequisites for the definition of entropy and temperature definition" in Chapter 9).

The section "Heat supply" below shows how to use (11.5) in order to calculate the entropy change of a system when the initial and final states are equilibrium states. No assumption must be made about the macrostates that occur in the course of the process.

The statements (11.2) and (11.7) partially overlap. For example, from (11.5) the entropy increase $\Delta S > 0$ can be calculated that arises from a temperature equalization in the overall closed system (Chapter 12). However, the statement (11.2) is only partially contained in (11.7) because (11.7) requires a defined temperature.

A heat transfer $đQ \neq 0$ can take place by contact with a heat bath. The transfer is only quasi-static if the temperature T of the heat bath is nearly equal to that of the system. A small temperature difference is required so that a heat exchange takes place at all. The ideal limiting case "quasi-static" is only realized in the limit of a vanishing temperature difference (and thus an infinitely long process duration). All real processes contain at least small non quasi-static (and therefore irreversible) components.

General case

For a general process, the statements (11.2) and (11.5) can be summarized as follows. The heat exchange takes place by contact to a heat with temperature T_W. At the point of contact, the system adopts locally the temperature T_W. The heat bath now quasi-statically transfers the heat $đQ_{q.s.}$. According to (11.5), the corresponding entropy change is $d_e S = đQ/T_W$; the index "e" stands for the contact with the *external* heat bath. Parallel to this, time dependent heat conducting processes take place within the system (index "i" for internal). For the corresponding internal entropy change equation (11.2) yields $d_i S \geq 0$. Overall, this gives the entropy change $dS = d_e S + d_i S$ or

$$dS = \frac{đQ_{q.s.}}{T_W} + d_i S, \qquad d_i S \geq 0 \qquad \text{(2nd law of TD)} \qquad (11.8)$$

To discuss this formulation, let us consider the heat supply to the Earth through the solar radiation. The heat radiation is characterized by the surface temperature $T_W = T_\odot$ of the Sun (about $6000\,\text{K}$, Chapter 33). We could (in principle) use this energy with an absorber of the temperature $T_{\text{absorber}} = T_\odot$. In doing so, the entropy initially increases by $d_e S = đQ/T_W$. Between the absorber and the Earth's surface (with $T_E \approx 300$ K), the temperature can now be equalized by heat conduction. Then

$d_i S = đQ(1/T_E - 1/T_W)$ is maximum and one gets a total of $dS = đQ/T_E$. (The entropy change due to temperature equalization will be discussed in detail in the next chapter). The heat transport from the higher level T_W to the lower level T_E can, however, also be used to convert part of the heat quantity $đQ$ into work. This occurs in nature, for example, through conversion into wind energy and technically in any kind of solar power plant. In this case, $dS < đQ/T_E$.

The example shows that contact with the heat bath (here the absorber) does not necessarily have to be the edge of the system (such as the outer layer of the atmosphere); rather, the contact can also be inside the system.

The following should also be noted about the absorber in this example. If the absorber has exactly the temperature of the radiation, then it emits as much energy as it absorbs. In practice, it must therefore have a have at least a slightly lower temperature, $T_{absorber} = T_\odot - \Delta T$. As discussed at the end of the last section, the process is quasi-static only in limiting $\Delta T \to 0$.

In a solar furnace, the Sun's rays are focused using concave mirrors. In a real solar furnace, thereby one might achieve temperatures of about 4000 to 5000 K; this is significantly below the theoretical upper limit $T_W = T_\odot$. The entropy transfer $dS = đQ/T_{absorber}$ already contains a portion $d_i S$, corresponding to the heat transport from level T_W to $T_{absorber}$.

Various formulations

Various formulations of the 2nd law can be found in the literature. The historical formulations by Clausius (1850), Kelvin (1851) and Carnot (1824) are often cited:

1. Clausius: There is no process which only transfers heat from a system with lower temperature to a system with higher temperature.

2. Kelvin: There is no thermodynamic process that only converts heat into work. Or: There is no perpetuum mobile of the second kind.

3. Carnot: There is no heat engine that is more efficient than a Carnot process.

The process excluded by point 1 would lead to $\Delta S = \Delta Q(1/T_> - 1/T_<) < 0$ for the closed system. The other points are further discussed in Chapter 19 on heat engines. Many authors[1] proceed from these formulations.

Today, an equation or inequality is usually referred to as the 2nd law of TD. Some authors[2] refer to the statement (11.2) as the 2nd law of TD. Other authors place the statement (11.7) in the center. Reif [6] refers to the statements (11.2) and (11.5) together as the 2nd law of TD. The formulation (11.8) is used by Brenig [6]) and in a similar form by Schmutzer[3]. In representations, in which (11.2) is referred

---

[1]For example K. Huang, *Statistical Mechanics*, 2nd ed., John Wiley, 1987

[2]For example Landau-Lifschitz [8] or F. Mandl, *Statistical Physics*, 2nd ed., John Wiley, 1988

[3]E. Schmutzer, *Fundamentals of Theoretical Physics*, Part 1, B.I.-Wissenschaftsverlag, Zürich 1989

Table 11.1 The specific heat c_P of water and copper (at normal pressure and room temperature).

System	specific heat	melting temperature	latent heat (melting)
Water	4.2 J/(g K)	273.15 K	334 J/g
Copper	0.38 J/(g K)	1356.6 K	205 J/g
Iron	0.45 J/(g K)	1808 K	277 J/g

to as the 2nd law of TD, of course also $dS = đQ_{\text{q.s.}}/T$ is introduced. Conversely, also $\Delta S \geq 0$ for closed systems is found in any textbook.

For the development of thermodynamics (Part III), the equations (11.2) and (11.5) are sufficient. However, these two statements are not independent of each other. Using (11.5), the entropy increase $\Delta S > 0$ can be calculated for the temperature equalization in the overall closed system (Chapter 12).

Heat input

For arbitrary processes $a \to b$ between two equilibrium states, the entropy change $\Delta S = S_b - S_a$ can be calculated with (11.5) if a quasi-static path from state a to state b is constructed. Along this constructed path, the system then passes through a sequence of equilibrium states, so that $T = T(E, x)$ in

$$\Delta S = \int_a^b \frac{đQ_{\text{q.s.}}}{T} \tag{11.9}$$

is defined. In addition to the heat transfer, also quasi-static work can be admitted. By such a construction, any two equilibrium states a and b can be connected.

In a quasi-static process, the temperature of a system is increased from T to $T + dT$ by the of heat input $dQ_{\text{q.s.}}$. The ratio

$$C = \frac{đQ_{\text{q.s.}}}{dT} \tag{11.10}$$

is defined as the *heat capacity* of the system. The heat capacity per mass, $c = C/M$, is referred to as the *specific heat*. Table 11.1 lists some experimental values. The heat transfer must be specified by stating which macroscopic parameters (such as V or P) are kept constant; these conditions are specified by the respective index in C_V and C_P. If the volume is kept constant, there is no work transfer associated with the process (since $đW_{\text{q.s.}} = -P\,dV = 0$). The differences between various heat capacities (in particular $C_P - C_V$) will be discussed later; for solids and liquids they are usually small. The heat capacity and the specific heat are functions of the macroscopic state variables, i.e. $C = C(T, P,...)$.

In the following, we assume that the heat capacity of the system under consideration is constant, $C = \text{const}$. Then (11.9) and (11.10) yield

$$\Delta S = \int_a^b \frac{đQ_{\text{q.s.}}}{T} = \int_a^b \frac{C\, dT}{T} = C \ln\left(\frac{T_b}{T_a}\right) \tag{11.11}$$

and

$$\Delta Q = \int_a^b đQ_{\text{q.s.}} = C\,(T_b - T_a) \tag{11.12}$$

To calculate the integral in (11.11), we assumed a quasi-static path. The condition "quasi-static" is fulfilled by a sufficiently slow process (slow heat input and possibly slow change of external parameters). Then, at each stage of the process $a \to b$ a temperature is defined.

If the amount of heat ΔQ is supplied to the system at constant external parameters (no work), then (11.11) and (11.12) apply to a non quasi-static process, too. In this case, the initial state $a = (E_a, x_a)$ and ΔQ determine the final state b, namely by $E_b = E_a + \Delta Q$ and $x_b = x_a$. Then the temperatures T_a and T_b and the entropy change $\Delta S = S_b - S_a$ are uniquely determined.

A non quasi-static input of heat could be achieved by heating a piece of matter locally (for example with a Bunsen burner or on a hotplate) and then thermally insulated. In the system (the piece of matter), time and position dependent heat conduction processes take place, i.e. non quasi-static processes. Eventually, however, an equilibrium state is restored.

If matter is heated at a constant pressure, it usually expands; the external parameters are therefore not constant. In many cases, this expansion will occur so slowly that the respective work is quasi-static (i.e. equal to $-\Delta W_{\text{q.s.}} = \int P\, dV$). In this case, the final state is independent of whether the amount of heat is actually supplied quasi-statically. Therefore, one can use ΔS from (11.11) (but now with $C = C_P$), even if the heat is supplied locally and the process is not quasi-static.

Third law of TD

Quantum mechanical systems usually have exactly one state with the lowest possible energy E_0, the ground state. This state is separated from the first excited state by a finite energy gap $E_1 - E_0$. For $E \to E_0$ there is only one state in the interval $[E, E + \delta E]$, i.e.

$$\Omega(E) = 1 \quad \text{for } E \to E_0 \tag{11.13}$$

and

$$S(E) = k_{\text{B}} \ln \Omega(E) \xrightarrow{E \to E_0} 0 \tag{11.14}$$

In contrast to this, we have so far considered systems and energies for which a large number of states lie in the δE interval.

The energy $E - E_0$ is available for the excitation of degrees of freedom. Thus (6.21) becomes

$$\Omega(E) \propto (E - E_0)^{\gamma f} \tag{11.15}$$

From this follows

$$\frac{1}{k_{\mathrm{B}} T} = \frac{\partial \ln \Omega}{\partial E} \propto \frac{\gamma f}{E - E_0} \xrightarrow{E \to E_0} \infty \tag{11.16}$$

Therefore, $E \to E_0$ means as much as $T \to 0$. This reflects the physical meaning of temperature as the (available) energy per degree of freedom. For the discrete quantum mechanical states (with E_0, E_1, ...) the relation between $E \to E_0$ and $T \to 0$ is justified in more detail in Part IV. From (11.14) and (11.16) we obtain

$$\boxed{\quad S \xrightarrow{T \to 0} 0 \qquad \text{3rd law of TD} \quad} \tag{11.17}$$

This is the 3rd law of TD, also called Nernst's theorem or *Nernst heat theorem*.

Experimentally, $T = 0$ cannot be achieved exactly; feasible are values down to $T \approx 10^{-7}$ K. In practical applications, $T \to 10^{-3}$ K is often considered equivalent to $T \to 0$. At such temperatures, almost all degrees of freedom are frozen in; and the system adopts the ground state. However, there is an exception, namely the setting of the spins of atomic nuclei. Atomic nuclei with an odd number of nucleons always have a finite spin. The nuclear spins interact with the environment via its small magnetic moment μ_{nucl}. This magnetic moment is of the same size as the magnetic moment of a nucleon; it is about a factor of 10^3 smaller than that of an electron. The interaction of the nuclear spins with each other and with the environment is extremely weak. For these nuclear spins, a temperature of $T = 10^{-3}$ K can then be a *high* temperature at which all possible spin states are equally probable. For N nuclei with spin $1/2$ this means that all $\Omega_0 = 2^N$ spin settings (5.6) are equally probable. In this case, the entropy goes towards the value $S_0 = k_{\mathrm{B}} \ln \Omega_0 = N k_{\mathrm{B}} \ln 2$ for small temperatures. The value S_0 depends only on the kind of atomic nuclei, but not on the energy or on external parameters of the system. In this sense, the 3rd law of TD may be also written in the form

$$S \xrightarrow{T \to 0^+} S_0 \qquad \text{(3rd law of TD)} \tag{11.18}$$

Here, $T \to 0^+$ means the approach the to a practically achievable, low temperature; and S_0 is a constant. In principle, however, (11.17) also applies to systems with nuclear spin; because the nuclear spins will also align in some way at a sufficiently low small temperature. This spin ordering sets in when $k_{\mathrm{B}} T$ is comparable to the strength of the interaction of the nuclear spins with each other or with residual magnetic fields (for example, in the solid).

Supplements

The laws of thermodynamic provide the basis for the relations between macroscopic quantities. Thereby, thermodynamics mostly refers to systems in equilibrium. Non-equilibrium states are discussed only qualitatively, based on the empirical fact that "isolated systems move towards equilibrium". The equilibrium condition "$T_1 = T_2$" for heat exchange leads to the qualitative statement that a temperature difference "$T_1 \neq T_2$" initiates a heat transfer, and that this heat transfer eventually result in the equilibrium state.

In an axiomatic sense, the three law of TD are not complete. For the definition of temperature as an observable we need in particular the statement that equilibrium under heat exchange is equivalent to equal temperature. This is sometimes formulated as the zeroth law of TD [6]. For practical application, we also need the empirical fact that closed systems actually move (in finite time) to equilibrium. Compared to this requirement, the 3rd law $\Delta S \geq 0$ is a weak statement only.

Exercises

11.1 Entropy change during heat exchange I

A steel vessel with the mass 2 kg contains half a liter of water; the system has adopted room temperature. Now an ice cube of 100 g is added to the water and the system is thermally insulated. Which water temperature will be finally reached? Calculate the entropy change.

11.2 Entropy change during heat exchange II

A kilogram of water with temperature 10 °C is brought into thermal contact with a heat reservoir with the temperature 90 °C. After some time, a new equilibrium is established. For this process, calculate the entropy change of the water, the heat bath and the entire system.

11.3 Entropy of a rubber band

A rubber band is modeled as a system with the one external parameter, its length L. The dependence of the entropy $S(E, L)$ on L shall be calculated in the following model: The rubber band is simulated by a chain of N links of length d. Each link lies parallel or antiparallel on a straight line. The neighbor of a specific link continues the chain to the right or to the left with equal probability. (If all links show to the right, the chain has the length $L = Nd$.) Determine the number Ω_L of the configurations for fixed L and the respective entropy

$$S(E, L) - S(E, 0) = -\frac{k_{\mathrm{B}} L^2}{2 N d^2} \qquad \text{for} \quad L \ll N d$$

First express Ω_L by $m = n_+ - n_- = L/d \ll N$ where $n_\pm$ is the number of links pointing to the right (left). Calculate the force f with which the rubber band is tensioned.

12 Reversibility

Based on the second law of TD, we classify processes as reversible or irreversible. We then investigate some processes (heat exchange and expansion) with respect to this classification. We discuss the microscopic basis of the irreversibility of the free expansion.

Closed system

Based on (11.2), we distinguish for closed systems

$$\Delta S = 0 \qquad \text{reversible process} \tag{12.1}$$

$$\Delta S > 0 \qquad \text{irreversible process} \tag{12.2}$$

Examples of an irreversible process are in particular the temperature equalization and free expansion, Figure 12.1. These examples will be discussed in detail below.

Reversible processes are ideal limit cases. Real processes can at best be *almost* reversible. A reversible process can be reversed by an arbitrarily small external effort. As an example we will consider the quasi-static expansion.

Open system

Based on (11.7), we distinguish for open systems

$$dS = \frac{\dd Q}{T} \qquad \text{reversible process} \tag{12.3}$$

$$dS > \frac{\dd Q}{T} \qquad \text{irreversible process} \tag{12.4}$$

For a quasi-static process the equal sign applies; such a process is reversible. Therefore, the index "rev" instead of "q.s." in (11.4) is common:

$$dS = \frac{\dd Q_{\text{q.s.}}}{T} = \frac{\dd Q_{\text{rev}}}{T} \qquad \text{(alternative notations)} \tag{12.5}$$

If the equals sign in (11.7) does not apply, the process is is not quasi-static; according to (12.4) it is irreversible. A quasi-static or reversible process consists of a sequence of equilibrium states. The term "quasi-static" emphasizes the experimental condition ($\tau_{\text{exp}}/\tau_{\text{relax}} \to \infty$), the term "reversible" emphasizes the invertibility of the process[1].

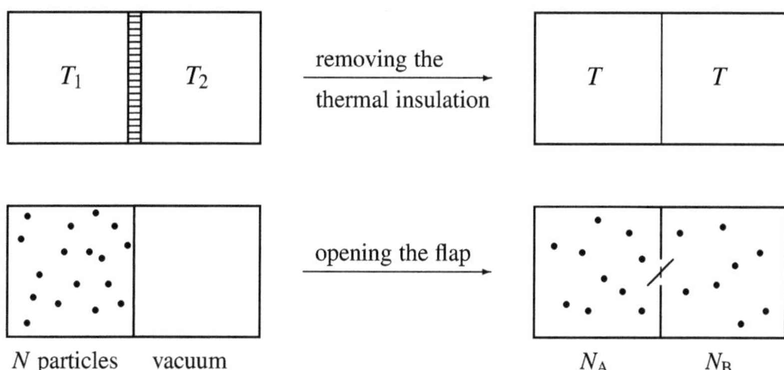

Figure 12.1   After removal of constraints, more microstates are accessible than before. In the upper part, a process is initiated by the removal of a thermal isolation between two subsystems. The temperature then equalizes. In the lower part, a process is initiated by opening a flap. The gas then expands to the entire available volume. In the closed system, none of the processes can take place in the opposite direction; they are irreversible.

A quasi-static process in an open system is *reversible* in the literal sense: A change of external parameters $x_a \rightarrow x_b$ and/or a heat transfer causes the quasi-static process $a \rightarrow b$. The occurring equilibrium states $a,...,$ intermediate states,..., b can now be passed through in the opposite direction if the change in the external parameters is reversed ($dx_i \rightarrow -dx_i$, this implies $đW_{\text{q.s.}} \rightarrow -đW_{\text{q.s.}}$) and the sign of the heat transfer is changed ($đQ_{\text{q.s.}} \rightarrow -đQ_{\text{q.s.}}$). The process defined in this way leads from E_b, x_b back to the initial state E_a, x_a. Then $đQ_{\text{q.s.}} \rightarrow -đQ_{\text{q.s.}}$ implies $dS \rightarrow -dS$ for the individual infinitesimal sections of the process. After reversion S_b becomes S_a again.

Deviations from the minimum entropy change $dS = đQ_{\text{q.s.}}/T$ are always positive. Therefore, such contributions cannot cancel each other in the reversed process. If such deviations occur, then the process is irreversible.

As an example of an irreversible process, consider again the gas in Figure 8.3, which is enclosed by an oscillating piston. After a full oscillation of the piston, the volume of the gas returns to its original value; the change of the external parameter has been reversed. The energy, however, reaches a value that is higher, $\Delta E = \Delta E_{\text{irrev}}^{(1)}$, (8.26). The reversible limiting case $\Delta E_{\text{irrev}} \rightarrow 0$ requires that the piston velocity goes to zero. Every real process takes place with a finite speed and therefore leads to some irreversible components.

---

[1]We have defined the condition *quasi-static* by *sequence of equilibrium states*. Then "quasi-static" is equivalent to "reversible". Occasionally [7], the condition "quasi-static" is defined by "sequence of partial equilibrium states". Example: If the heat transfer in Figure 12.1 top is very slow, then the subvolumes each for itself go through equilibrium states. According to this alternative definition the heat transfer is called quasi-static (it may indeed be very slow), but it remains of course non-reversible with $\Delta S > 0$ for the combined system.

To summarize, we can state that the reversible processes (12.3) as well as (12.1) are ideal limit cases. In reality, there are only approximately reversible processes, i.e. in which the irreversible parts are very small.

Temperature equalization

We consider the temperature equalization between two subsystems which initially have different temperatures and which are then brought into thermal contact (Figure 12.1 top). The thermal contact leads to an equalization of the temperatures; the closed overall system is in equilibrium. We have already discussed this process in Chapter 9, where the focus was on statistical considerations. We briefly recall this point. After that, we deal with the process from a thermodynamic point of view.

The heat exchange considered in Figure 9.1 leads to the temperature equalization. Figure 9.3 makes it clear that at equilibrium ($T_1 = T_2$) the number Ω_b of the accessible microstates is much larger than in the initial state a, i.e.

$$\Omega_b \gg \Omega_a \tag{12.6}$$

This is the microscopic cause of non reversibility: In the final state b, all Ω_b microstates are equally probable. This also includes the Ω_a of the initial state. However, because of (12.6), they have a vanishingly small statistical weight; their occurrence is extremely unlikely. The example of free expansion (next section) illustrates in detail that "improbable" here means "practically impossible".

First we treat the irreversible temperature equalization thermodynamically. The entropy changes of the subsystems can be calculated using (12.3), i.e. under the assumption that the subsystems each pass through a sequence of equilibrium states. As described in the section "Heat supply" in Chapter 11, the entropy change ΔS calculated in this way is also correct if the heat exchange is not slow. The entropy change $\Delta S = S_b - S_a$ depends only on the initial and final states. These states are determined by the initial and final temperatures of the two systems.

As an example, let us consider the cooling of a filled beer bottle B in a lake L. For both systems, we assume constant heat capacities, C_B and C_L. The beer bottle will lose the amount of heat

$$\Delta Q_B = C_B \left(T_b - T_a\right) \tag{12.7}$$

by contact with the lake; here T_a is the initial temperature and T_b is the final temperature of the beer bottle. If the beer bottle is cooled, $T_b < T_a$ and $\Delta Q < 0$; however, (12.7) and the following equations also apply for $T_b > T_a$. In the closed system "beer bottle plus lake" the energy is constant

$$\Delta E = \Delta E_B + \Delta E_L = \Delta Q_B + \Delta Q_L = 0 \tag{12.8}$$

There is no work transfer. From $\Delta Q_L = -\Delta Q_B$ follows for the temperature change ΔT_L of the lake

$$|\Delta T_L| = \frac{|\Delta Q_L|}{C_L} = \frac{|\Delta Q_B|}{C_L} = \frac{C_B}{C_L} |T_b - T_a| \ll |T_b - T_a| \tag{12.9}$$

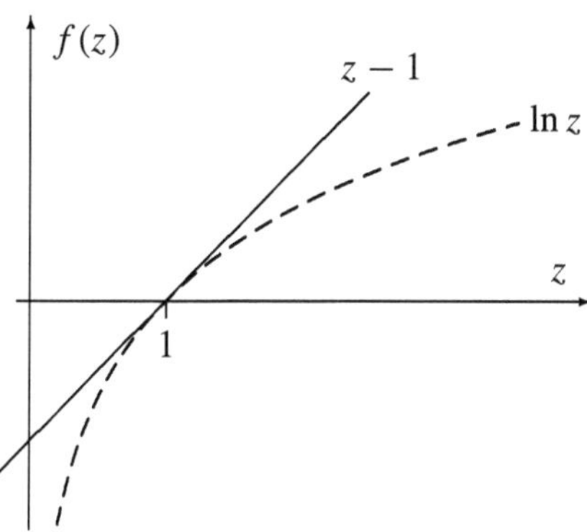

Figure 12.2 Graph of the functions $\ln z$ and $z - 1$. Because of $z - 1 \geq \ln z$, ΔS from (12.13) is always larger than or equal to zero.

The lake plays the role of a *heat bath* (or heat reservoir): Because of $C_L \gg C_B$, the temperature change of the lake (of the heat bath) can be neglected, i.e.

$$T_L = \text{const.}, \qquad T_b = T_L \tag{12.10}$$

In the equilibrium state, the temperatures of the subsystems are equal. In the final state b, the beer bottle and the lake have therefore the same temperature T_b.

We determine the entropy changes of the two subsystems:

$$\Delta S_L = \int_a^b \frac{dQ_{L,\,\text{q.s.}}}{T} = \frac{\Delta Q_L}{T_b} = -\frac{\Delta Q_B}{T_b} = C_B \frac{T_a - T_b}{T_b} \tag{12.11}$$

$$\Delta S_B = \int_a^b \frac{dQ_{B,\,\text{q.s.}}}{T} = \int_{T_a}^{T_b} \frac{C_B\, dT}{T} = C_B \ln\left(\frac{T_b}{T_a}\right) \tag{12.12}$$

This gives us the entropy change of the overall system:

$$\Delta S = \Delta S_L + \Delta S_B = C_B\,(z - 1 - \ln z) \quad \text{with} \quad z = T_a/T_b \tag{12.13}$$

Figure 12.2 immediately shows that

$$\ln z \leq z - 1 \tag{12.14}$$

From the last two equations follows

$$\Delta S \geq 0 \qquad \text{(beer bottle + lake)} \tag{12.15}$$

The equals sign only applies to $T_b = T_a$. With a finite temperature difference, the temperature equalization (cooling or warming of the beer bottle in the lake) is associated with an increase in entropy in the closed system. Such a temperature equalization is irreversible.

For the calculation of ΔS, a quasi-static process *assumed*. This means that the heat exchange should take place so slowly that the system beer bottle has a defined temperature at all times. The results, in particular ΔS_L and ΔS_B, are also valid if only the initial an final states are equilibrium states (see section "heat supply" in Chapter 11). In the real experiment, time and position dependent temperatures will occur in the beer bottle; the bottle is then temporarily in non-equilibrium states.

Free expansion

A volume V is divided into two equal parts by a wall. In one part there is a gas, the other contains nothing (i.e. a vacuum). Now a rotary flap in the wall is opened (Figure 12.1 bottom right). The opening of the flap takes place without work, $đW = 0$. The system as a whole is thermally insulated, i.e. $đQ = 0$. This means that the energy E remains unchanged, $dE = đQ + đW = 0$; also the number of particles N is unchanged. By opening the flap, however, the accessible volume is doubled, $V_b = 2V_a$. According to (6.16) $\Omega \propto V^N$ holds. This means that for the considered process in the closed system we get

$$\Omega_b = 2^N \, \Omega_a \gg \Omega_a \tag{12.16}$$

Because of $\Omega_b \gg \Omega_a$ (i.e. $\Delta S = S_b - S_a > 0$), the process $a \to b$ is irreversible. In the following we discuss the statistical basis of this irreversibility.

Before the flap opens, the system is in an equilibrium state a. After opening, the gas spreads rapidly over the entire volume; the timescale for this is L/c_s, where L is the length scale of the volume and c_s is the speed of sound. This rough estimate applies for a sufficiently large flap. After the flap is opened, the density distribution is time dependent, and the system it runs through non equilibrium states. After a few L/c_s the new equilibrium state b is reached.

All Ω_b microstates of the final state b are equally probable. These states also include the Ω_a microstates of the initial state a; these states are still accessible states in the sense of the fundamental postulate. Within the Ω_b states these Ω_a states have a vanishing statistical weight. The probability to find one of the Ω_a states in the ensemble of the final states b is given by

$$P = \frac{\Omega_a}{\Omega_b} = 2^{-N} \approx 10^{-2 \cdot 10^{23}} \qquad \left(N = 6 \cdot 10^{23} \right) \tag{12.17}$$

For the numerical value, we have assumed that there is one mole of the gas in considered volume. This value of P means in fact the impossibility of finding one of the Ω_a microstates in the final state b. We explain this "impossibility" in more detail.

We refer to a classical, ideal gas with the microstates (5.8),

$$r = (r_1, ..., r_N, p_1, ..., p_N) \tag{12.18}$$

At a specific time, a certain state r is present, in the next moment another one. Equilibrium means that all states r occur with equal probability.

We divide the microstates in the ensemble of the final state b into classes, where a class k indicates the division of the particles into the subvolumes. Through

$$k = (+, +, -, -, -, -, +, +, +, -, +, ...) \tag{12.19}$$

we specify whether the particle 1, 2, ..., N is in the left $(+)$ or right $(-)$ partial volume. There are 2^N different such divisions or classes, $k = 1, 2, ..., 2^N$; for the sake of simplicity, we assume distinguishable particles. Each class comprises the same number of microstates, namely $\Omega_b/2^N = \Omega_a$ states.

The initial state $a = (+, +, +, +, +, +, +, \dots)$ represents one of the 2^N classes. To clarify the question of whether and when the state a is reached again, we investigate after what time Δt the system changes from one class to another.

At room temperature, air molecules move with a speed $\overline{v} \approx 400$ m/s. In an ideal gas each particle collides after a time $\mathcal{O}(L/\overline{v})$ to the intermediate wall; here L is the linear extension perpendicular to the dividing wall. If the flap opening accounts for the 10^{-3}-th part of the area of the partition wall, a particle changes from one partial volume to the other on average after the time

$$\tau \sim \frac{L}{\overline{v}}\, 10^3 \approx 1\,\mathrm{s} \tag{12.20}$$

For the numerical value, we have the volume $V = L^3 = 22.4 \cdot 10^3\,\mathrm{cm}^3$ of a mole under normal conditions. (The actual processes in a real gas are different due to collisions. But an order of magnitude more or less does not matter in our estimate.) After the time τ approximately all $N = 6 \cdot 10^{23}$ atoms have changed sides once. Thereby, the system has changed from one class (12.19) to another N times. The system therefore changes the class after the time

$$\Delta t = \frac{\tau}{N} \sim 10^{-24}\,\mathrm{s} \tag{12.21}$$

The time in which *all* 2^N classes (12.19) are passed through just once is

$$T \sim 2^N\, \Delta t \tag{12.22}$$

This is roughly the time, in which the initial state a is also passed through once; with a little bit of luck, this may already happen after the time $T/100$. We estimate T numerically:

$$T \sim 2^{6\cdot 10^{23}}\, 10^{-24}\,\mathrm{s} \approx 10^{2\cdot 10^{23}-24}\,\mathrm{s} = 10^{2\cdot 10^{23}-24-18}\, t_{\mathrm{world}} \tag{12.23}$$

The time T is so large that we can say that the initial state a will *never* be reached again. Even if T is taken as a multiple of the world age $t_{\mathrm{world}} \approx 10^{10}$ years, this changes the exponent of the prefactor only slightly. It is true that the microstates of the ensemble a are contained in the ensemble of b; however, they have vanishing statistical weight. The improbability of their occurrence becomes a de facto impossibility.

We specify the probability $W(N_A)$ that N_A particles are present in the original volume (we refer to Figure 12.1, bottom right. i.e. to situation after the flap opening)). With equally large subvolumes, each individual particle is with probability $p = 1/2$ on the left, and with $q = 1/2$ on the right. In Chapter 3, the probability $W_N(n)$ was calculated that $n = N_A$ particles are in a subvolume. Because of $Npq \gg 1$, $W_N(n)$ can be approximated by a Gaussian function,

$$W(N_A) = W_N(N_A) \overset{(3.15)}{=} \frac{1}{\sqrt{2\pi}\,\Delta N_A} \exp\left(-\frac{\left(N_A - \overline{N_A}\right)^2}{2\,\Delta N_A^2}\right) \tag{12.24}$$

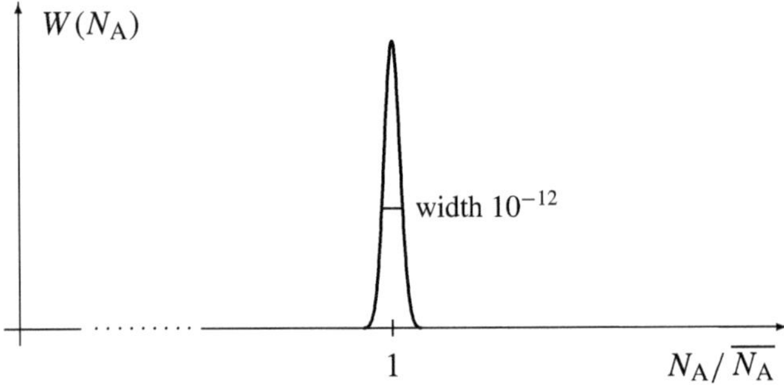

Figure 12.3 The N atoms of an ideal gas are equally likely positioned everywhere in the accessible volume V (Figure 12.1 below right). The figure shows the probability $W(N_A)$ that just N_A particles are in the left, original subvolume. The distribution shown is extremely sharp: The finite width shown in the figure implies that the abscissa value 2 (all particles in the initial volume) lies about 1 million kilometers to the right of the center of the distribution. If we mark the 2 on the abscissa in this figure, then the depicted distribution would be a line with a width of about 10^{-10} mm (i.e. it were invisible).

Here $\overline{N_A} = Np = N/2$ and $\Delta N_A = \sqrt{Npq} = \sqrt{N}/2$. This distribution is sketched in Figure 12.3. For a macroscopic system (e.g. with $N = 10^{24}$) it is extraordinarily sharp,

$$\frac{\Delta N_A}{\overline{N_A}} = \frac{1}{\sqrt{N}} = 10^{-12} \tag{12.25}$$

This sharpness means that almost all of the 2^N classes are in the immediate vicinity of $\overline{N_A} = N/2$, i.e. at the maximum of the probability distribution. The equilibrium can therefore be characterized by

$$\text{equilibrium:}\quad W(N_A) = \text{maximum} \tag{12.26}$$

The free expansion discussed in this section is instructive for the following reasons:

1. The connection with the statistical foundations (random walk in Chapter 3) is particularly close and obvious.

2. The example is characteristic of other distributions in statistical equilibrium. If we use $\xi = N_A$ in the general distribution $W(\xi)$ from (10.6), we obtain (12.24). If we use $\xi = E_A$, we obtain the corresponding distribution for heat exchange (Chapter 9). The statement (12.26) corresponds to "$S = \text{maximum}$".

3. This example illustrates in a particularly simple and transparent way the cause of the irreversibility of a process with $\Delta S > 0$ (or $\Omega_b \gg \Omega_a$) in a closed system.

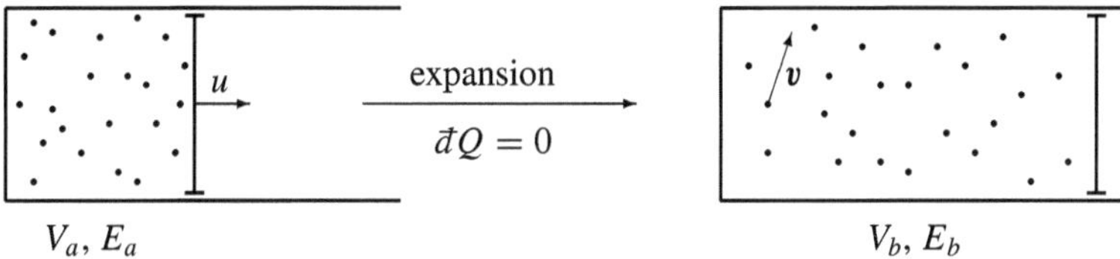

Figure 12.4 A gas is expanded adiabatically (no heat transfer). The average piston velocity is u, and the mean particle velocity is $\overline{v}$. In the quasi-static limit ($u/\overline{v} \to 0$), the work done is $\Delta W = \Delta W_{\text{q.s.}} = -\int dV\, P < 0$; the expansion is reversible only in this limit. If the piston is pulled out very quickly ($u \gg \overline{v}$), then $\Delta W = 0$ (free expansion). If the piston oscillates for a while on its way from V_a to V_b, the gas can be heated up. Therefore, there is no upper limit for ΔW for or a general expansion process. In all cases, $E_b - E_a = \Delta W$ holds.

Adiabatic expansion

We now consider various adiabatic expansion processes (Figure 12.4). In the last section of Chapter 8, we had found:

$$\Delta W = \begin{cases} \Delta W_{\text{q.s.}} = -\int dV\, P & \text{quasi-static expansion} \\ 0 & \text{free expansion} \\ \text{no upper limit} & \text{arbitrary expansion} \end{cases} \tag{12.27}$$

Quasi-static case

In the quasi-static case, $\Delta W = \Delta W_{\text{q.s.}}$ holds, and the equal sign in (11.6) and (11.7) applies. The quasi-static expansion is reversible.

Using the example of a monatomic ideal gas, we discuss the number of microstates in a quasi-static adiabatic expansion. According to (6.20) the following applies

$$\ln \Omega(E, V, N) = \frac{3N}{2} \ln\left(\frac{E}{N}\right) + N \ln\left(\frac{V}{N}\right) + N \ln c \tag{12.28}$$

The energy change for an adiabatic and quasi-static volume change by dV results from

$$dE = đQ_{\text{q.s.}} + đW_{\text{q.s.}} = đW_{\text{q.s.}} \stackrel{(8.14)}{=} -\frac{2}{3}\frac{E}{V}\, dV \tag{12.29}$$

This gives us

$$d\ln \Omega = \frac{3N}{2}\frac{dE}{E} + \frac{N}{V}\, dV = \frac{3N}{2}\frac{-2}{3V}\, dV + \frac{N}{V}\, dV = 0 \tag{12.30}$$

From this follows $d\Omega/dV = 0$ and

$$\Omega_b = \Omega_a \tag{12.31}$$

The quasi-static adiabatic expansion fulfills (12.3) and is therefore reversible. One may also construct a closed system leading to a reversibility according to (12.1). To do this, one may add a storage for the work performed by the gas; this could be achieved, for example, by lifting a weight via a gearbox. For the closed system consisting of the gas and the storage, $\Omega_b = \Omega_a$ or $\Delta S = 0$, because the few degrees of freedom of the work storage can be neglected when calculating the number of microstates. This results in an (almost) reversible process in a closed system; because the stored work can be used to compress the gas again. In reality, reversing the process requires an (arbitrarily) small external intervention.

Non quasi-static case

In the non quasi-static case, $\Delta W > \Delta W_{\text{q.s.}}$ applies. This implies the greater sign in (11.6) and (12.4). The non quasi-static expansion is an irreversible process.

In the last section, we specifically discussed free expansion. From $E_b = E_a$ and $V_b = 2\,V_a$ follows $\Omega_b \gg \Omega_a$. This is the statistical cause of irreversibility.

The last line in (12.27) was explained in Chapter 8 using the example of a gas enclosed by an oscillating piston. The piston motion (with finite speed) irreversibly transfers energy to the gas; this energy is proportional to the duration of the oscillations and is therefore not limited. In a numerical example (8.27), we have obtained $E_b = 1.01\,E_a$ for 100 piston oscillations, whereas $V_b = V_a$. Then $\Omega \propto E^{3N/2}$ implies again $\Omega_b \gg \Omega_a$.

In both cases (free expansion and oscillating piston) we have $\Delta S > 0$ whereby $\Delta Q = 0$. Therefore (12.4) applies; the processes are irreversible. The example of the oscillating piston makes it clear that the reversible limit can only be only be achieved if piston speed goes to zero.

Final remarks

For any given process, $đQ/T$ is a lower bound for the entropy increase. Irreversible processes lead to an additional (often undesirable) entropy increase.

Work $đW$ can be converted into heat $đQ$ at any time (electric heater), but heat can only be converted into work to a limited extent (heat engine, Chapter 19). In order not to waste the more valuable energy form "work", as little work as possible should be transferred to the system. According to (11.5) this means staying as close as possible to the quasi-static (i.e. reversible) limit. From (11.6) follows (11.7) and vice versa; the difference to the lower bound $đW_{\text{q.s.}}$ for the work is equal to that of $T\,dS$ to the lower bound $đQ$. The irreversible parts of the process that lead to a difference from these lower bounds (or optimal values).

Exercises

12.1   Discussion of the function $f(x) = x - 1 - \ln x$

Discuss the function $f(x) = x - 1 - \ln x$. Determine the extremes and sketch the graph of the function.

A stone with the heat capacity C has the temperature T_1. It is thrown into a swimming pool with the (constant) temperature T_2 and then assumes the temperature T_2. Determine the entropy change of the system (stone plus pool) and establish the connection with the discussed function.

12.2   Entropy change during heat exchange III

A body A with the initial temperature T_A is brought into thermal contact with a body B with the initial temperature T_B. The transferred heats are given by

$$\mathrm{d}Q_{A,\,\mathrm{q.s.}} = C_A \, dT, \qquad \mathrm{d}Q_{B,\,\mathrm{q.s.}} = C_B \, dT$$

with approximately constant heat capacities C_A and C_B.

Determine the temperature of the equilibrium state. Calculate the change in entropy of the overall system ΔS. Show $\Delta S \geq 0$.

13 Statistical physics and thermodynamics

We summarize the basics of statistical physics as described in Chapters 5 – 12. In doing so, we formulate the main tasks of statistical physics and thermodynamics.

Statistical physics

Statistical physics is based on the microscopic structure of the considered system. It treats the microscopic degrees of freedom statistically. This leads to statements about macroscopic quantities and their relations. We largely restrict ourselves to equilibrium states.

The derivation of the macroscopic properties from its microscopic structure of a system can be described by the scheme:

$$H(x) \xrightarrow{\;1.\;} E_r(x) \xrightarrow{\;2.\;} \Omega(E,x) \xrightarrow{\;3.\;} S(E,x),\; T(E,x),\; X(E,x) \tag{13.1}$$

We explain the individual steps in general and specifically for the ideal gas:

1. Starting point is the Hamilton operator H of the system. The parameters on which H depends are denoted by $x = (x_1,..., x_n)$. These parameters can be, for example, the number of particles, the volume or an external magnetic field. The first step (first arrow in (13.1)) consists of the determination of the eigenvalues $E_r(x)$ of $H(x)$. As a result, the

$$\text{microstates } r \text{ with the energy } E_r(x) \tag{13.2}$$

   of the system are defined. For the many-body systems the determination of the eigenvalues of H is a complex and in general unsolvable problem. Therefore, one often uses model Hamilton operators or approximate solutions.

2. The second step consists of calculating the partition function

$$\Omega(E,x) = \sum_{r\,:\,E - \delta E \,\le\, E_r(x) \,\le\, E} 1 \tag{13.3}$$

   This is a non-trivial step, too. Keep in mind that the sum over the microstates r implies f sums over the quantum numbers n_i in $r = (n_1,..., n_f)$. For $f = \mathcal{O}(10^{24})$ these summations are only feasible in particularly simple cases.

3. The third step is based on the fundamental postulate, i.e.

$$
P_r = \begin{cases} \dfrac{1}{\Omega(E, x)} & E - \delta E \le E_r(x) \le E \\ 0 & \text{otherwise} \end{cases}
\tag{13.4}
$$

Hereby, we restrict ourselves to equilibrium states. These macrostates are defined by the energy E and the external parameters. All macroscopic quantities can be expressed by E and x. These macroscopic quantities include in particular the entropy, the temperature and the generalized forces:

$$
S(E, x) = k_B \ln \Omega(E, x)
\tag{13.5}
$$

$$
\frac{1}{T} = \frac{\partial S(E, x)}{\partial E}, \qquad \frac{X_i}{T} = \frac{\partial S(E, x)}{\partial x_i}
\tag{13.6}
$$

The third step in (13.1) consists in writing down (13.5) and in evaluating the partial derivatives (13.6). It implies non-trivial preconditions such as the description of equilibrium states by the fundamental postulate. This step connects the microscopic level (Ω) with the macroscopic (S) level. It is therefore of crucial conceptual importance. As a step to be carried out, however, it is trivial in contrast to the first two steps: All one has to do is identify S with $k_B \ln \Omega$ and perform the partial derivatives.

The scheme (13.1) can also be applied to classical systems. In this case, H is the Hamilton function, and for the counting (13.3) it must be taken into account that each microstate corresponds to a finite volume in phase space.

For specific systems, we will later use the canonical or grand canonical partition function (Part IV) instead of the microcanonical partition function Ω. Also these other partition functions yield all relevant macroscopic quantities.

For an ideal, monatomic gas we summarize the outlined steps:

1. The Hamilton operator $H(x)$ is given by (6.2); the external parameters are $x = (V, N)$. The microstates r and their energy E_r are

$$
r = (n_1, \ldots, n_{3N}), \qquad E_r(V, N) = \sum_{k=1}^{3N} \frac{\pi^2 \hbar^2}{2m L^2} n_k^2
\tag{13.7}
$$

2. The evaluation of the partition function with these E_r results in (6.20), i.e.

$$
\ln \Omega(E, V, N) = \frac{3}{2} N \ln \left(\frac{E}{N} \right) + N \ln \left(\frac{V}{N} \right) + N \ln c
\tag{13.8}
$$

3. The entropy follows from this

$$
S(E, V, N) = \frac{3}{2} N k_B \ln \left(\frac{E}{N} \right) + N k_B \ln \left(\frac{V}{N} \right) + N k_B \ln c
\tag{13.9}
$$

The first part of (13.6) gives the caloric equation of state

$$E = \frac{3}{2} N k_\mathrm{B} T \tag{13.10}$$

For $x = V$ and $X = P$, the second part of (13.6) gives the thermal equation of state

$$PV = N k_\mathrm{B} T \tag{13.11}$$

As outlined here in general and specifically for the ideal gas, statistical physics derives the relations for the macroscopic variables from the microscopic structure of the system. In Part V we will apply this procedure to numerous systems.

Thermodynamics

Thermodynamics (TD) deals solely with macroscopic quantities and their relations with each other; it does not refer to the underlying microscopic structure. Historically, it was developed as a discipline before the connection with the microscopic structure was known. This connection can be characterized by the central equation $S = k_\mathrm{B} \ln \Omega$. This relation was established by Boltzmann at a time when S was only defined macroscopically (as a observable).

Laws of TD

Statistical physics is based on the Hamilton operator of the considered system and on the fundamental postulate. In contrast, thermodynamics (TD) is based on the following laws:

$$dE = đQ + đW \qquad \text{1st law of TD} \tag{13.12}$$

$$\begin{aligned} &\Delta S \geq 0 \qquad\quad \text{(closed system)} \\ &dS = đQ_\mathrm{q.s.}/T \quad \text{(quasi-static process)} \end{aligned} \qquad \text{2nd law of TD} \tag{13.13}$$

$$S \xrightarrow{T \to 0} 0 \qquad \text{3rd law of TD} \tag{13.14}$$

These statements can be understood as the basic laws of thermodynamics like the Newtonian axioms for mechanics or Maxwell's equations for electrodynamics. Such laws of nature cannot be derived. They are either postulated or established as a generalization of key experiments.

In the previous chapters, we have discussed these laws in connection to the underlying microscopic structure. In doing so, we have made them partially plausible. For example, we discussed the improbability of finding all the atoms of an ideal gas in a partial volume (Chapter 12). This example illuminated the statistical basis of the factual impossibility of a process with $\Delta S < 0$ in a closed system. In addition, we have invoked as a fact of experience that a closed system strives to equilibrium. And we have defined specific experimental conditions (adiabatic, no change of external parameters). This lead to the classification of possible energy transfers, i.e. to the 1st law of TD.

Observables

In the previous chapters, we have introduced macroscopic quantities on a microscopic basis; this applies in particular to the entropy and the temperature. From this we derived properties of these quantities that allow to link them to observables. A central property of temperature was (9.17):

$$T_A = T_B \qquad \text{(equilibrium at thermal contact)} \qquad (13.15)$$

In thermodynamics, the temperature is introduced from the outset as a observable, whereby (13.15) is assumed.

Equations of state

The laws of TD do not refer to specific systems. The statements that follow from them therefore apply in general. One such statement is, for example, the efficiency of an ideal heat engine (Chapter 19). More detailed statements can be obtained if the equations of state of the considered system are used. In statistical physics, these equations of state are determined according to the scheme (13.1); in thermodynamics they are introduced as phenomenological assumptions.

We will mainly consider *homogeneous systems in equilibrium*. A homogeneous system is of the same structure everywhere in the range of the volume V. This applies, for example, to a volume of gas or a copper rod, but not for the system "gas plus box" or for a silver-plated copper rod. Such composite systems can possibly be treated as two homogeneous systems. For homogeneous systems, one can often restrict the external parameters to the volume V and to the particle number N. The particle number can be replaced by the mass M of the system.

If only the external parameters V and N occur, then the state of equilibrium is determined by E, V and N. This means that

$$T = T(E, V, N) \quad \text{and} \quad P = P(E, V, N) \qquad (13.16)$$

are fixed. The first relationship may be resolved to E,

$$E = E(T, V, N) \qquad \text{caloric equation of state} \qquad (13.17)$$

If we insert this into the second relation, we get

$$P = P(T, V, N) \qquad \text{thermal equation of state} \qquad (13.18)$$

As we will see in thermodynamics (Part III), these two equations of state are not completely independent of each other. In statistical physics, the equations of state are derived from the microscopic structure (such as (13.10) and (13.11)). In thermodynamics, they are introduced as phenomenological *assumptions*.

Equilibrium states

According to (13.1), the equilibrium state of a system is defined by the energy E and the parameters $x = (x_1,..., x_n)$. All other macroscopic quantities such as S, T, P or X are also fixed in the equilibrium state; they are functions of E and $x_1,..., x_n$. The macroscopic quantities that are given in an equilibrium state are called (thermodynamic) *state variables*. To determine a thermodynamic state, one must specify $n + 1$ state variables. Possible choices are:

$$
\text{equilibrium state} =
\begin{cases}
E, x_1, \ldots, x_n \\
\quad\text{or} \\
y_1, \ldots, y_{n+1}
\end{cases}
=
\begin{cases}
E, V \\
T, V \\
T, P \\
V, P \\
S, V \\
\vdots
\end{cases}
\tag{13.19}
$$

The last row refers to the frequent case where the external parameters are V and $N = \text{const}$. The dependence on N is omitted.

In statistical physics, we start with the energy eigenvalues $E_r(x_1,..., x_n)$. The primary state variables are therefore $E, x_1,..., x_n$. However, one can decide to use other state variables $y_i = f_i(E, x_1,..., x_n)$. In thermodynamics, initially all possible sets of state variables are of equal right. Which of these sets is used depends on pragmatic considerations. Many calculations in thermodynamics concern the transition between different variables.

Exercises

13.1   Magnetization of an ideal spin system

For a system of N spin $1/2$ particles in an external magnetic field B, the partition function Ω is given by (6.24). Determine the energy $E = E(T, B)$. Establish the relation between the magnetization M (magnetic moment per volume) and the generalized force associated with B. Calculate the magnetization $M(T, B)$ and sketch it as a function of B/T.

13.2   Entropy and temperature in a two-level system

A system consists of a large number N of distinguishable, independent particles. Each particle has the choice between the two energy levels $\varepsilon_1 = 0$ and $\varepsilon_2 = \varepsilon$. What is the number of states Ω_n in which exactly n particles are in the (excited) level ε_2? Determine the number of states $\Omega(E)$ in the interval $(E, E - \varepsilon)$. Calculate the entropy $S(E)$ and the temperature T of the equilibrium state. Determine the ratio $n/(N - n)$ of the mean occupation numbers and the energy E as functions of the temperature. What follows from the condition $T \geq 0$ for the ratio of the occupation numbers? What energy results for $T \to \infty$?

14 Measurement of macroscopic observables

We discuss the measurability of macroscopic quantities. In particular, the initially microscopically introduced entropy $S = k_B \ln \Omega$ and temperature (given by $1/T = \partial S/\partial E$) are defined as measurable quantities.

Energy

We consider two equilibrium states a and b. The definition of work as an observable is assumed to be known. One can then measure the work W_{ab} required to move from a to b. Due to the conservation of energy, the following applies

$$E_b - E_a = A_{ab} = \text{work done on the system} \tag{14.1}$$

This defines the measurement of energy differences. The unit of the energy is

$$[E] = J = Nm = VAs \tag{14.2}$$

i.e. joule, newton-meter or volt-ampere-second. By (14.1) the energy of an arbitrary system state is fixed except for a constant. This constant is left open; alternatively, a specific equilibrium state a can be chosen as the energy zero point.

Using the example of a gas, we explain how work can lead from a state a to any other state b. An equilibrium state of a gas can be specified by the pressure P and the volume V; let the number of particles N be constant. Then two arbitrary states a and b are defined by P_a, V_a and P_b, V_b, i.e. by two points in the P-V- diagram in Figure 14.1.

First, we expand (or compress) the thermally isolated gas quasi-statically from V_a to V_b. In the process, the energy $W_{\text{mech}} = \Delta W_{\text{q.s.}} = -\int P\, dV$ is supplied to the gas; depending on the sign of $V_a - V_b$, the work W_{mech} is positive or negative. The gas reaches a new equilibrium state P_b', V_b. Then, for constant volume V_b the gas is heating by an electric resistor adding the energy W_{electr}. This is continued until the desired higher pressure P_b is reached. Since the external parameters (i.e. the volume) remain constant in this step, the work W_{electr} is equal to the heat absorbed by the system, $\Delta Q = W_{\text{electr}} > 0$. The 1st law of TD yields the energy balance (14.1),

$$E_b - E_a = \Delta E = \Delta Q + \Delta W = W_{\text{electr}} + W_{\text{mech}} \tag{14.3}$$

With the specified method, two arbitrary states P_a, V_a and P_b, V_b can be connected with each other. The direction of the process ($a \to b$ or $b \to a$) must be selected such that $W_{\text{electr}} > 0$.

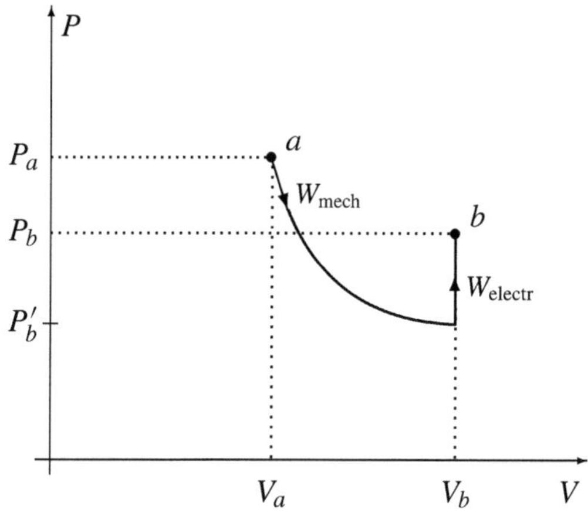

Figure 14.1   The energy difference between two equilibrium states a and b can be determined by measuring the work supplied.

Heat

For any process $a \to b$ we can apply the 1st law of TD

$$\Delta Q = \Delta E - \Delta W \tag{14.4}$$

Here ΔQ is the amount of heat absorbed by the system, $\Delta E = E_b - E_a$ is the change in energy and ΔW the work done on the system. The previous section explained the measurement of ΔE, and that of ΔW is assumed to be known. Then (14.4) defines heat ΔQ as an observable. The amount of heat is measured in joules, just like the energy:

$$[Q] = \mathrm{J} = \mathrm{Nm} = \mathrm{VAs} \tag{14.5}$$

In the past, the unit calorie was used. A calorie is the amount of heat needed to raise the temperature of one gram of water from $14.5\,^\circ\mathrm{C}$ to $15.5\,^\circ\mathrm{C}$ (normal pressure). The specific heat c_P of water is

$$c_P = 1\,\frac{\mathrm{cal}}{\mathrm{g\,K}} \approx 4.1868\,\frac{\mathrm{J}}{\mathrm{g\,K}} \qquad (\text{water, } P = 1\,\mathrm{atm},\ T \approx 15\,^\circ\mathrm{C}) \tag{14.6}$$

This results in the conversion $1\,\mathrm{cal} \approx 4.2\,\mathrm{J}$ between calorie and joule. The unit K stands for the kelvin to be defined below.

Generalized forces

The external parameters $x = (x_1, ..., x_n)$ are macroscopic observables; examples are the volume V, the number of particles N (equivalent to mass) and an external magnetic field B. If the i-th parameter is changed quasi-statically, the work

$$đW_\mathrm{q.s.} = -X_i\,dx_i \qquad (\text{quasi-static, } dx_{j \neq i} = 0) \tag{14.7}$$

is supplied to the system. The quantities $đW_\mathrm{q.s.}$ and dx_i can be measured. Thereby the generalized forces X_i are defined as measurable quantities.

If the parameter is the volume, $x_i = V$, then the generalized force is the pressure, $X_i = P$. According to (14.7) it has the dimension energy per volume or force per area. The unit of pressure is

$$[P] = \text{Pa} = \text{pascal} = \frac{\text{N}}{\text{m}^2} = 10^{-5}\,\text{bar} \tag{14.8}$$

Other old-fashioned units are $1\,\text{at} = 1\,\text{kp/cm}^2 \approx 0.98\,\text{bar}$ (technical atmosphere, kp = kilopond) and $1\,\text{atm} = 760\,\text{Torr} \approx 1.013\,\text{bar}$ (standard atmosphere).

For the parameter $x_i = N$, the generalized force is equal to the negative chemical potential $X_i = -\mu$ (Chapter 20). For the magnetic field $x_i = B$, the generalized force (8.5) is equal to the magnetic moment $X_i = VM$.

Temperature

We now come to the central topic of this chapter, the measurement of the temperature. For the temperature introduced in Chapter 9 the following holds

$$T_A = T_M \qquad \begin{array}{c}\text{(equilibrium between A and}\\ \text{M with thermal contact)}\end{array} \tag{14.9}$$

We have also derived the equation of state (13.11) for an ideal gas:

$$PV = Nk_B T \tag{14.10}$$

We consider two macroscopic equilibrium systems A and B with the (unknown) temperatures T_A and T_B. For the temperature measurement we use a third system M, which consists of a small amount of an ideal gas in a fixed volume. The system M is first brought into contact with A, then with B. As the amount of gas is small, the temperature of the system to be measured does not change noticeably. The contact with A leads to the temperature $T_M = T_A$, and the contact with B to $T_M = T_B$. According to (14.10), the gas then has the pressure $P_A = Nk_B T_A/V$ or $P_B = Nk_B T_B/V$. From this follows

$$\frac{T_A}{T_B} = \frac{P_A}{P_B} \tag{14.11}$$

The pressures P_A and P_B of the gas can be measured. This defines the measurement of the ratio T_A/T_B of any two systems.

A real gas approaches the ideal gas for low density. Therefore, we may replace (14.11) by

$$\left(\frac{P_A}{P_B}\right)_{\text{real}} \xrightarrow{N/V \to 0} \left(\frac{P_A}{P_B}\right)_{\text{ideal}} = \frac{T_A}{T_B} \tag{14.12}$$

The pressure of a real gas in contact with A or B is measured for different densities. Then one plots P_A/P_B as a function of the density N/V. From this plot, one may read off plot limiting value for $N/V \to 0$. This method is, however, not suitable for low temperatures, as ultimately all real gases condense into liquids. For this

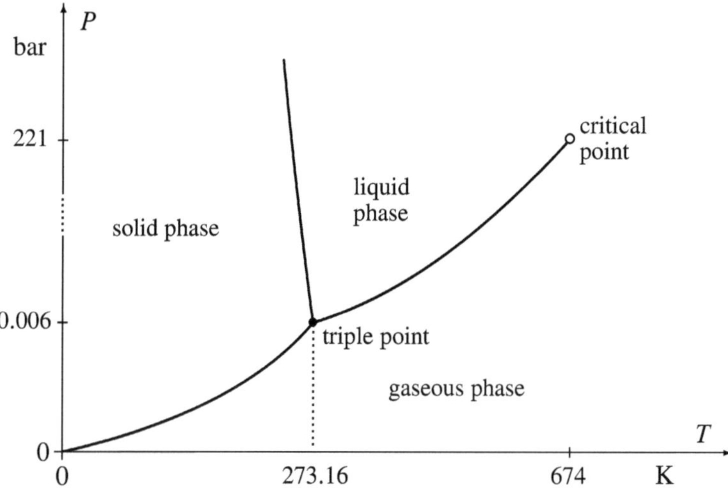

Figure 14.2 Qualitative sketch of the state diagram of water. By definition, the temperature of the triple point is given the *exact* the value 273.16 K. This defines the unit kelvin. The sketch also shows the critical point above which the transition gaseous ↔ liquid becomes continuous.

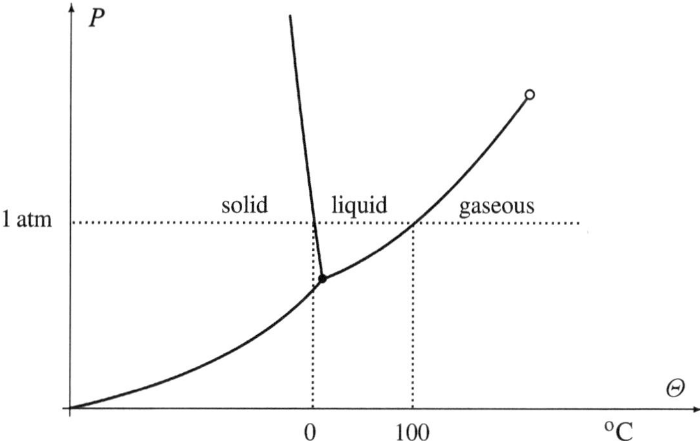

Figure 14.3 The Celsius scale $\Theta_{\text{historical}}$ was defined by the melting and boiling point of water at normal pressure (1 atm ≈ 1.013 bar). The temperature values 0 °C and 100 °C are assigned to these points. The sketch is not to scale; for example, the triple point lies at 0.01 °C, i.e. in the given scale practically at the same temperature as the melting point.

Table 14.1 Comparison of the Kelvin scale T and of the Celsius scale Θ. The triple, freezing and boiling points refer to water (at normal pressure). If no "$\approx$" sign is shown, the numerical value applies exactly. The numerical value in $T_{\text{triple}} = 273.16\,\text{K}$ was chosen such that one gets a simple conversion, namely $\Delta T = 1\,\text{K} = 1\,°\text{C} = \Delta\Theta$.

	T	$\Theta_{\text{historical}}$	Θ
Zero point	$0\,\text{K}$	$\approx -273.15\,°\text{C}$	$-273.15\,°\text{C}$
Triple point	$273.16\,\text{K}$	$\approx 0.01\,°\text{C}$	$0.01\,°\text{C}$
Freezing point	$\approx 273.15\,\text{K}$	$0\,°\text{C}$	$\approx 0\,°\text{C}$
Boiling point	$\approx 373.15\,\text{K}$	$100\,°\text{C}$	$\approx 100\,°\text{C}$

purpose equation (14.10) must be replaced by another relationship which contains the T together with other measurable variables.

If we arbitrarily fix the value of T_{B} for a specific system B, then (14.11) determines the value of T_{A} of all other systems A. As a standard system for the definition of the fixed point one chooses pure water in which the three phases, water vapor, water and ice, are in equilibrium. For this triple point of water, the following convention is agreed upon:

$$T_{\text{triple}} \overset{\text{def}}{=} 273.16\,\text{K} \qquad (\text{K} = \text{kelvin}) \tag{14.13}$$

Exactly this numerical value, i.e. $273.160000\ldots$, is used. This means that $1\,\text{K}$ is the $1/273.16$ th part of the temperature range between $T = 0$ and T_{triple}. This is comparable with the convention that 1 degree is the 360 th part of a full circle. Because of this analogy, one could also use "degrees kelvin"; however, this squiggling is only used for the Celsius scale (yet to be introduced).

At this point, we add a few comments on the state diagram of water shown in Figure 14.2. The equilibrium state of a certain amount of water can be determined by two variables. One often uses the pressure P and the temperature T because they are relatively easy to measure. In the P-T diagram, each point corresponds to a specific state equilibrium state. In different P-T ranges water occurs as a liquid, as a solid (ice) or as a gas (water vapor). Two of these *phases* are generally separated from each other by a curve in the P-T diagram, Figure 14.2. For P-T values on this curve, the two adjacent phases are in equilibrium with each other. For example, if pieces of ice are floating in a thermally insulated bucket of water, the temperature of the water and the ice is then a function of the pressure. At a certain point in the state diagram (Figure 14.2), the three curves that separate two phases meet. A system in which water, ice and water vapor are in equilibrium has a well defined temperature and is therefore suitable for fixing the temperature scale.

Celsius scale

In addition, we define the temperature Θ of the Celsius scale by

$$\Theta \overset{\text{def}}{=} \left(\frac{T}{\text{K}} - 273.15 \right)\,{}^{\circ}\text{C} \tag{14.14}$$

This means that a temperature difference of 1 kelvin is exactly equal to 1 degree celsius:

$$\Delta T = 1\,\text{K} = 1\,{}^{\circ}\text{C} = \Delta\Theta \tag{14.15}$$

In (14.14) we used different symbols, T and Θ, in order point out the different units. In practical use the symbol T is also used for the Celsius scale, for example $T = 15\,{}^{\circ}\text{C}$.

Historically, the Celsius scale was defined by the freezing and boiling point of water at normal pressure, see Figure 14.3 and Table 14.1. The choice of the numerical value in (14.13) together with the definition (14.14) was made in such a way that

$$\Theta \approx \Theta_{\text{historical}} \tag{14.16}$$

Table 14.1 compares the Kelvin scale, the historical scale and the current Celsius scale with each other. The differences between the two Celsius scales are of the size $10^{-3}\,{}^{\circ}\text{C}$.

At the time of the introduction of the Celsius scale, the existence of a lower temperature limit, the absolute zero point $T = 0$, was not known. On the historical Celsius scale, the zero temperature is therefore to be determined experimentally. The temperature T, which was introduced later, implies the existence of the zero point and is therefore also called the "absolute" temperature. We will not use this term.

Thermometer

The method (14.10)–(14.12) defines T as the measurable quantity. For a practical measurement, this method is suitable only to a limited extent. We we discuss the question of suitable measuring devices for T. As such *thermometer*, all macroscopic systems M are suited that fulfill two conditions:

1. A macroscopic parameter ϑ changes by suitable amounts when M is brought into thermal contact with other systems.

2. The system M must be small compared to the system to be measured.

The first condition implies that the respective change of ϑ is easy to be read off. The second condition shall guarantee that the disturbance of the system to be measured is negligibly small. Whether a system M is suitable depends on the kind of application, in particular on the temperature range to be measured, on the systems to be examined and on the desired accuracy. Examples of such systems M are

- Mercury thermometer: $\vartheta = h =$ height of the column.

- Gas at constant pressure: $\vartheta = V =$ volume.

- Electrical resistance: $\vartheta = R =$ resistance.

Such a thermometer is brought into contact with systems whose temperature T is known. Thereby a measured ϑ–value can be assigned to the known T–value:

$$T = T_{\mathrm{M}}(\vartheta) \qquad \text{(calibration)} \tag{14.17}$$

By this calibration, a scale for T can be attached to the thermometer. Two different calibrated thermometers M and M$'$ then show the same temperature.

Entropy

According to the 2nd law of TD, the entropy difference between two equilibrium states a and b is given by

$$S_b - S_a = \int_a^b \frac{\mathdj Q_{\mathrm{q.s.}}}{T} \tag{14.18}$$

This reduces the entropy measurement to a combined heat and temperature measurements (which were discussed above). The absolute value of entropy can be determined by the 3rd law of TD.

The unit of entropy is joule per kelvin,

$$[S] = \frac{\mathrm{J}}{\mathrm{K}} \tag{14.19}$$

Boltzmann constant

The definition (14.13) fixes the Boltzmann constant k_{B}. An ideal gas with N atoms in a volume V has at T_{triple} has a certain pressure P. If one insert the measured values for P, V and N, and the definition $T_{\mathrm{triple}} = 273.16\,\mathrm{K}$ in (14.10), then one obtains[1] for k_{B}:

$$k_{\mathrm{B}} = (1.380\,648\,52 \pm 0.000\,000\,79) \cdot 10^{-23}\,\frac{\mathrm{J}}{\mathrm{K}} \qquad \text{(Boltzmann constant)} \tag{14.20}$$

The energy unit electron volt (eV) is often more practical than $\mathrm{J} = \mathrm{C\,V}$; in eV, the symbol e denotes an elementary charge. It has the value $\mathrm{e} \approx 1.6 \cdot 10^{-19}\,\mathrm{C}$; where C is the unit coulomb. For estimations, we note that room temperature corresponds to approximately one fortieth of an electron volt:

$$300\,k_{\mathrm{B}}\mathrm{K} \approx \frac{1}{40}\,\mathrm{eV} \tag{14.21}$$

---

[1] *2014 CODATA recommended values* at http://physics.nist.gov/constants

We have first defined the temperature scale here; form this follows the value of k_B. The microscopic definition of the temperature in Chapter 9 showed that $k_B T$ (with initially arbitrary k_B) is proportional to the energy per degree of freedom. From a theoretical point of view, it would be natural to set $k_B = 1$ and measure the temperature in to measure the temperature in energy units:

$$k_B = 1 \quad \longrightarrow \quad [T] = \text{Joule} \tag{14.22}$$

With this choice, the temperature of the triple point of water would $T_{\text{triple}} \approx 3.77 \cdot 10^{-21}$ J. A unit such as kelvin or degrees celsius is therefore superfluous; in $k_B T$ it is truncated anyway. These units are historical ballast and a hindrance to the physical understanding of the temperature.

Loschmidt constant

The mass m of an atom is essentially proportional to the number A of nucleons in the atomic nucleus. Therefore, A grams of different monatomic gases contain approximately the same number of particles; this number is called *Loschmidt constant* (also Loschmidt's number or Avogadro constant) L_0. We note that there is a deviating definition which is also quite common: The Loschmidt constant is the number of gas particles of an ideal gas in one cubic meter under normal conditions.

Compared to the simple proportionality to atomic number A, there are corrections due to the different masses of neutrons and protons, and due to the mass defect (binding energy in the nucleus), and due to the contribution of the electrons, and due to possible isotope mixtures. One *defines* therefore L_0 as the number of atoms in 12 g pure ^{12}C. The L_0 defined in this way may be determined experimentally[1]:

$$L_0 = (6.022\,140\,857 \pm 0.000\,000\,074) \cdot 10^{23} \tag{14.23}$$

The amount of substance that L_0 molecules of a certain type of substance X (for example H_2O or O_2) is now defined as one mole or 1 mol. The following applies

$$1\,\text{mol X} \,\hat{=}\, L_0\, m_x\, \text{X} \qquad (m_x = \text{mass of the molecule}) \tag{14.24}$$

One mole of water is therefore an amount of water with a mass of 18 g, and one mole of oxygen gas (O_2) corresponds to about 32 g of oxygen. We also use the Loschmidt constant in the form

$$L = \frac{L_0}{\text{mol}} \approx \frac{6 \cdot 10^{23}}{\text{mol}} \tag{14.25}$$

The amount of a substance can be expressed by the *mass* ($M = N m_x$, in units of g or kg) or as *amount of substance ν* (in units of mole). The following applies to this substance amount

$$\nu = \frac{M}{L_0\, m_x}\,\text{mol} = \frac{N}{L_0}\,\text{mol} = \frac{N}{L} \tag{14.26}$$

In the MKSA system, the units mole and kg are regarded as independent units. Deviating from this, 1 mol carbon could also be defined as 12 g carbon; an equal sign would then be used in (14.24).

We insert $\nu = N/L$ into the ideal gas law (14.10):

$$PV = Nk_{\mathrm{B}}T = \nu L k_{\mathrm{B}}T = \nu RT \tag{14.27}$$

Here k_{B} and L have been combined to the historically older *gas constant R*,

$$R = L k_{\mathrm{B}} \approx 8.3145 \, \frac{\mathrm{J}}{\mathrm{K\,mol}} \tag{14.28}$$

The gas constant is alternatively called molar gas constant, universal gas constant or ideal gas constant. The quantity R refers to one mole, just as k_{B} refers to one particle. These quantities have the same dimension as entropy. As a rule, the following applies

$$S = N \, \mathcal{O}(k_{\mathrm{B}}) = \nu \, \mathcal{O}(R) \tag{14.29}$$

In the microscopic description, we prefer to relate the quantities to a particle and use k_{B}, whereas in thermodynamics, on the other hand, we often refer to the number of moles and use R. For the ideal gas law, these alternatives are

$$\begin{aligned} Pv &= k_{\mathrm{B}}T \qquad \text{for } v = V/N \\ Pv &= RT \qquad \text{for } v = V/\nu \end{aligned} \tag{14.30}$$

We use the same symbol v for V/N and V/ν; the respective meaning follows from the context. The volume per mole is called molar volume .

Standard conditions

The state of homogeneous systems can often be defined by its pressure and temperature. Then material constants (such as density, specific heat, compressibility, conductivity) are functions of P and T. For example, if one says that the density is so many kilograms per cubic meter, one must –in principle– also specify the values for P and T. If such a specification is missing, one usually refers to *standard conditions* for P and T. This is understood to mean

$$\begin{aligned} P &= 101\,325\,\mathrm{Pa} \approx 1\,\mathrm{bar} \\ T &= 0\,{}^{\circ}\mathrm{C} = 273.15\,\mathrm{K} \end{aligned} \qquad \text{standard conditions} \tag{14.31}$$

Somewhat deviating from this are the *normal conditions* which mean $T = 20\,{}^{\circ}\mathrm{C}$ (and the same pressure). These are approximately the conditions in the laboratory if no special precautions are taken. If the pressure P, the temperature T and the number of particles N of a gas are given, then the volume V is fixed. We calculate the volume of one mole of an ideal gas:

$$\left.\begin{aligned} PV &= Nk_{\mathrm{B}}T, \quad N = L_0 \\ &\text{normal conditions} \end{aligned}\right\} \quad \longrightarrow \quad V \approx 22.4\,\mathrm{L} \tag{14.32}$$

This can often be used as an estimate for common gases. For example, 22 liters (unit sign L, $1\,\mathrm{L} = 10^{-3}\,\mathrm{m}^3$) of auditorium air has a mass of about 30 grams.

III Thermodynamics

15 State variables

The tasks and goals of thermodynamics (outlined in Chapter 13) are investigated in detail. The thermodynamics (TD) deals with the macroscopic properties of various systems without referring to the underlying microscopic structure. The basis of the investigations are the laws of TD. They may be supplemented by assumptions about the equations of state for specific systems.

In this chapter, we consider the state variables and general relationships between them. State variables (or state functions, or state quantities) are the macroscopic quantities that are fixed in a thermodynamic state.

State variables

In thermodynamics, the considered *states* are mostly equilibrium states. All considered processes begin and end in a state of equilibrium. We will also investigate quasi-static processes in which all intermediate states are equilibrium states. If we consider non quasi-static processes, we do not make specific statements about the intermediate states (which are then non-equilibrium states).

We restrict ourselves initially to homogeneous systems whose only external parameters are V and N; the number of particles N can also be replaced by the number of moles ν, (14.26). The equilibrium state is then determined by the energy E, the volume V and the particle number N. These three quantities may be replaced by three other macroscopic quantities. The quantities defining the state are called *state variables* or state functions. We denote them summarily with y. Possible state variables for homogeneous systems are

$$\text{state variables:}\quad y = (E, V, N),\ (T, V, N),\ (T, P, N),\ (S, V, N),\ldots \quad (15.1)$$

The entropy S, the temperature T and the pressure P have been introduced in Part II. Other designations are U for the (internal) energy instead of E, and p instead of P for the pressure.

Homogeneous systems can be gases, liquids or solids. A phase equilibrium (such as water and ice), on the other hand, is a system that is composed of two or more homogeneous systems.

State variables are physical quantities that are are defined in a state of equilibrium, i.e. all quantities that are a (unique) function of the state variables:

$$\text{state variables:} \quad f = f(y) \overset{\text{e.g.}}{=} f(T, P, N) \tag{15.2}$$

For homogeneous systems, the state variables are either *extensive* or *intensive* quantities. If the homogeneous system is divided into two parts, A and B, the following applies to the state variables,

$$\begin{aligned} f \text{ is extensive if} \quad f = f_A + f_B \\ f \text{ is intensive if} \quad f = f_A = f_B \end{aligned} \tag{15.3}$$

Examples of extensive quantities are energy, mass (or number of particles), heat capacity, entropy and volume. In contrast, energy density, particle density, specific heat, temperature and pressure are intensive quantities. Since the division into subsystems A and B is arbitrary and since the system is homogeneous, extensive quantities are proportional to N, whereas intensive quantities do not depend on N. For the state variables T, P, N the following holds

$$f = \begin{cases} N\,g(T, P) & \text{extensive} \\ f(T, P) & \text{intensive} \end{cases} \tag{15.4}$$

Using the variables E, V and N, an extensive quantity is of the form $f = N\,g(E/N, V/N)$.

We will initially restrict ourselves to processes in which N is constant. Then (15.1) becomes

$$\text{state variables:} \quad y = (E, V),\ (T, V),\ (T, P),\ (S, V),\ \ldots \tag{15.5}$$

If we choose the state variables T and V, for example, then the functions $S = S(T, V)$, $P = P(T, V)$ or $E = E(T, V)$ are state variables, too (in addition to T and V itself). Which quantities are chosen as variables is a question of expediency. In a microscopic treatment (as for the ideal gas in Chapter 6) E and V are the preferred variables. Experimentally it is on the other hand, the more easily accessible variables are T and P. Depending on the variable to be calculated, thermodynamics often switches from one pair of variables to another.

Partial derivatives

State functions like $S(E, V)$ depend in general on several variables. In thermodynamic processes, changes of state variables are investigated. For this purpose, a partial derivative describes how the considered quantity behaves as a function of one state variable when the other state variables are held fixed.

Physical notation

We explain the specific notation commonly used in thermodynamics for the functions (15.2) and their partial derivatives. As an example, we consider the entropy S as a state function depending on the variables T, V or on E, V. The mathematical (left) and physical (right) notations for this are:

$$S = f(T, V) = g(E, V) \quad \text{or} \quad S = S(T, V) = S(E, V) \qquad (15.6)$$

If the entropy S is written as a function of T, V or E, V, this results in *different* functions. This is shown in the left part by the different designations (f and g). However, $f(T, V)$ and $g(E, V)$ stand for the same physical quantity and therefore have the same function value (for a given state). It is therefore common in physics to use the same letter, like S on the right part of Equation (15.6).

   This physical notation is dangerous insofar as we designate different functions with the same letter; the substitution of expressions into the arguments can then lead to errors. For example, when working with dimensionless variables, $S(3, 4)$ is obviously ambiguous: One may relate this to $S(E, V)$ and to $S(T, V)$ yielding $S(3, 4) \neq S(3, 4)$. On the other hand, we will usually work with at least four different sets of variables. If we then (formally correctly) denote different functions differently, we would have to introduce four different letters for the entropy. So we have the choice between dangerous notation and a confusing variety of designations. It is common to opt for the former.

   Let us note: For a state variable we always use the same letter (e.g. S), even if we switch to a different set of variable and thus to a different function. If the variables are characterized by letters, such as $S(E, V)$ or $S(T, V)$, the conventions for the variable designation show which function is meant.

   For the two possibilities $S(E, V)$ and $S(T, V)$ we write on the exact differentials

$$dS = \frac{\partial S(E, V)}{\partial E} dE + \frac{\partial S(E, V)}{\partial V} dV = \frac{\partial S(T, V)}{\partial T} dT + \frac{\partial S(T, V)}{\partial V} dV \quad (15.7)$$

The value S in a certain state is independent of whether it is described by E, V or T, V. Therefore, this also applies for the entropy difference dS between two infinitesimally neighboring states. In thermodynamics, the partial derivatives are commonly written in the form

$$\left(\frac{\partial S}{\partial E} \right)_V \equiv \frac{\partial S(E, V)}{\partial E} \qquad (15.8)$$

Then the first form of dS given in (15.7) becomes

$$dS = \left(\frac{\partial S}{\partial E} \right)_V dE + \left(\frac{\partial S}{\partial V} \right)_E dV \qquad (15.9)$$

For a quasi-static process, $dE = đQ_{\text{q.s.}} + đW_{\text{q.s.}}$ (1st law of TD), $T dS = đQ_{\text{q.s.}}$ (2nd law of TD) and $đW_{\text{q.s.}} = -P dV$ lead to the differential

$$dS = \frac{1}{T} dE + \frac{P}{T} dV \qquad (15.10)$$

By (15.9) or (15.10), dS is given as a function of the variables E and V. Therefore, the coefficients of the differentials must agree, i.e.

$$\frac{1}{T} = \left(\frac{\partial S}{\partial E}\right)_V, \qquad \frac{P}{T} = \left(\frac{\partial S}{\partial V}\right)_E \qquad (15.11)$$

This contains the following statements:

1. The state variables are E and V. The entropy is a function of these variables, $S = S(E, V)$.

2. The right-hand sides $\partial S(E, V)/\partial E$ and $\partial S(E, V)/\partial V$ are functions of E and V, too. Therefore, T and P/T are functions of E and V, i.e. $T = T(E, V)$ and $P = P(E, V)$.

Heat capacity

A special partial derivative is the heat capacity. For its definition, we assume that T is one of the state variables i.e. $y = (T, z)$. We now consider a process in which the other variables z are fixed and in which heat is supplied quasi-statically. Due to the heat transfer, the state and thus the temperature must change. The ratio between supplied heat $dQ_{\text{q.s.}}$ and the temperature change dT is defined as the *heat capacity* C_z of the system:

$$C_z = \left.\frac{dQ_{\text{q.s.}}}{dT}\right|_{z=\text{const.}} = \lim_{\Delta T \to 0} \left.\frac{\Delta Q_{\text{q.s.}}}{\Delta T}\right|_{z=\text{const.}} \qquad (15.12)$$

With the specification "$z = $ const." the quotient $dQ_{\text{q.s.}}/dT$ is uniquely defined. This quotient cannot not be written as a partial derivative, because Q is not a state variable (there is no function $Q(T, z)$). Using the 2nd law of TD, $\Delta Q_{\text{q.s.}} = T\,\Delta S$, we can write (15.12) as a partial derivative of the entropy:

$$C_z = \lim_{\Delta T \to 0} \left.\frac{T\,\Delta S}{\Delta T}\right|_{z=\text{const.}} = T\left(\frac{\partial S}{\partial T}\right)_z \qquad (15.13)$$

With $S(T, z)$, also $C_z(T, z)$ is a state variable.

The heat capacity of a homogeneous system is given by C_z. For such systems, we have distinguished in (15.3) between extensive and intensive quantities. The heat capacity is an extensive quantity. The ratio

$$c_z = \frac{C_z}{N}, \qquad c_z = \frac{C_z}{M} \quad \text{oder} \quad c_z = \frac{C_z}{\nu} \qquad (15.14)$$

is then denoted as the *specific heat* of the considered substance; this is an intensive quantity. The heat capacity can be related to the number of particles N or to the mass M (in grams) or the amount of substance ν (number of moles); we do not introduce separate symbols for the resulting specific heat.

The state variables T, z are in particular T, V, N or T, P, N. For this, the specific heat (per particle) is

$$
\begin{aligned}
c_P &= c_P(T, P) &= \frac{T}{N}\frac{\partial S(T, P, N)}{\partial T} \\[2mm]
c_V &= c_V(T, V/N) &= \frac{T}{N}\frac{\partial S(T, V, N)}{\partial T}
\end{aligned}
\tag{15.15}
$$

In the arguments of c_P and c_V we have taken into account that these are intensive quantities. In the index z of c_z, the particle number N is not listed because usually $N = \text{const.}$ is assumed.

Exact differential

To formulate some general mathematical statements we consider a function $f(x, y)$ that depends on two variables. In the applications, x, y and f are state variables.

We assume that the function $f(x, y)$ is defined for all occurring variable values and that it is twice differentiable. The differentiability of a function of two variables is equivalent to one of the following two statements: (i) In the neighborhood of a specific point, the surface $z = f(x, y)$ can be approximated by a tangential plane. (ii) The partial derivatives exist and are continuous. The mere existence of the partial derivative is not sufficient. For a function of several variables, differentiability is a much stronger condition than for the function of one variable.

The exact (or complete) differential of the function $f(x, y)$ is

$$
\begin{aligned}
df &= \frac{\partial f(x, y)}{\partial x}\,dx + \frac{\partial f(x, y)}{\partial y}\,dy = \left(\frac{\partial f}{\partial x}\right)_y dx + \left(\frac{\partial f}{\partial y}\right)_x dy \\[2mm]
&= A(x, y)\,dx + B(x, y)\,dy
\end{aligned}
\tag{15.16}
$$

In the following, we examine the integration of df to $f(x, y)$ and some relations that arise from (15.16).

Integration

The quantity df is the difference between $f(x + dx, y + dy)$ and $f(x, y)$. By summing up the differences df one gets the finite difference

$$
f(x, y) - f(x_0, y_0) = \int_{x_0, y_0}^{x, y} df
\tag{15.17}
$$

One may consider two different paths that lead from x_0, y_0 to x, y. Inverting the direction of one of the path, one may construct a closed path for which

$$
\oint df = 0
\tag{15.18}
$$

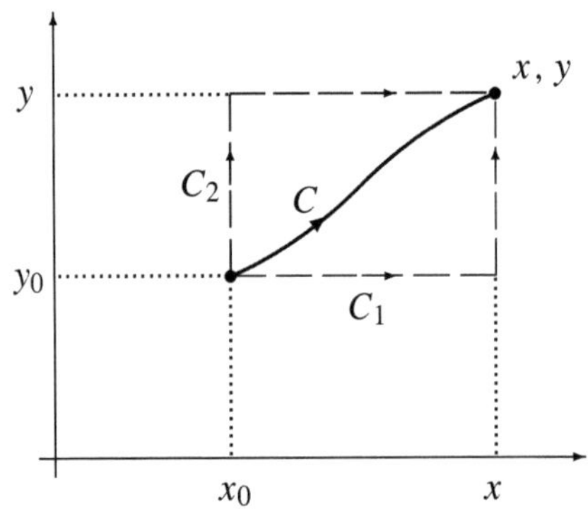

Figure 15.1 The exact differential df of a function $f(x, y)$ can be calculated by using an arbitrary way C. The calculation along the paths C_1 or C_2 is particularly simple.

holds. This applies to any closed path.

The integration (15.17) can be carried out along any path which leads from (x_0, y_0) to (x, y) in the x-y-plane. The result is independent of the chosen path, since in the summation of the differences df only the values at the end points survive. Figure 15.1 shows three possible paths including

$$
\begin{aligned}
C_1 &: \quad (x_0 \to x, \ y = y_0) \quad \text{and} \quad (x = \text{const.}, \ y_0 \to y) \\
C_2 &: \quad (x = x_0, \ y_0 \to y) \quad \text{and} \quad (x_0 \to x, \ y = \text{const.})
\end{aligned}
\tag{15.19}
$$

For the path C_1, the integral (15.17) becomes

$$
f(x, y) - f(x_0, y_0) = \int_{x_0}^{x} dx' \, \frac{\partial f(x', y_0)}{\partial x'} + \int_{y_0}^{y} dy' \, \frac{\partial f(x, y')}{\partial y'}
\tag{15.20}
$$

This means that the function $f(x, y)$ can be calculated from its partial derivatives.

Alternatively, one can proceed as follows: If the function $f(x, y)$ has the partial derivative $\partial f / \partial x = A(x, y)$, then we can set

$$
f(x, y) = \int^{x} dx' \, A(x', y) + F_1(y) = \int dx \, A(x, y) + F_1(y)
\tag{15.21}
$$

This is an indefinite integral; it can also be written as a integral with the upper limit x and a constant lower limit. The occurring integration constant is a function of the fixed variable y. Similarly, from the known partial derivative $\partial f / \partial y = B(x, y)$ we get

$$
f(x, y) = \int^{y} dy' \, B(x, y') + F_2(x) = \int dy \, B(x, y) + F_2(x)
\tag{15.22}
$$

The unknown function $F_1(y)$ must be part of the indefinite integral in (15.22); correspondingly, $F_2(x)$ is contained in the integral in (15.21). From both forms together, $f(x, y)$ can be determined up to a constant.

Relationships between partial derivatives

From the twofold differentiability of $f(x, y)$ follows

$$\frac{\partial^2 f(x, y)}{\partial x\, \partial y} = \frac{\partial^2 f(x, y)}{\partial y\, \partial x} \tag{15.23}$$

For (15.16) this means

$$\left(\frac{\partial A}{\partial y}\right)_x = \left(\frac{\partial B}{\partial x}\right)_y \tag{15.24}$$

For an expression of the form $A\, dx + B\, dy$ with arbitrary functions $A(x, y)$ and $B(x, y)$ the following applies

$$\begin{array}{c} A(x, y)\, dx + B(x, y)\, dy \\ \text{is an exact differential} \end{array} \quad \longleftrightarrow \quad \left(\frac{\partial A}{\partial y}\right)_x = \left(\frac{\partial B}{\partial x}\right)_y \tag{15.25}$$

Starting from (15.16) we deduced (15.24). The other direction of deduction is done in Exercise 15.1.

From the expression for the exact differential (15.16), the ratio of dx and dy at *constant f* can be read off. For $df = 0$ we get

$$\left(\frac{\partial x}{\partial y}\right)_f = -\frac{\left(\dfrac{\partial f}{\partial y}\right)_x}{\left(\dfrac{\partial f}{\partial x}\right)_y} \tag{15.26}$$

For constant y, i.e. for $dy = 0$, (15.16) yields the relation

$$\left(\frac{\partial x}{\partial f}\right)_y = \frac{1}{\left(\dfrac{\partial f}{\partial x}\right)_y} \tag{15.27}$$

The exact differential (15.16) can be written on for all state variables. Then (15.24), (15.26) and (15.27) result in a large number of relations between partial derivatives.

Exercises

15.1 Path integral and exact differential

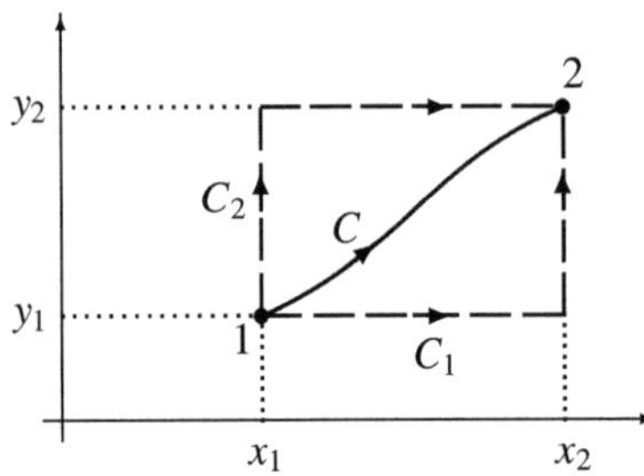

The path integral

$$I = \int_{1,C}^{2} \left[\, A(x, y)\, dx + B(x, y)\, dy \,\right]$$

with the fixed end points 1 and 2 can be calculated for different ways C. Show that I is independent of the path if and only if $\partial A/\partial y = \partial B/\partial x$ holds.

Hint: Write the integral in the form $I = \int_C d\boldsymbol{r} \cdot \boldsymbol{V}$ with $d\boldsymbol{r} = dx\, \boldsymbol{e}_x + dy\, \boldsymbol{e}_y$. Consider two different paths C_1 and C_2 (as in the picture) that have the same endpoints end points, and use Stokes' theorem.

16 Ideal gas

For the ideal gas, we perform some exemplary thermodynamic calculations. We show that the energy of the ideal gas does not depend on the volume and evaluate the difference $c_P - c_V$. Eventually we consider the isothermal and adiabatic P-V-relations, and the entropy of the ideal gas.

Equation of state

In thermodynamics, we make *assumptions* about the equations of state for specific systems. The most frequently considered approaches for the thermal equation of state for a gas are:

$$P = \frac{\nu R T}{V} = \frac{R T}{v} \qquad \text{(ideal gas)} \qquad (16.1)$$

$$P = -\frac{a}{v^2} + \frac{R T}{v - b} \qquad \text{(van der Waals gas)} \qquad (16.2)$$

Here $v = V/\nu$ is the volume per number of moles. In thermodynamics (TD), both equations are empirical approaches. In contrast to this, the derivation of (16.1) in Chapter 13 was based on the microscopic structure. The van der Waals equation is a more general approach that describes deviations of real gases from (16.1) with the help of two parameters a and b. It is quoted here to make it clear that (16.1) is only an approximation for the behavior of real gases; ultimately, this also applies to (16.2). In Chapter 28, the equation of state (16.2) will be justified microscopically.

Energy

We show that the energy of the ideal gas is independent of volume. The quantity of substance under consideration shall be constant throughout the following, $\nu = $ const. We can therefore restrict ourselves to two state variables, such as T, V or T, P.

From $dE = đQ_{\text{q.s.}} + đW_{\text{q.s.}}$ (1st law of TD), $T dS = đQ_{\text{q.s.}}$ (2nd law of TD) and $đW_{\text{q.s.}} = -P dV$ follows:

$$\boxed{dS = \frac{1}{T} dE + \frac{P}{T} dV} \qquad (16.3)$$

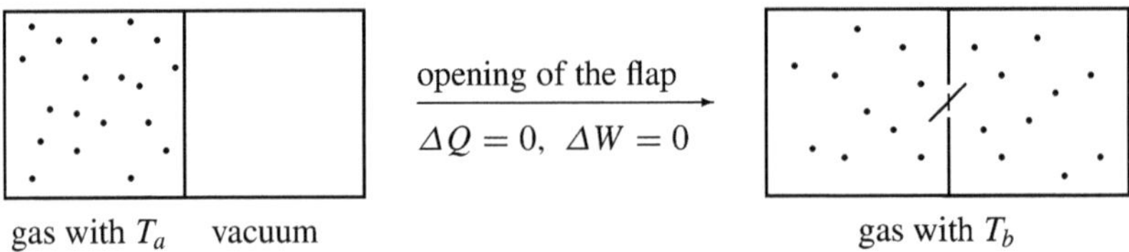

Figure 16.1 Sketch of the Gay-Lussac experiment: At the beginning, the gas is in a partial volume and has the temperature T_a, the other partial volume is empty (vacuum). The system is thermally insulated. By opening a flap between the two volumes, the gas expands into the whole volume without doing any work. In the final state, the temperature T_b is measured. For an ideal gas, $T_b = T_a$ holds.

In thermodynamics, we will often start from this fundamental relationship. We insert (16.1), $P/T = \nu R/V$, and the exact differential dE for $E(T, V)$:

$$dS = \underbrace{\frac{1}{T}\left(\frac{\partial E}{\partial T}\right)_V dT}_{(\partial S/\partial T)_V} + \underbrace{\left[\frac{1}{T}\left(\frac{\partial E}{\partial V}\right)_T + \frac{\nu R}{V}\right] dV}_{(\partial S/\partial V)_T} \qquad (16.4)$$

The identification with the partial derivatives of $S(T, V)$ is obtained if (16.4) is compared with the exact differential dS for $S(T, V)$. The second derivative $\partial^2 S/\partial T \partial V$ may be obtained from the first coefficient as well as from the second one:

$$\frac{\partial^2 S(T, V)}{\partial T \partial V} = \begin{cases} \dfrac{\partial}{\partial V}\left(\dfrac{\partial S}{\partial T}\right)_V = \dfrac{1}{T}\dfrac{\partial^2 E(T, V)}{\partial V \partial T} \\[2em] \dfrac{\partial}{\partial T}\left(\dfrac{\partial S}{\partial V}\right)_T = \dfrac{1}{T}\dfrac{\partial^2 E(T, V)}{\partial T \partial V} - \dfrac{1}{T^2}\dfrac{\partial E(T, V)}{\partial V} \end{cases} \qquad (16.5)$$

The equality of both expressions implies

$$\left(\frac{\partial E}{\partial V}\right)_T = 0 \qquad \text{(ideal gas)} \qquad (16.6)$$

The energy of the ideal gas can therefore be written as a function of T alone:

$$E = E(T, V) = E(T) \qquad \text{(ideal gas)} \qquad (16.7)$$

Figure 16.1 shows a simple arrangement that can be used to check whether a real gas behaves in this way. For this purpose the gas is expanded adiabatically ($đQ = 0$) and without work ($đW = 0$) from the initial volume V_a to the final volume V_b. This *free* expansion is characterized by $dE = đQ + đW = 0$, so it takes place without any change in energy:

$$E_b - E_a = E(T_b, V_b) - E(T_a, V_a) = 0 \qquad \text{(free expansion)} \qquad (16.8)$$

From (16.7) then follows

$$E(T_b) - E(T_a) = 0, \quad \text{also } T_b = T_a \qquad \begin{array}{c}\text{(free expansion}\\ \text{of the ideal gas)}\end{array} \qquad (16.9)$$

The volume independence of the energy implies that the temperature does not change by a free expansion. A real gas, however, will cool down during free expansion (Chapter 18).

Microscopically, the result $T_b = T_a$ for a free expansion is easy to understand. For an ideal gas, E is the sum of the kinetic energies of the atoms. The velocities of the particles do not change when they fly into the other subvolume. The average energy per particle and therefore the temperature remain the same.

Specific heats

We calculate the difference $c_P - c_V$ of the specific heats for an ideal gas. The heat capacity was defined in (15.13) by

$$C_z = T \left(\frac{\partial S}{\partial T} \right)_z \qquad (16.10)$$

One of the state variables is the temperature T; the other is $z = V$ or $z = P$. From (16.3) follows

$$T \, dS = dE \quad \text{for } V = \text{const.} \qquad (16.11)$$

This gives us

$$C_V = T \left(\frac{\partial S}{\partial T} \right)_V = \left(\frac{\partial E}{\partial T} \right)_V \qquad (16.12)$$

This holds in general because the equation of state was not used.

We now insert (16.12) and (16.6) into (16.4):

$$dS = \frac{C_V(T)}{T} \, dT + \frac{\nu R}{V} \, dV \qquad \text{(ideal gas)} \qquad (16.13)$$

Because of (16.7), the heat capacity $C_V(T) = (\partial E / \partial T)_V$ is a function of T only.

To determine C_P we express dS by dT and dP. To do this we write the equation of state of the ideal gas as $V = V(T, P) = \nu R T / P$ and form the exact differential

$$dV = \frac{\nu R}{P} \, dT - \frac{\nu R T}{P^2} \, dP \qquad (16.14)$$

We use this in (16.13):

$$dS = \left(\frac{C_V}{T} + \frac{\nu^2 R^2}{PV} \right) dT - \frac{\nu^2 R^2 T}{V P^2} \, dP \overset{(16.1)}{=} \frac{C_V + \nu R}{T} \, dT - \frac{\nu R}{P} \, dP$$

$$(16.15)$$

Table 16.1 The table compares the specific heat $c_P = 5/2\,R$ of the monatomic ideal gas with that of some real substances. If the specific heat is related to the number of particles, $c_P = C_P/N$, then the gas constant R had to be replaced by the Boltzmann constant k_B. The values apply to normal pressure and room temperature.

System	$c_P = C_P/\nu$
Ideal gas (monatomic)	$5/2\,R$
Noble gases	$2.5\,R$
Hydrogen gas	$3.5\,R$
Water	$9.0\,R$
Benzene	$16.4\,R$
Copper	$2.9\,R$
Diamond	$0.7\,R$

Table 16.2 The table shows the experimental values for c_P and $\gamma = c_P/c_V$ at normal conditions. Using the experimental value for c_P, a value γ_{theor} (last column) was calculated according to (16.17).

Gas	Symbol	c_P/R	γ_{exp}	γ_{theor}
Helium	He	2.50	1.667	1.667
Oxygen	O_2	3.521	1.397	1.397
Carbon dioxide	CO_2	4.312	1.301	1.302
Ethanol	C_2H_5OH	7.419	1.16	1.156

From this we can read off C_P:

$$C_P = C_V + \nu R \quad \text{or} \quad c_P - c_V = R \qquad \text{(ideal gas)} \qquad (16.16)$$

The difference $(C_P - C_V)\, dT$ is equal to the expansion work $P\, dV = \nu R\, dT$, which the gas performs when the temperature is increased under constant pressure.

The result $c_P = c_V + R$ follows from the laws of TD and the ideal gas law. From the microscopic treatment we had obtained $E(T, V) = 3\nu R T/2$, (13.10), for the monatomic ideal gas. This yields the specific heats $c_V = 3R/2$ and $c_P = 5R/2$.

Table 16.1 shows the specific heats c_P for some substances. Typical values of c_P are a few R (per mol) or a few k_B (per particle). Diamond has a particularly low specific heat. Substances with polyatomic molecules usually have higher values, because they have more degrees of freedom that can be excited. The specific heat c_P is itself a state variable, i.e. a function of T and P. The specified values apply to room temperature and normal pressure.

The equation of state (16.1) does not determine the absolute values of c_V and c_P, but only their difference. To compare the result (16.16) with the experiment we form the dimensionless ratio

$$\gamma \equiv \frac{c_P}{c_V} \overset{\text{(ideal gas)}}{=} \frac{1}{1 - R/c_P} \qquad (16.17)$$

Table 16.2 compares this quantity with the experimental values γ_exp for real gases; the experimental values may be obtained from the direct measurement of c_V and c_P, or from the speed of sound, Exercise 16.1. The comparison with γ_exp shows that the relation $c_P - c_V = R$ (found for ideal gases) is often also well fulfilled for real gases. However, this no longer holds if one approaches the vapor pressure curve.

Isothermal and adiabatic curves

An *isotherm* or *isothermal curve* is defined by $T = \text{const.}$ in the P-V diagram. For the ideal gas we get

$$PV = \text{const.} \qquad \text{(isotherm)} \qquad (16.18)$$

This statement is also called *Boyle-Mariotte law.*

In contrast to the isotherms we consider now the adiabats (adiabatic curves), which are obtained for a quasi-static *and* adiabatic volume change. Because $dS = \tilde{d}Q_\mathrm{q.s.} = 0$ these are also curves of constant entropy (isentrope or isentropic curve).

In the following, we assume that the heat capacity C_V is temperature independent,

$$C_V(T) = \text{const.} \qquad (16.19)$$

This applies to an ideal monatomic gas with $C_V = 3Nk_\mathrm{B}/2$. For polyatomic gases, (16.19) can be a sensible approximation in limited temperature ranges.

For $dS = 0$, (16.13) becomes

$$C_V \frac{dT}{T} + (C_P - C_V) \frac{dV}{V} = 0 \qquad (16.20)$$

We have used $C_P = C_V + \nu R$. We divide this equation by C_V. Because of (16.19) we can integrate:

$$\ln T + (\gamma - 1) \ln V = \text{const.} \tag{16.21}$$

Here $\gamma \equiv c_P/c_V$. We can write the left-hand side as $\ln(T\, V^{\gamma-1})$. This yields

$$T\, V^{\gamma-1} = \text{const.} \tag{16.22}$$

Similarly, (16.15) with $dS = 0$ results in

$$C_P \frac{dT}{T} - \left(C_P - C_V\right) \frac{dP}{P} = 0 \tag{16.23}$$

After division by C_P and integration we get from this

$$T\, P^{1/\gamma - 1} = \text{const.} \tag{16.24}$$

The elimination of T from (16.22) and (16.24) finally results in

$$P\, V^{\gamma} = \text{const.} \qquad \text{(adiabat)} \tag{16.25}$$

This *adiabat* (or adiabatic curve) in the P-V diagram may be contrasted with the isotherm (16.18). We note, that the adiabatic curve is steeper than the isothermal one in the P-V. The statements (16.22), (16.24) and (16.25) are also referred to as Poisson equations.

Entropy

We calculate the entropy $S(V, T)$ of an ideal gas. The exact differential dS for the entropy $S(T, V)$ of the ideal gas is given in (16.13). We integrate this according to (15.21) and (15.22):

$$S(T, V) = \int dT' \, \frac{C_V(T')}{T'} + F_1(V) \tag{16.26}$$

$$S(T, V) = \int dV' \, \frac{\nu R}{V'} + F_2(T) \tag{16.27}$$

The unknown functions $F_1(V)$ and $F_2(T)$ are given by the integral in the other expression (except for a constant). This gives us the entropy of the ideal gas,

$$S(T, V) = \nu \left(\int^T dT' \, \frac{c_V(T')}{T'} + R \ln V + \text{const.} \right) \tag{16.28}$$

where $c_V = C_V/\nu$. For fixed integral limits we get

$$S(T, V) - S(T_0, V_0) = \nu \int_{T_0}^T dT' \, \frac{c_V(T')}{T'} + \nu R \ln \frac{V}{V_0} \tag{16.29}$$

If the specific heat is temperature independent, the following holds

$$S(T, V) - S(T_0, V_0) = \nu \, c_V \, \ln \frac{T}{T_0} + \nu R \, \ln \frac{V}{V_0} \qquad (c_V = \text{const.}) \qquad (16.30)$$

We specify the particle number dependence of the entropy. Instead of $c_V = C_V/\nu$ we use $c_V = C_V/N$ (without introducing a new symbol); then $\nu \, c_V$ is replaced by $N c_V$. Together with $\nu R = N k_{\mathrm{B}}$, (16.28) becomes

$$S(T, V, N) = N \left(\int^{T} dT' \, \frac{c_V(T')}{T'} + k_{\mathrm{B}} \, \ln V + g(N) \right) \qquad (16.31)$$

Since $N = \text{const.}$ was previously assumed, the constant in (16.28) may depend on N. As an extensive quantity, S is of the form $S = N \, s(T, V/N)$; from this follows $g(N) = -k_{\mathrm{B}} \ln N + \text{const.}$, thus

$$S(T, V, N) = N \left(\int^{T} dT' \, \frac{c_V(T')}{T'} + k_{\mathrm{B}} \, \ln \frac{V}{N} + \text{const.} \right) \qquad (16.32)$$

Exercises

16.1   Compressibility and sound velocity

Determine the adiabatic compressibility and the speed of sound of an ideal gas:

$$\kappa_S = -\frac{1}{V}\left(\frac{\partial V}{\partial P}\right)_S, \qquad c_S = \sqrt{\left(\frac{\partial P}{\partial \varrho}\right)_S} \tag{16.33}$$

Here ϱ is the mass density of the gas. Estimate c_S for air ($\gamma \equiv c_P/c_V \approx 1.4$) under normal conditions numerically. Measuring the sound velocity is a simple method for determining γ.

16.2   Special volume-pressure relation

The change in volume of an ideal gas shall take place under the condition

$$\frac{dP}{P} = a\,\frac{dV}{V} \tag{16.34}$$

where a is a constant. Determine $P = P(V)$, $V = V(T)$ and the heat capacity $C_a = đQ/dT$. What results in the limiting cases $a = 0$ and $a \to \infty$?

16.3   Entropy of the ideal gas

From the ideal gas law $PV = Nk_{\mathrm{B}}T$ follows

$$S(T, V, N) = N\left(\int dT\,\frac{c_V(T)}{T} + k_{\mathrm{B}}\ln\frac{V}{N} + \text{const.}\right) \tag{16.35}$$

for the entropy. From $S = k_{\mathrm{B}}\ln\Omega$ with $\Omega(E, V, N)$ from (6.20), another expression is obtained $S(E, V, N)$ for the entropy of an ideal gas. How are these two expressions related to each other?

16.4   Mixing of a gas

A closed volume is divided by a wall into two subvolumes V_1 and V_2. In each partial volume there are N particles of the same monatomic, ideal gas. The temperatures T_1 and T_2 are chosen so that the pressure in both subvolumes is the same ($P_1 = P_2 = P_0$).

The wall is now pulled out sideways. Calculate the temperature and the pressure of the resulting state. Determine the change in entropy as a function of T_1, T_2 and N. What results for $T_1 = T_2$?

17 Thermodynamic potentials

We introduce the enthalpy H, the free energy F and the free enthalpy G. Together with the energy E, these quantities are referred to as thermodynamic potentials[1]. We show that all thermodynamic relations follow from one of the potentials. i.e. each thermodynamic potential contains the complete thermodynamic information.

We consider homogeneous systems whose equilibrium states are characterized by two state variables (15.5). We restrict ourselves here to the following pairs of variables:

$$\text{state variables: } (S, V), \ (T, V), \ (S, P) \ \text{ or } \ (T, P) \tag{17.1}$$

The *thermodynamic potentials* to be introduced have the following properties:

1. They are state variables with the dimension of an energy.

2. Their partial derivatives are simple expressions.

To show the analogy to the potential of mechanics, we consider the two-dimensional motion of a particle. The position of the particle is determined by the variables (x, y). If the particle moves in the potential $U = U(x, y)$, the partial derivatives

$$K_x = -\frac{\partial U(x, y)}{\partial x} \ , \qquad K_y = -\frac{\partial U(x, y)}{\partial y} \tag{17.2}$$

are the forces that push the particle in the x- or y-direction. From the laws of TD ($dE = \text{\dj}Q_{\text{q.s.}} + \text{\dj}W_{\text{q.s.}}$, $\text{\dj}Q_{\text{q.s.}} = T\,dS$) and from $\text{\dj}W_{\text{q.s.}} = -P\,dV$ follows

$$dE = T\,dS - P\,dV \tag{17.3}$$

From this we read off

$$T = \frac{\partial E(S, V)}{\partial S} \ , \qquad P = -\frac{\partial E(S, V)}{\partial V} \tag{17.4}$$

---

[1]**Designations:** We summarize deviating designations found in the literature. What we call energy E (the energy of the considered system) is often named internal energy U. The next symbols F, G and H are nearly always used. The free energy F is also called Helmholtz free energy. The name enthalpy H is the standard designation. The free enthalpy H is also called Gibbs free energy. If the particle number N is used as a state variable, a further "grand canonical potential" J will be introduced (Chapter 22); for this other designation like "Landau potential" Ω or Φ_G may be found; we chose the letter J in continuation of E, F, H,.... — At this point we mention that the lower case letter p is often used for the pressure (instead of our P). As intensive state variables both, temperature and pressure, are to be seen in parallel; and for the temperature a capital letter is uniquely used.

The temperature T is the driving force for heat or entropy exchange, and the pressure P is the driving force for volume exchange (Chapters 9 and 10). In this respect, T and P *thermodynamic forces*, which acts towards a change of the system state (S, V), just like the forces K_x and K_y act towards a change of the position (x, y).

The analogy to mechanics does not exist in an essential point: In mechanics, the state of a particle is defined by the positions (x, y) *and* by the velocities $(\dot{x}, \dot{y})$; changes of the mechanical state are therefore *dynamic* processes. A thermodynamic state is already defined by the variables (17.1) which correspond to the positions of the particle. The speed of the variable changes plays no role in thermodynamics. This is due to the fact that by (17.3) we restrict ourselves to *quasi-static* processes in (17.3). For this reason, the name *thermostatics* instead of *thermodynamics* would be appropriate.

Regardless of this limitation, we consider non quasi-static processes that start from an (equilibrium) state a via unspecified intermediate states to the state b. In this case, we restrict ourselves to statements about the initial and final states.

In the following, we consider the four thermodynamic potentials that correspond to the four pairs of variables in (17.1).

Definitions

By

$$
\begin{array}{lll}
\text{Energy:} & E & \\
\text{Free energy:} & F = E - TS & \\
\text{Enthalpy:} & H = E + PV & \text{(17.5)} \\
\text{Free enthalpy:} & G = E - TS + PV &
\end{array}
$$

four state variables are defined that have the dimension of an energy. As stated in the footnote[1], other common designations are *internal energy* for the energy and *Gibbs potential* for the free enthalpy.

Differentials

From (17.3) and

$$
d(TS) = T\,dS + S\,dT \quad \text{and} \quad d(PV) = P\,dV + V\,dP \tag{17.6}
$$

we get in the exact differentials of the quantities (17.5),

$$
\begin{array}{rcll}
dE & = & T\,dS - P\,dV & \text{(17.7)} \\
dF & = & -S\,dT - P\,dV & \text{(17.8)} \\
dH & = & T\,dS + V\,dP & \text{(17.9)} \\
dG & = & -S\,dT + V\,dP & \text{(17.10)}
\end{array}
$$

Instead of *exact differential* one sometimes finds (less common) designations complete, full or perfect differential. The transitions between the potentials are done according to the scheme:

$$dA = \ldots \pm X\,dY \ldots \quad \Longrightarrow \quad dB = d(A \mp XY) = \ldots \mp Y\,dX \ldots \qquad (17.11)$$

Such a transition is called *Legendre transformation*.

Natural variables

The state variables E, F, H and G listed in (17.5) are called G are called *thermodynamic potentials* if they are written as functions of their *natural variables*. These are the variables whose differentials occur on the right-hand side of (17.7)–(17.10). The thermodynamic potentials are therefore of the form

$$\begin{aligned}
E &= E(S, V) &\text{(energy)} &\qquad (17.12)\\
F &= F(T, V) &\text{(free energy)} &\qquad (17.13)\\
H &= H(S, P) &\text{(enthalpy)} &\qquad (17.14)\\
G &= G(T, P) &\text{(free enthalpy)} &\qquad (17.15)
\end{aligned}$$

To obtain the thermodynamic potential of the free energy, for example, the corresponding variables must be used in the definition $F = E - TS$, i.e.

$$F = F(T, V) = E(T, V) - T\,S(T, V) \qquad (17.16)$$

Thermodynamic forces

For each function in (17.12)–(17.15) we write down the exact differential. The comparison with (17.7)–(17.10) then results in simple expressions for the partial derivatives with respect to a natural variable:

$$\left(\frac{\partial E}{\partial S}\right)_V = T, \qquad \left(\frac{\partial E}{\partial V}\right)_S = -P \qquad (17.17)$$

$$\left(\frac{\partial F}{\partial T}\right)_V = -S, \qquad \left(\frac{\partial F}{\partial V}\right)_T = -P \qquad (17.18)$$

$$\left(\frac{\partial H}{\partial S}\right)_P = T, \qquad \left(\frac{\partial H}{\partial P}\right)_S = V \qquad (17.19)$$

$$\left(\frac{\partial G}{\partial T}\right)_P = -S, \qquad \left(\frac{\partial G}{\partial P}\right)_T = V \qquad (17.20)$$

The results are state variables which may be considered as *thermodynamic forces*. For example, the pressure is the thermodynamic force acting towards a volume change. Thereby it must be specified which other state variable is kept constant during the volume change.

Maxwell relations

For each exact differential (15.16) we obtain the relation (15.24). We write on (15.16) for the differentials (17.17)–(17.20):

$$\left(\frac{\partial T}{\partial V}\right)_S = -\left(\frac{\partial P}{\partial S}\right)_V \qquad \text{(from } dE) \qquad (17.21)$$

$$\left(\frac{\partial S}{\partial V}\right)_T = \left(\frac{\partial P}{\partial T}\right)_V \qquad \text{(from } dF) \qquad (17.22)$$

$$\left(\frac{\partial T}{\partial P}\right)_S = \left(\frac{\partial V}{\partial S}\right)_P \qquad \text{(from } dH) \qquad (17.23)$$

$$-\left(\frac{\partial S}{\partial P}\right)_T = \left(\frac{\partial V}{\partial T}\right)_P \qquad \text{(from } dG) \qquad (17.24)$$

For the thermodynamic potentials, these relations are called *Maxwell relations*.

Generalizations

The presented relations can be applied to the case of several external parameters $x = (x_1,..., x_n)$. For a quasi-static process, then $đW_{\text{q.s.}} = -\sum X_i \, dx_i$. From the laws of TD ($dE = đQ_{\text{q.s.}} + đW_{\text{q.s.}}$ and $T\,dS = đQ_{\text{q.s.}}$) then follows

$$\boxed{dE = T\,dS - \sum_{i=1}^{n} X_i \, dx_i} \qquad (17.25)$$

This is the generalization of (17.7); the differentials of the other thermodynamic potentials also follow from this. For example, the number of particles N or a magnetic field B may occur as external parameters. In this case the energy would be of the form $E(S, V, B, N)$ with the differential

$$dE = T\,dS - P\,dV - VM\,dB + \mu\,dN \qquad (17.26)$$

The new generalized forces are the magnetic moment VM and the chemical potential μ (Chapters 20–21). Overall, we obtain the following thermodynamic forces

$$\left(\frac{\partial E}{\partial S}\right)_{V,B,N} = T\,, \qquad \left(\frac{\partial E}{\partial V}\right)_{S,B,N} = -P$$

$$\left(\frac{\partial E}{\partial B}\right)_{S,V,N} = -VM\,, \qquad \left(\frac{\partial E}{\partial N}\right)_{S,V,B} = \mu \qquad (17.27)$$

This results in six Maxwell relations. By Legendre transformations (17.1), additional thermodynamic potentials can be defined.

Complete thermodynamic information

From the knowledge of one of the thermodynamic potentials (as a function of its natural variables) all other potentials and the thermal and caloric equation of state can be determined. In this sense, a thermodynamic potential contains the complete thermodynamic information about the considered system.

We determine the thermodynamic relations from the thermodynamic potential $E(S, V)$. For this, we start with the partial derivatives of the potential. From

$$T = \left(\frac{\partial E}{\partial S}\right)_V = T(S, V) \longrightarrow S = S(T, V) \tag{17.28}$$

follows the entropy $S(T, V)$. By inserting it into $E(S, V)$, the caloric equation of state equation of state $E = E(T, V)$ is obtained:

$$E = E(S(T, V), V) = E(T, V) \tag{17.29}$$

The thermal equation of state $P = P(T, V)$ follows from

$$P = -\left(\frac{\partial E}{\partial V}\right)_S = P(S, V) = P(S(T, V), V) = P(T, V) \tag{17.30}$$

It should be remembered that in the physical notation the same letter is used for different functions (which result from a change of variables). With $S(T, V)$ from (17.28) one also gets the free energy $F(T, V) = E(T, V) - T\, S(T, V)$ as a function of the natural variables. The reader may find out for himself how to determine $G(T, P)$ and $H(S, P)$ from $E(S, V)$, and how to obtain the equations of state from $G(T, P)$ or $H(S, P)$.

In the microscopic treatment, $\Omega(E, V)$ is calculated, or generally $\Omega(E, x) = \Omega(E, x_1, ..., x_n)$. On may now resolve $S = k_B \ln \Omega = S(E, x)$ for E, yielding the thermodynamic potential $E(S, x)$. Therefore, the partition function $\Omega(E, x)$ or the entropy $S(E, x)$ contain the complete information, too.

We have seen that all thermodynamic relations can be derived from one of the thermodynamic potentials or from $S(E, x)$. Conversely, one can start from the experimentally more easily accessible equations of state. For a homogeneous system with the external parameter V, the complete information requires the thermal equation of state equation of state and the heat capacity C_V as a function of T for a fixed volume V_0:

$$P(T, V) \quad \text{and} \quad C_V(T, V_0) \qquad \text{(complete information)} \tag{17.31}$$

We show that all thermodynamic relations follow from these relations, too. To do this, we apply the derivative $(\partial/\partial T)_V$ to the Maxwell relation (17.22):

$$\left(\frac{\partial C_V}{\partial V}\right)_T = T \left(\frac{\partial^2 P}{\partial T^2}\right)_V \tag{17.32}$$

From this and (17.31), the function $C_V(T, V)$ can be determined:

$$C_V(T, V) = C_V(T, V_0) + T \int_{V_0}^{V} dV' \, \frac{\partial^2 P(T, V')}{\partial T^2} \tag{17.33}$$

With the Maxwell relation (17.22) one also gets

$$dS = \left(\frac{\partial S}{\partial T}\right)_V dT + \left(\frac{\partial S}{\partial V}\right)_T dV = \frac{C_V}{T} dT + \left(\frac{\partial P}{\partial T}\right)_V dV \tag{17.34}$$

and thus

$$dE = T \, dS - P \, dV = C_V \, dT + \left[T \left(\frac{\partial P}{\partial T}\right)_V - P \right] dV \tag{17.35}$$

Therefore, the differentials dS and dE are determined by (17.31). From them the functions $S(T, V)$ and $E(T, V)$ can be calculated (using (15.20)). The elimination of T from $S(T, V)$ and $E(T, V)$ yields the thermodynamic potential $E = E(S, V)$.

We summarize the results of this section:

$$\left.\begin{array}{l} E(S, V) \\ F(T, V) \\ H(S, P) \\ G(T, P) \\ S(E, V) \\ \Omega(E, V) \\ P(T, V), \ C_V(T, V_0) \end{array}\right\} \quad \text{—} \left| \begin{array}{l} \text{each line contains} \\ \text{the complete thermo-} \\ \text{dynamic information} \end{array} \right. \tag{17.36}$$

Extremal principles

The equilibrium condition "$S = $ maximal" applies for a closed system with given values for the energy E and the external parameters x. We supplement this extremal condition by

$$S = \text{maximal} \quad \text{for given } E, V \tag{17.37}$$

$$F = \text{minimal} \quad \text{for given } T, V \tag{17.38}$$

$$G = \text{minimal} \quad \text{for given } T, P \tag{17.39}$$

Here we consider the volume V as the only external parameter; the number of particles N shall be constant. For deriving (17.38) and (17.39), we consider two macroscopic systems A and B that can exchange heat and/or volume, see Figure 9.1 or Figure 10.1. We assume that the system A is much smaller than B:

$$E_A \ll E = E_A + E_B, \qquad V_A \ll V = V_A + V_B \tag{17.40}$$

Then the contact with B means the specification of the temperature and/or of the pressure for system A.

If only heat exchange is allowed (volumes constant), then the entropy S of the entire system is of the form

$$S(E_A) = S_A(E_A, V_A) + S_B(E - E_A, V_B) = \text{maximal} \qquad (17.41)$$

Since the system A is small, we can expand S_B into powers of E_A:

$$S = S_A + S_B(E, V_B) - \frac{\partial S_B(E_B, V_B)}{\partial E_B} E_A = \text{const.} + S_A - \frac{E_A}{T} \qquad (17.42)$$

Here $T = T_B$ is the temperature of the heat bath B. From $S = \text{maximal}$ it follows for the free energy $F_A = E_A - T S_A$ of the subsystem A:

$$F_A = \text{minimal} \qquad (T, V \text{ given}) \qquad (17.43)$$

The condition of maximal entropy for the closed overall system implies the condition of minimal free energy for the subsystem A.

For a minimal $F = E - TS$, the energy should be small and the entropy should be large. These are opposing goals, since a smaller energy goes in the direction of more order, but a larger entropy goes in the direction of more disorder. The influence of entropy increases with increasing temperature, as T appears as a coefficient of S. This interplay between order and disorder will be studied in the ideal spin system (Chapter 26).

We repeat the derivation (17.41)–(17.43) for the case that the two systems can exchange heat and volume as in Figure 10.1. In

$$S(E_A, V_A) = S_A(E_A, V_A) + S_B(E - E_A, V - V_A) = \text{maximal} \qquad (17.44)$$

we expand S_B into powers of E_A and V_A:

$$\begin{aligned}
S &= S_A + S_B(E, V) - \frac{\partial S_B(E_B, V_B)}{\partial E_B} E_A - \frac{\partial S_B(E_B, V_B)}{\partial V_B} V_A \\[2mm]
&= \text{const.} + S_A - \frac{E_A}{T} - \frac{P V_A}{T}
\end{aligned} \qquad (17.45)$$

Here $P = P_B$ is the pressure that is given by the large system B. From $S = \text{maximal}$ follows for the free enthalpy $G_A = E_A - T S_A + P V$ of the subsystem A:

$$G_A = \text{minimal} \qquad (T, P \text{ given}) \qquad (17.46)$$

The specification of temperature and pressure usually corresponds to the experimental situation.

The statement "$S = \text{maximum}$ in equilibrium" also implies that $\Delta S \geq 0$ holds for an arbitrary process in a closed system. During the transition from a non-equilibrium state to an equilibrium state, the entropy grows, it "strives towards a maximum". Given the temperature and pressure, the free enthalpy strives towards a minimum.

Exercises

17.1   Equation of state for volume-independent energy

Derive the general form of the thermal equation of state for a material that satisfies the relation $(\partial E/\partial V)_T = 0$.

17.2   Special equation of state

The following information is given for a gas (with $N = \text{const.}$):

$$P = \frac{f(T)}{V} \quad \text{and} \quad \left(\frac{\partial E}{\partial V}\right)_T = bP \qquad (b = \text{const.})$$

Determine the function $f(T)$ from this. It is advisable to solve exercise 17.1 first.

17.3   Energy density of the photon gas

The energy E and the pressure P of a photon gas in a cavity with the volume V are of the form

$$\frac{E(T, V)}{V} = U(T), \qquad P(T, V) = \frac{U(T)}{3} \tag{17.47}$$

From this, determine the T-dependence of the energy density $U(T)$. Calculate the entropy S and the thermodynamic potentials $E(S, V)$, $F(T, V)$, $G(T, P)$ and $H(S, P)$. It is advisable to solve problem 17.1 first.

17.4   Thermodynamic relations from free enthalpy

The free enthalpy $G(T, P) = E - TS + PV$ shall be given (and $N = \text{const.}$). How can one obtain the equation of state $P(V, T)$ and the heat capacity $C_V(V, T)$?

17.5   Thermodynamic relations from enthalpy

The enthalpy $H(S, P) = E + PV$ shall be given as a function of the natural variables ($N = \text{const.}$). How does one obtain the heat capacity $C_P(T, P)$ and the isothermal compressibility $\kappa_T(T, P) = -(\partial V/\partial P)_T/V$?

17.6   Extremal condition for the enthalpy

Show that $H(S, P)$ is minimal for a given S and P.

17.7   Density profile of the earth's atmosphere

For a column of air (base area A) in a homogeneous gravity field $\boldsymbol{g} = g\,\boldsymbol{e}_z$ the particle density $n(z) = N/V$ shall be determined. The air is assumed to be an ideal gas with a constant specific heat $c_V(T) = \text{const.}$.

1. *Convective equilibrium:* The entropy shall be constant, $S(z) = $ const. Calculate $n(z)$ from the condition of minimum energy. First determine the relationship between the temperature T and the density $n = N/V$ for $dS = 0$, and the energy density E/V as a function of z and $n(z)$. Then determine the energy $E[n]$ as a function of $n(z)$.

2. *Barometric formula:* The temperature shall be constant, $T(z) = $ const. Determine $n(z)$ from the condition minimum free energy. To do this, determine the free energy $F[n]$ as a functional of $n(z)$.

17.8   Entropy, heat capacity and equation of state

Derive the following relations:

$$dS = \frac{C_P}{T}\, dT - \left(\frac{\partial V}{\partial T}\right)_P dP \tag{17.48}$$

$$dS = \frac{C_V}{T}\, dT + \left(\frac{\partial P}{\partial T}\right)_V dV \tag{17.49}$$

17.9   Difference $C_P - C_V$ for a solid

For a solid, the thermal equation of state

$$V = V_0 - A P + B T \tag{17.50}$$

and the heat capacity $C_P = C$ at constant pressure are given; here A, B and C are material dependent constants. Calculate the heat capacity C_V at constant volume and the internal energy E.

17.10   Difference $C_P - C_V$ for the van der Waals gas

The van der Waals gas satisfies the equation of state

$$P = \frac{RT}{v - b} - \frac{a}{v^2}$$

where $v = V/v$ is the volume per mole. Calculate the difference $C_P - C_V$ of the heat capacities and determine the leading correction term to the ideal gas with $C_P - C_V = v R$. Estimate the relative size of the correction term for carbon dioxide under normal conditions. For this, the parameter $a = 27(R T_{cr})^2/(64 P_{cr})$ is given by the critical values $P_{cr} = 71.5\,\text{bar}$ and $T_{cr} = 304.2\,\text{K}$; this relationship will be derived later in Exercise 37.1.

18 State changes

We investigate the changes of thermodynamic state for heat supply and for volume change. A general expression for the difference $C_P - C_V$ of the heat capacities is derived. The temperature changes for the free expansion, for the quasi-static, adiabatic expansion and for the Joule-Thomson process are calculated.

We consider homogeneous systems with two state variables as in (17.1). For a volume change (like expansion), a condition for the second variable must be specified. This applies accordingly for a process with heat supply. This specification usually consists in keeping a specific other variable constant. We investigate the heat supply for constant pressure or constant volume, and the expansion for constant energy, entropy or enthalpy.

Heat supply

The heat capacities

$$C_P = \left. \frac{đQ_{\text{q.s.}}}{dT} \right|_{P=\text{const.}} = T \left(\frac{\partial S}{\partial T} \right)_P \tag{18.1}$$

and

$$C_V = \left. \frac{đQ_{\text{q.s.}}}{dT} \right|_{V=\text{const.}} = T \left(\frac{\partial S}{\partial T} \right)_V \tag{18.2}$$

determine the temperature increase when heat is added. Usually, when heat is added, either the pressure or the volume is kept constant. For the ideal gas, we had obtained $C_P - C_V = \nu R$ for the difference of the heat capacities. Now we evaluate $C_P - C_V$ in a form that is suitable for an arbitrary equation of state $P = P(T, V)$.
We start from (17.34) for dS:

$$dS = \frac{C_V}{T} dT + \left(\frac{\partial P}{\partial T} \right)_V dV \tag{18.3}$$

For C_P we need dS as a function of dT and dP. For this we insert dV for $V = V(T, P)$ in (18.3):

$$dS = \frac{C_V}{T} dT + \left(\frac{\partial P}{\partial T} \right)_V \left[\left(\frac{\partial V}{\partial T} \right)_P dT + \left(\frac{\partial V}{\partial P} \right)_T dP \right] \tag{18.4}$$

From this we read off

$$C_P = C_V + T \left(\frac{\partial P}{\partial T}\right)_V \left(\frac{\partial V}{\partial T}\right)_P \tag{18.5}$$

We introduce the expansion coefficient α,

$$\alpha = \frac{1}{V}\left(\frac{\partial V}{\partial T}\right)_P \tag{18.6}$$

and the isothermal compressibility κ_T :

$$\kappa_T = -\frac{1}{V}\left(\frac{\partial V}{\partial P}\right)_T \tag{18.7}$$

This gives us

$$\left(\frac{\partial P}{\partial T}\right)_V \overset{(15.26)}{=} -\left(\frac{\partial V}{\partial T}\right)_P \Big/ \left(\frac{\partial V}{\partial P}\right)_T = \frac{\alpha}{\kappa_T} \tag{18.8}$$

We use the last three equations in (18.5):

$$\boxed{C_P - C_V = \frac{V T \alpha^2}{\kappa_T}} \tag{18.9}$$

For gases, the quantities C_P, C_V, α and κ_T can be measured easily. For solids or liquids, the condition $P = $ const. is easier to realize than $V = $ const.; the preferred experimental heat capacity is therefore C_P. In a microscopic calculation, on the other hand, V usually appears as a given external parameter; therefore, C_V is theoretically easier to access.

For a stable system,

$$\kappa_T > 0 \qquad \text{(stability condition)} \tag{18.10}$$

must hold. If the volume becomes smaller when the pressure is reduced, the system would become smaller and smaller by itself leading to an implosion. The sign of the expansion coefficient, on the other hand, is not fixed,

$$\text{usually: } \alpha > 0, \text{ but possible: } \alpha \le 0 \tag{18.11}$$

The best-known example of $\alpha < 0$ is water in the range between 0 and $4\,°\text{C}$ at normal pressure. A rubber band can also contract when the temperature is increased. From (18.9) and (18.10) follows

$$C_P \ge C_V \tag{18.12}$$

The following applies to the ideal gas

$$\alpha = \frac{1}{V}\left(\frac{\partial V}{\partial T}\right)_P = \frac{1}{T}, \qquad \kappa_T = -\frac{1}{V}\left(\frac{\partial V}{\partial P}\right)_T = \frac{1}{P} \qquad \text{(ideal gas)} \tag{18.13}$$

Table 18.1 Expansion coefficient, compressibility and specific heat for the monoatomic ideal gas and for copper under normal conditions. The values for the ideal gas apply in good approximation for noble gases.

Physical quantity	Ideal gas (noble gas)	Copper
α	$3.4 \cdot 10^{-3}\ \mathrm{K^{-1}}$	$5.0 \cdot 10^{-5}\ \mathrm{K^{-1}}$
κ_T	$1.0\ \mathrm{bar^{-1}}$	$7.4 \cdot 10^{-7}\ \mathrm{bar^{-1}}$
c_P	$21\ \mathrm{J\,K^{-1}\,mol^{-1}}$	$24.5\ \mathrm{J\,K^{-1}\,mol^{-1}}$
$c_P - c_V$	$8.3\ \mathrm{J\,K^{-1}\,mol^{-1}}$	$0.7\ \mathrm{J\,K^{-1}\,mol^{-1}}$
$\gamma = c_P/c_V$	$5/3$	1.03

We insert this into (18.9) and use $PV = \nu RT$:

$$C_P - C_V = \nu R \quad \text{or} \quad c_P - c_V = R \qquad \text{(ideal gas)} \qquad (18.14)$$

This result is known from (16.16). According to Table 16.2, many real gases approximately satisfy this relation.

For an ideal gas, the difference $(C_P - C_V)\,dT$ can be interpreted as the expansion work $P\,dV = \nu R\,dT$ to be performed when the temperature increases under constant pressure. However, this interpretation is not generally valid; this follows from the possibility $\alpha < 0$.

The general relations (18.1)–(18.12) apply for gases, liquids and solids; it is only assumed that the system homogeneous. Table 18.1 compares the values for an ideal gas at normal conditions with those for copper. The expansion coefficient for copper is two orders of magnitude smaller, the compressibility is six orders of magnitude smaller. In general, α and κ_T are much smaller for solids and liquids than for gases. For solids and liquids, the relative difference $(c_P - c_V)/c_P$ is mostly small compared to 1.

Expansion

We consider the expansion (compression) under the following conditions (Figure 18.1):

1. Free expansion: $E = \text{const.}$

2. Quasi-static, adiabatic expansion: $S = \text{const.}$

3. Joule-Thomson expansion: $H = \text{const.}$

Since the states of the considered homogeneous systems are defined by two variables, the specification of one constant state variable is sufficient to define the kind of expansion. The associated experimental setups are shown in Figure 18.1.

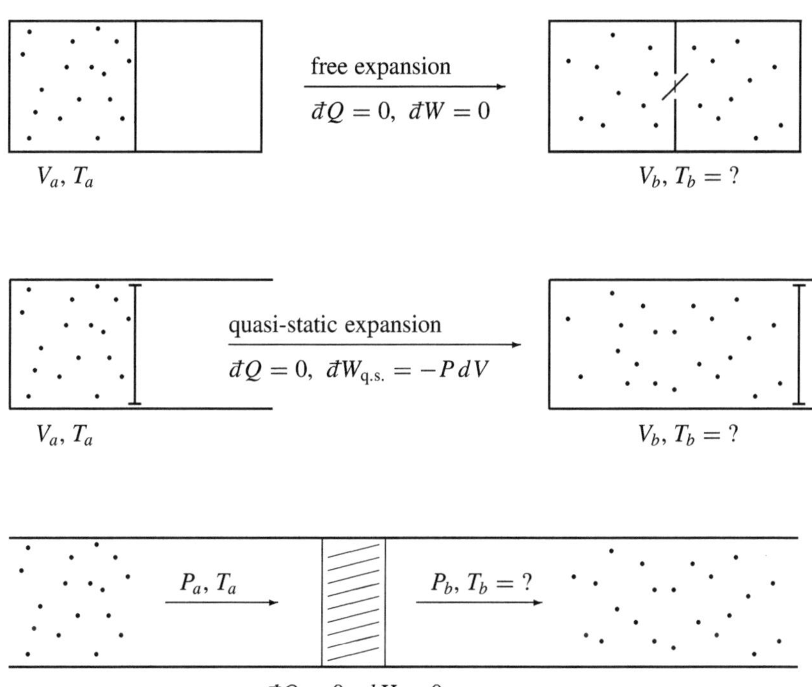

Figure 18.1 Different expansion processes are compared: In the Gay-Lussac experiment (top), the gas expands adiabatically and without work output. In the middle, the gas performs the work $P\,dV$ during a quasi-static, adiabatic expansion. In the Joule-Thomson expansion (bottom), the gas is pressed through a porous plug. The respective temperature changes can be calculated and measured.

We want to calculate the temperature change associated with the expansion, i.e. the partial derivatives

$$\left(\frac{\partial T}{\partial V}\right)_E, \qquad \left(\frac{\partial T}{\partial V}\right)_S, \qquad \left(\frac{\partial T}{\partial P}\right)_H \tag{18.15}$$

In the Joule-Thomson experiment, the temperature change is a function of the pressure difference (which depends on the structure of the porous plug). The wanted partial derivatives can be read from the exact differential of the constant variable quantity A:

$$dA = B\,dT + C\,dV \quad \Longrightarrow \quad \left(\frac{\partial T}{\partial V}\right)_A = -\frac{C}{B} \tag{18.16}$$

For the Joule-Thomson experiment, V must be replaced by P. The result $-C/B$ is given by the respective partial derivatives. These partial derivatives can be expressed by experimentally easily accessible quantities. The results may be further evaluated for specific thermal equations of state; for this purpose, we consider (16.1) or (16.2).

Free expansion

The arrangement for the free expansion (Figure 18.1 above) is also known as the Gay-Lussac experiment. The system consists of two subvolumes, one filled with gas, the other is empty (vacuum). The system is thermally isolated ($đQ = 0$). Then a throttle valve (or a flap) is opened (without work), and the gas distributes over the whole volume without performing any work. Therefore

$$dE = đQ + đW = 0 \tag{18.17}$$

holds for this process. As discussed in Chapter 12, $\Omega_b \gg \Omega_a$ implies $S_b > S_a$; the process is irreversible. The process runs through non-equilibrium states; it is not quasi-static. The initial state (T_a, V_a) and the final state final state (T_b, V_b) shall be equilibrium states again. Then the 1st law of TD and (18.17) imply

$$E(T_b, V_b) = E(T_a, V_a) \tag{18.18}$$

This condition determines the amount by which the temperature changes. This temperature change is given by the first expression in (18.15). According to the scheme (18.16) we express dE by dT and dV:

$$dE \overset{(17.35)}{=} C_V\,dT + \left[T\left(\frac{\partial P}{\partial T}\right)_V - P\right]dV \tag{18.19}$$

This yields the desired temperature change

$$\left(\frac{\partial T}{\partial V}\right)_E = \frac{1}{C_V}\left[P - T\left(\frac{\partial P}{\partial T}\right)_V\right] \tag{18.20}$$

We evaluate this for (16.1) and (16.2):

$$\left(\frac{\partial T}{\partial v}\right)_E = v \left(\frac{\partial T}{\partial V}\right)_E = \begin{cases} 0 & \text{(ideal gas)} \\[2mm] -\dfrac{a}{c_V\, v^2} & \text{(van der Waals gas)} \end{cases} \tag{18.21}$$

For an ideal gas, we already have obtained this result in (16.7). For an ideal gas, the energy is the sum of the kinetic energies of the atoms, and the kinetic energies of the atoms do not change as they pass through the open throttle valve.

In the real gas, on the other hand, there is cooling. The term $-a/v^2$ in (16.2) describes the reduction in pressure due to the attractive part of the interaction. At lower gas density the particles feel less of this attraction on the average, so that the potential energy increases. Since the energy is conserved during free expansion, the kinetic energy of the particles must then decrease, causing the temperature to fall. The strongly repulsive, short-range part of the interaction reduces the effectively available volume (term with b in (16.2)), but has no influence on the energy. In the context of thermodynamics, (16.2) is an empirical approach. Accordingly, the explanation of the temperature reduction given here is only qualitatively.

Quasi-static, adiabatic expansion

The thermally insulated ($đQ = 0$) gas is expanded by slowly pulling out a piston, Figure 18.1 center. Thereby the gas performs the work $-đW_{\mathrm{q.s.}} = P\, dV \neq 0$. For this quasi-static and adiabatic process, the 2nd law of TD says

$$dS = \frac{đQ_{\mathrm{q.s.}}}{T} = 0 \tag{18.22}$$

This gives us for the final and initial state

$$S(T_b, V_b) = S(T_a, V_a) \tag{18.23}$$

This condition determines the amount ΔT by which the temperature changes when the volume changes by ΔV. For an infinitesimal change, this relationship is given by the second expression in (18.15). According to the scheme (18.16) we express dS by dT and dV:

$$dS \overset{(17.34)}{=} \frac{c_V}{T}\, dT + \left(\frac{\partial P}{\partial T}\right)_V dV \tag{18.24}$$

This results in the desired temperature change:

$$\left(\frac{\partial T}{\partial V}\right)_S = -\frac{T}{c_V}\left(\frac{\partial P}{\partial T}\right)_V \overset{(18.8)}{=} -\frac{T\,\alpha}{c_V\,\kappa_T} \tag{18.25}$$

For an ideal gas, $\alpha = 1/T$ and $\kappa_T = 1/P$ we get

$$\left(\frac{\partial T}{\partial V}\right)_S = -\frac{P}{c_V} \qquad \text{(ideal gas)} \tag{18.26}$$

For the van der Waals gas, the temperature change shall be calculated in Exercise 18.2.

For a gas, $\alpha > 0$ always applies, so that $(\partial T/\partial V)_S < 0$. The gas therefore cools down during quasi-static, adiabatic expansion. This cooling can be used to achieve lower temperatures. The principle of a *refrigeration* (cooling engine) based on this process is the following:

1. The gas initially has the ambient temperature T_1. It is now compressed quasi-statically and adiabatically. Thereby it heats up.

2. By thermal contact with the environment, it cools down again to the ambient temperature T_1.

3. Now it is expanded quasi-statically and adiabatically to its original volume. In doing so, it reaches the lower temperature $T_2 < T_1$.

Once a sufficiently large cold reservoir with T_2 has been created, another cycle can begin with T_2 as the new ambient temperature. The temperatures that can be achieved by such a process are practically limited by the fact that real gases condense at low temperatures. Nitrogen condenses under normal pressure at around 77 K, whereas helium only condenses at around 5 K. Liquid nitrogen or liquid helium often serve as a cold reservoir in the laboratory.

The third step for could also be a free expansion for a real gas. However, the temperature reduction achieved in this way is generally small, because the term a/v^2 in the van der Waals equation is a correction term only, $a/v^2 \ll P$.

Joule-Thomson expansion

In this expansion, the pressure P_a forces a gas through a porous, Figure 18.1 bottom. Behind the plug, the gas has the lower pressure P_b. The pressure difference and the resulting flow velocity depend on the structure of the porous plug. The system shall be thermally insulated, i.e.

$$\text{đ}Q = 0 \tag{18.27}$$

The system as a whole has no defined temperature; it is not an equilibrium state. Therefore, it is also not a quasi-static process, and $dS = 0$ does not follow from $\text{đ}Q = 0$. We may, however, assume separate equilibrium states for the gases on the left and right side of the plug.

By forcing the gas through the plug, the gas is compressed before the plug and expanded afterwards (corresponding to the lower pressure P_b). If one mole of the gas before the plug has the volume V_a, the work

$$P_a V_a \tag{18.28}$$

must be performed to move this amount of gas forward. Behind the plug the gas must then do the work

$$P_b V_b \tag{18.29}$$

to push the same amount of gas (1 mol) further. On the whole, the work

$$\Delta W = P_a \, V_a - P_b \, V_b \tag{18.30}$$

is supplied to the gas. From the 1st law of TD follows

$$\Delta E = E_b - E_a = \Delta Q + \Delta W = \Delta W \tag{18.31}$$

and thus

$$E_b + P_b \, V_b = E_a + P_a \, V_a \quad \text{or} \quad H_b = H_a \tag{18.32}$$

This means that the enthalpy $H = E + PV$ is constant in this process. We write the enthalpy as a function of T and P:

$$H(T_b, \, P_b) = H(T_a, \, P_a) \tag{18.33}$$

This condition determines the amount by which the temperature changes due to the pressure change. For an infinitesimal change, this relationship is given by the third expression in (18.15). According to the scheme (18.16) we express dH by dT and dP:

$$dH = T \, dS + V \, dP = T \left[\left(\frac{\partial S}{\partial T} \right)_P dT + \left(\frac{\partial S}{\partial P} \right)_T dP \right] + V \, dP$$

$$= C_P \, dT - T \left(\frac{\partial V}{\partial T} \right)_P dP + V \, dP = C_P \, dT + (V - V T \alpha) \, dP \tag{18.34}$$

We have used the Maxwell relation for $dG = -S \, dT + V \, dP$ and introduced the expansion coefficient α. We note that the heat capacity C_P is linked to $H(T, P)$ in a simple way:

$$C_P = T \left(\frac{\partial S}{\partial T} \right)_P = \left(\frac{\partial H}{\partial T} \right)_P \tag{18.35}$$

This can be compared to (16.12), $C_V = \partial E(T, V)/\partial T$.

From (18.34) we read off the temperature change:

$$\mu_{\mathrm{JT}} \equiv \left(\frac{\partial T}{\partial P} \right)_H = \frac{V}{C_P} \, (T \alpha - 1) \tag{18.36}$$

The quantity μ_{JT} is called Joule-Thomson coefficient. For the ideal gas this coefficient disappears because $\alpha = 1/T$. For a real gas, μ_{JT} can be positive or negative. For nitrogen gas under normal conditions, μ_{JT} is positive; the Joule-Thomson process (with $dP < 0$) therefore leads to a cooling ($dT < 0$). For lower temperatures μ_{JT} increases, so that the cooling effect becomes stronger. In Exercise 18.4, the Joule-Thomson coefficient is calculated for the van der Waals gas.

As a continuous process, the Joule-Thomson expansion is particularly easy to realize technically. At low temperatures the efficiency of a Joule-Thomson cooling machine may be comparable with that of a quasi-static adiabatic expansion.

Exercises

18.1 Isothermal quasi-static expansion

A mole of an ideal gas expands isothermally and quasi-statically from the volume V_a to $V_b = 2.7\,V_a$. Determine the heat ΔQ absorbed during the process. Calculate ΔQ in joules if the process is carried out at room temperature.

18.2 Adiabatic expansion of the van der Waals gas

Determine the temperature change of the van der Waals gas during a quasi-static, adiabatic expansion.

18.3 Expansion coefficient of the van der Waals gas

Calculate the expansion coefficient $\alpha = (\partial V/\partial T)_P / V$ as a function of P and v for the van der Waals gas.

18.4 Inversion curve in the Joule-Thomson process

The sign of the Joule-Thomson coefficient

$$\mu_{\mathrm{JT}} \equiv \left(\frac{\partial T}{\partial P}\right)_H = \frac{V}{C_P}\left(T\alpha - 1\right)$$

determines whether cooling or heating occurs in a Joule-Thomson process. Determine the *inversion curves* $T = T_{\mathrm{i}}(v)$ and $P = P_{\mathrm{i}}(T)$ defined by $\mu_{\mathrm{JT}} = 0$ for the van der Waals gas (16.2). Sketch and discuss the curve $P_{\mathrm{i}}(T)$.

19 Heat engines

We investigate processes in which heat is converted into work. A perpetuum mobile of the 2nd kind contradicts the 2nd law of TD. From the laws of TD follows the maximum achievable efficiency of a heat engine or a heat pump.

We consider engines that work cyclically. This means that the engine E after each cycle returns to the same state again and again (such as a complete cycle of a combustion engine). This includes continuously operating engines; for them, the cycle span can be selected at will.

Perpetuum mobile of the second kind

An ideal heat engine would be one that, in one cycle, extracts the heat q from a heat reservoir and converts it into the work $w = q$. Such a hypothetical apparatus is referred to as a perpetuum mobile of the second kind. The schematic of such a process is shown in Figure 19.1. The heat reservoir should be so large that its temperature T practically does not change during the heat extraction, for example a lake.

Such a engine cannot be realized because it contradicts the 2nd law of TD. After one cycle, the engine E in Figure 19.1 is in the same state, so $\Delta E_{\mathrm{E}} = 0$ holds. During a cycle, it absorbs the heat q and performs the work w. We insert $\Delta Q = q$ and $\Delta W = -w$ into the 1st law of TD:

$$\Delta E_{\mathrm{E}} = q - w = 0 \qquad \text{(1 cycle)} \tag{19.1}$$

thus

$$w = q \tag{19.2}$$

The conversion of the amount of heat q into work w is compatible with the first law of TD (the energy conservation). The 1st law would contradict a device that generates work out of nothing; such a hypothetical engine is called a perpetuum mobile of the 1st kind.

For a closed system, the 2nd law of TD states $\Delta S \geq 0$. To obtain a closed system, we supplement the arrangement in Figure 19.1 by a storage S for the work w. For example, the work could be stored in a spring; this is a system with one degree of freedom ($f = 1$). Another possibility would be a weight that is lifted in the gravitational field; in concrete terms, this kind of storage is achieved by a water

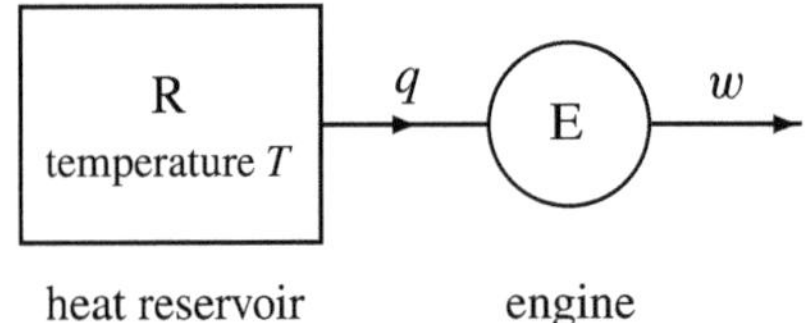

Figure 19.1 Scheme of a perpetuum mobile of the second kind, which converts heat q into work $w = q$.

pumping station with two reservoirs at different heights. For the closed system consisting of heat reservoir R, engine E and energy storage S, the following applies

$$\Delta S = \Delta S_R + \Delta S_E + \Delta S_S \geq 0 \qquad \text{(2nd law of TD)} \qquad (19.3)$$

The energy storage with one degree of freedom has an entropy of size $S_S = \mathcal{O}(k_B)$. This is negligible compared to the entropy of the heat reservoir (much larger than $10^{24}\, k_B$). Therefore we can use

$$\Delta S_S \approx 0 \qquad \text{(storage, } f = 1) \qquad (19.4)$$

The engine E is in the same macro state after each cycle, i.e.

$$\Delta S_E = 0 \qquad \text{(after one cycle)} \qquad (19.5)$$

For the heat reservoir we obtain after one cycle

$$\Delta S_R = -\frac{q}{T} \qquad \text{(2nd law of TD)} \qquad (19.6)$$

For this $dS = đQ_{\text{q.s.}}/T$ of the 2nd law was used, assuming a quasi-static process. However, the so calculated entropy change is generally valid if the process consists exclusively of the heat transfer q (see discussion following (11.12)); therefore an index "q.s." in (19.6) is omitted. From the last three equations follows

$$\Delta S = \Delta S_R + \Delta S_E + \Delta S_S = -\frac{q}{T} < 0 \qquad (19.7)$$

In the closed system consisting of heat reservoir, machine and energy storage, the entropy decreases. This is in contradiction to the 2nd law. It is therefore impossible to build such a engine.

One idea for realizing a perpetuum mobile of the 2nd kind would be the following process:

1. In contact with the environment (heat reservoir), a box containing an ideal gas has the temperature T.

2. The box is thermally insulated. We wait until all particles happen to be in the left half of the box. At this point in time we insert a wall from the side without doing any work. After this the gas is enclosed in a smaller volume.

3. We let gas expand adiabatically and quasi-statically to its original volume. Thereby it performs the work w. The gas then has a lower temperature $T_<$.

4. In contact with the environment, the gas heats up again to T. Thereby the gas reaches its original state, i.e. $\Delta E = 0$. The gas absorbs the heat $q = w$ in this step. The cyclic process continues with the 2nd step.

This process is not feasible due to the duration of the second step. As we saw in Chapter 12, this time is unimaginably much larger than the world age. This also applies even if we only wait until the left half volume contains 1% more particles than the right half. For $N = \mathcal{O}(10^{24})$, a deviation of 1% from the mean value corresponds to about 10^{10} standard deviations and is therefore extremely improbable (practically impossible). The actually occurring fluctuations, $\Delta N/\overline{N} \approx 10^{-12}$, on the other hand, cannot be used by any physical wall for performing work.

Completely analogous to the example just given are two subsystems in thermal contact and waiting for one system to produce a usable deviation from its mean energy value. As we saw in Chapter 9, the fluctuation of the energy in one subsystem around the mean value is of the same relative size as in particle exchange. According to Chapter 10, this applies to any other macroscopic quantity, too.

A well-known thought experiment for the construction of a perpetuum mobile of the second kind is the *Maxwell demon*, which outwits the 2nd law in the following way: It sits at a small opening between two volumes of gas and can open or close it with a flap (like in Figure 16.1 right). It opens the flap if, by chance, a fast particle (faster than the average) flies from right to left or a slow particle wants to fly from left to right. Otherwise the flap remains closed. In this way, the left-hand volume gets gradually warmer than the right. The temperature difference is then be used to generate work.

However, Maxwell's demon has problems. Opening and closing the flap must not cost more work than can be gained later. The work w_{flap} to operate the flap must therefore be small compared to the excess energy of a fast particle. The energy of a particle is, according to Chapter 9, of the size $k_{\text{B}}T$. This means

$$w_{\text{flap}} \ll k_{\text{B}}T \tag{19.8}$$

The flap itself represents a system with one degree of freedom. After some time, when it is in equilibrium with its environment, the flap has the mean statistical energy

$$\overline{\varepsilon_{\text{flap}}} = \mathcal{O}(k_{\text{B}}T) \tag{19.9}$$

Due to $\overline{\varepsilon_{\text{flap}}} \gg w_{\text{flap}}$, the flap opens and closes uncontrollably. The hands of Maxwell's demon tremble so much that he cannot fulfill his task. The statements (19.8) and (19.9) and the conclusion drawn from them apply to any conceivable device for sorting the particles, i.e. also for an intelligent electronic solution.

The impossibility of building a periodic engine of the kind shown in Figure 19.1 is sometimes used as an alternative formulation to $\Delta S \geq 0$. This formulation

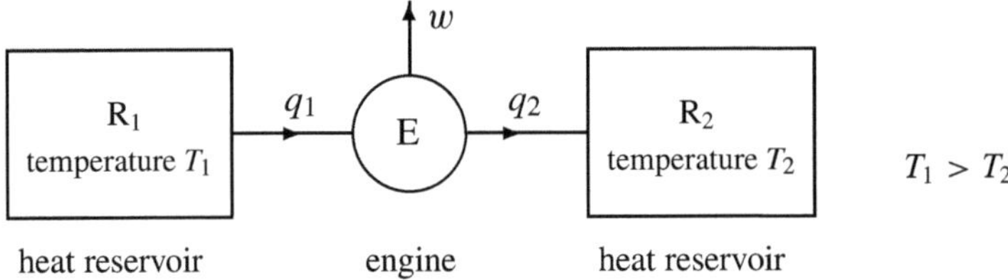

Figure 19.2 Scheme of a realizable heat engine. The engine E converts a part of the heat taken from a high temperature heat reservoir T_1 into work. The other part is passed on to a heat reservoir with a lower temperature T_2.

emphasizes that the form of energy "heat" (disordered motion) is inferior to the form of energy "work". The conversion $w \to q$ (for example by an electric heater) is possible without restriction. The conversion $q \to w$, on the other hand, is is only partially possible.

At this point, it should be noted that the 2nd law of TD defines a time direction; $\Delta S \geq 0$ means $dS/dt \geq 0$. Since the basic equations of mechanics and quantum mechanics are time-reversal invariant, they cannot by themselves justify $\Delta S \geq 0$. In Chapter 5, this time direction was introduced by the observation that a closed system strives to equilibrium by itself, i.e. to $S = $ maximal. This was not derived, but merely asserted as a fact of experience.

How a time direction can be distinguished on the basis mechanics or quantum mechanics has not been clarified finally; this question will be discussed in Chapter 41. The process of free expansion discussed in Chapter 12 may suffice as an obvious example. Here it is intuitively clear that a process $\Delta S < 0$ (with a $|\Delta S/\overline{S}| \gg N^{-1/2}$) is extremely unlikely. It should be noted that in other areas of physics, a time direction is distinguished. For example, in the scattering problem in quantum mechanics, an outgoing spherical wave is used, and in electrodynamics, the retarded potentials are preferred to the advanced ones..

Efficiency of a heat engine

A realizable heat engine converts heat into usable work. Thereby it utilizes the tendency of disordered energy to get evenly distributed. Heat flows by itself from the warmer system with the temperature T_1 to the colder one with T_2. The scheme of such a device is sketched in Figure 19.2. The specified energy quantities (q_1, q_2 and w) refer to one cycle of a periodically operating engine.

Such a engine is compatible with the laws of TD; however, these laws restrict the achievable efficiency. We first derive this restriction and then consider realizations of such engines.

After one cycle, the engine E is in the same state, so that $\Delta E_E = 0$. During a cycle, it absorbs the heat q_1, releases the heat q_2 and performs the work w. With

$\Delta Q = q_1 - q_2$ and $\Delta W = -w$, the 1st law of TD implies

$$\Delta E_{\mathrm{E}} = q_1 - q_2 - w = 0 \qquad \text{(after one cycle)} \qquad (19.10)$$

From this follows

$$w = q_1 - q_2 \qquad (19.11)$$

According to the 1st law, the difference $q_1 - q_2$ is converted into the work w.

We now apply the 2nd law of TD to the closed system of two heat reservoirs R_1 and R_2, the engine E and an energy storage S:

$$\Delta S = \Delta S_{R_1} + \Delta S_{R_2} + \Delta S_{\mathrm{E}} + \Delta S_{\mathrm{S}} \geq 0 \qquad \text{(2nd law of TD)} \qquad (19.12)$$

The storage (e.g. lifting an weight) shall have just one degree of freedom and yields a negligible contribution to entropy,

$$\Delta S_{\mathrm{S}} \approx 0 \qquad \text{(storage, } f = 1\text{)} \qquad (19.13)$$

The engine E is in the same state after each cycle,

$$\Delta S_{\mathrm{E}} = 0 \qquad \text{(after one cycle)} \qquad (19.14)$$

For the heat reservoirs we get

$$\Delta S_{R_1} = -\frac{q_1}{T_1}, \qquad \Delta S_{R_2} = \frac{q_2}{T_2} \qquad \text{(2nd law of TD)} \qquad (19.15)$$

We have used the 2nd law $dS = đQ_{\mathrm{q.s.}}/T$ assuming a quasi-static process. This result applies for a non quasi-static process, too, if the process consists exclusively of the heat transfer q (see discussion following (11.12)); therefore an index "q.s." in (19.15) is not necessary. We insert (19.13)–(19.15) into (19.12) into (19.12):

$$\Delta S = \frac{q_2}{T_2} - \frac{q_1}{T_1} \geq 0 \qquad (19.16)$$

We can also express this in the form

$$\frac{q_2}{q_1} \geq \frac{T_2}{T_1} \qquad (19.17)$$

In a power plant, the amount of heat extracted at a high temperature level from R_1 must be continuously replaced; this is done, for example, by burning oil or coal. The work w, on the other hand, is the usable energy. Therefore, we define *efficiency* (or thermal efficiency) by

$$\eta = \frac{\text{generated work}}{\text{expended heat}} = \frac{w}{q_1} = \frac{q_1 - q_2}{q_1} = 1 - \frac{q_2}{q_1} \qquad (19.18)$$

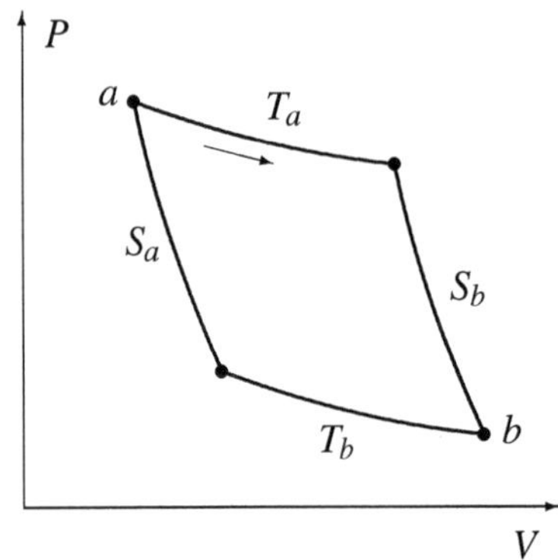

Figure 19.3 Scheme of the Carnot process. In the P-V diagram for the gas, one first chooses the two states a and b. Now one draw the isotherms ($T =$ const.) and the adiabats ($S =$ const.) through each of these two points. This creates a diamond (or rhombus) like shape. A process that runs clockwise along these lines can serve as a heat engine. A process running in the opposite direction may be used as a heat pump

The generated work might be saleable electric energy, the expended heat might be provided by burning fuel; so this ratio may be considered as the ratio benefit/cost. With (19.17) we obtain the following inequality for the efficiency

$$\eta \leq 1 - \frac{T_2}{T_1} \tag{19.19}$$

The maximum achievable efficiency is

$$\boxed{\eta_{\text{ideal}} = \frac{T_1 - T_2}{T_1} \qquad \begin{array}{l} \text{efficiency of an} \\ \text{ideal heat engine} \end{array}} \tag{19.20}$$

The standard example of a process with the ideal efficiency is the *Carnot process*. Here, a gas volume serves as the engine E. The gas volume runs through the cyclic process outlined in Figure 19.3, which consists of the following steps:

1. The gas expands quasi-statically from a to b in two partial steps:

   (i) isothermally from (T_a, S_a) to (T_a, S_b).

   (ii) adiabatically from (T_a, S_b) to (T_b, S_b).

2. The gas is compressed quasi-statically from b back to a in two partial steps:

   (i) isothermally from (T_b, S_b) to (T_b, S_a).

   (ii) adiabatically from (T_b, S_a) to (T_a, S_a).

For the isothermal steps, contact with a heat reservoir of the respective temperature is required. We denote the heat absorbed along the isotherms T_a and T_b by q_a and q_b, respectively. Then the total change in the entropy of the gas is

$$\Delta S_{\text{gas}} = \frac{q_a}{T_a} - \frac{q_b}{T_b} = 0 \tag{19.21}$$

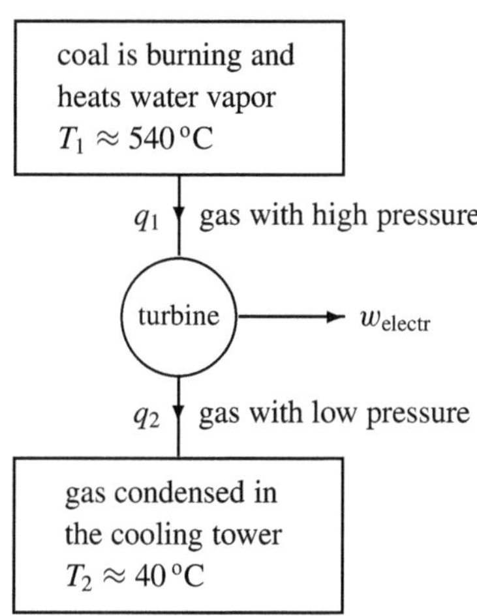

Figure 19.4 The scheme shown in Figure 19.2 can be realized by an oil or coal-fired power plant. An analogous scheme applies to a combustion engine: In this case gasoline is burned instead of coal, the turbine is replaced by the cylinder with its piston, mechanical work is generated and cooled gas is emitted through the exhaust.

Since the gas is returned to its initial state, $\Delta S_{\mathrm{gas}} = 0$ holds. For the same reason, $\Delta E_{\mathrm{gas}} = 0$ or

$$w = q_a - q_b \tag{19.22}$$

Since all steps are quasi-static, the work done by the gas is equal to $w = w_{\mathrm{q.s.}} = \oint P\,dV$; this equals the enclosed area in Figure 19.3. From (19.21) and (19.22) follows the efficiency of the Carnot process:

$$\eta_{\mathrm{Carnot}} = \frac{w}{q_a} = \frac{T_a - T_b}{T_a} = \eta_{\mathrm{ideal}} \tag{19.23}$$

We calculate the total entropy change of the closed system of heat system consisting of heat reservoirs, the gas and an energy storage. With $\Delta S_{\mathrm{R}_1} = -q_a/T_a$, $\Delta S_{\mathrm{R}_2} = q_b/T_b$, ΔS_{gas} from (19.21) and $\Delta S_{\mathrm{S}} \approx 0$ we obtain

$$\Delta S = \Delta S_{\mathrm{R}_1} + \Delta S_{\mathrm{R}_2} + \Delta S_{\mathrm{gas}} + \Delta S_{\mathrm{S}} = 0 \tag{19.24}$$

Because $\Delta S = 0$ the process is reversible. The Carnot process can also run in the opposite direction; it can then be used as a refrigerator or a heat pump.

We look at the diagram of a realistic thermal power plant sketched in Figure 19.4. For the temperatures shown in the figure temperatures shown in the figure, the ideal efficiency is

$$\eta_{\mathrm{ideal}} \approx \frac{813\,\mathrm{K} - 313\,\mathrm{K}}{813\,\mathrm{K}} \approx 62\% \tag{19.25}$$

In reality, a good steam turbine may achieve around 45 %. One liter of heating oil has a caloric value of about 40 000 kJ. This can be used to generate around 5 kWh of electrical energy:

$$1\ \mathrm{l\,oil} \,\widehat{=}\, 40\,000\ \mathrm{kJ} \approx 11\ \mathrm{kWh} \xrightarrow{\ \eta \approx 0.45\ } 5\ \mathrm{kWh} \tag{19.26}$$

At a price of 1 Euro (or USD) per liter of heating oil, this results in raw material costs of about 20 cents per kilowatt hour. This is a basis for the assessment of the household price for private consumers (at least for conventional power plants).

In order to maximize $\eta_{\text{ideal}} = 1 - T_2/T_1$, the temperature T_1 must be as large and T_2 as small as possible. Hereby T_2 is limited downwards by the ambient temperature. In order to increase T_1, combustion must be as concentrated as possible (in space and time). In a combustion engine, this requires heat-resistant material (such as ceramics). The actual processes in such an engine are of course more complicated than in the heat engines schemes considered here.

It would be desirable to generate as much high-quality energy w_{electr} as possible when burning coal or oil, and to use the unavoidable waste heat (q_2 in Figure 19.4) for heating (combined heat and power plant).

We can also reverse the process of the heat engine, Figure 19.5: By applying the work w we can pump heat from R_2 to R_1, i.e. from the lower temperature level to the higher one. As all variables q_1, q_2 and w are chosen positive, some of the signs in the above equations change. From the 1st law of TD follows

$$q_2 + w = q_1 \tag{19.27}$$

From the 2nd law of TD follows

$$\Delta S = \Delta S_{R_1} + \Delta S_{R_2} + \Delta S_E + \Delta S_S = \frac{q_1}{T_1} - \frac{q_2}{T_2} \geq 0 \tag{19.28}$$

For a *heat pump*, R_1 is the house to be heated and R_2 is the environment. The benefit is the heat q_1 pumped into the house, i.e.

$$\eta = \frac{q_1}{w}, \qquad \eta_{\text{ideal}} = \frac{T_1}{T_1 - T_2} \qquad \text{(heat pump)} \tag{19.29}$$

A heat pump works as a heating system by pumping heat from outside ($T_2 = 0\,°\text{C}$) into the house ($T_1 = 20\,°\text{C}$). We insert $T_1 \approx 293\,\text{K}$ and $T_2 \approx 273\,\text{K}$ into (19.29) and obtain

$$\eta_{\text{ideal}} \approx 15 \tag{19.30}$$

In principle, the electrical energy 1 kWh could be used to pump 15 kWh of heat into the house. In practice, one needs a higher temperature level $T_1 = 50\,°\text{C}$ in the house, for example for the domestic hot water. For this the ideal efficiency is $\eta_{\text{ideal}} \approx 6$; realistic values are $\eta \approx 3$.

From the physical principles, it is clear that the direct conversion of electrical energy into heat ($q = w_{\text{electr}}$ in the electric heater) is a waste, because one only achieves $\eta = q/w_{\text{electr}} = 1$ compared to $\eta \approx 3$ (realistic value) for a heat pump. Gas or oil central heating is also not an optimal energy utilization. One could use the oil to run a diesel engine, use the waste heat q_2 for heating and operate a heat pump with the generated work w. Even with modest and realizable efficiencies $\eta_{\text{engine}} = 0.3$ and $\eta_{\text{pump}} = 3$ it would be possible to double the heat generated for the house.

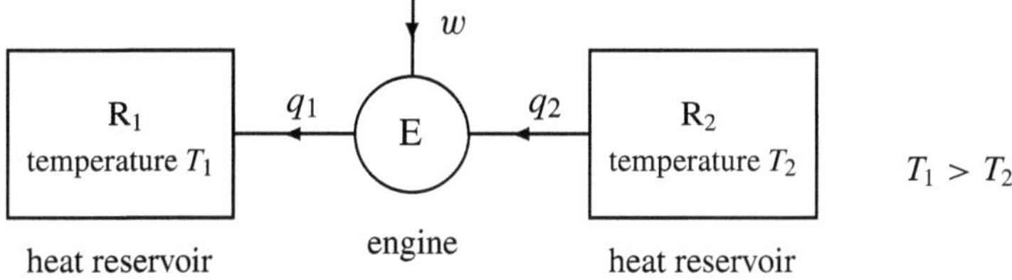

Figure 19.5   Diagram of a heat pump. With the expenditure of work w, heat can be transported from a lower temperature level to a higher one. It could be a refrigerator (T_2), from which heat is pumped into the environment (T_1). It could also be a heat pump that transfers heat from the surroundings (T_2) to the inside of the house (T_1).

We apply the scheme from Figure 19.5 to a refrigerator Then R_2 is the inside temperature and R_1 is the room in which the refrigerator is located. The benefit is the heat q_2 extracted from the refrigerator, i.e.

$$\eta = \frac{q_2}{w}, \qquad \eta_{\text{ideal}} = \frac{T_2}{T_1 - T_2} \qquad \text{(refrigerator)} \qquad (19.31)$$

Efficiencies far above 1 are possible here.

Exercises

19.1   Efficiency of a refrigerator

A refrigerator works at an internal temperature of $5\,^{\circ}\text{C}$. The heat that is pumped out of the refrigerator is transferred to a rear metal grate with temperature of $30\,^{\circ}\text{C}$. Someone covers the ventilation slots of the refrigerator; this enhances the temperature of the metal grate to $35\,^{\circ}\text{C}$. Estimate the percentage by which the energy consumption of the refrigerator increases.

19.2 Carnot process with ideal gas

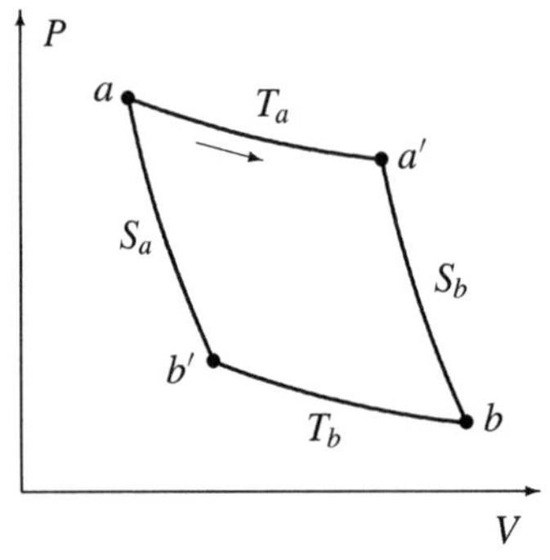

The cycle process $a \to a' \to b \to b' \to a$ takes place along isotherms and adiabats. It is carried out for one mole of a monatomic ideal gas. The pressure and volume are given for the states a and b, i.e. P_a, V_a, P_b and V_b. Determine the values T, V, P and S for all four states (a, a', b and b'). Calculate from this the heat amounts q_1 and q_3 absorbed in the first and third steps, and show $q_1/T_a + q_3/T_b = 0$.

19.3 Special cyclic process

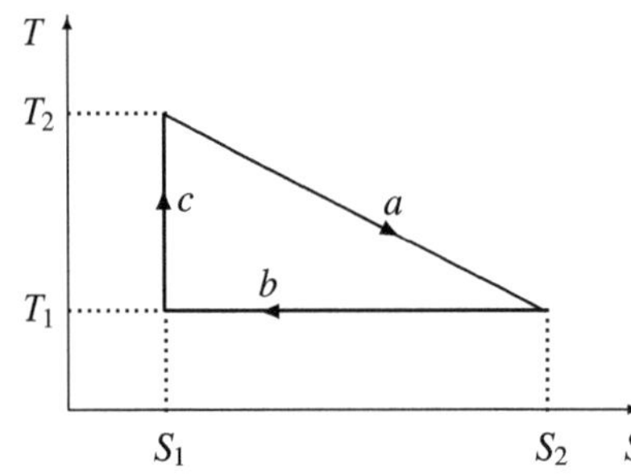

An ideal gas goes through the cyclic process described by the paths a, b and c. Calculate the individual work and heat contributions. What efficiency η results if the process is regarded as a heat engine?

19.4 Stirling process with ideal gas

We consider a monatomic ideal gas. A quasi-static cyclic process consists of the following steps 1, 2, 3 and 4:

1:	Isothermal expansion from V_1 to V_2	$T = T_1 = $ const.
2:	Isochoric cooling from T_1 to T_2	$V = V_2 = $ const.
3:	Isothermal compression from V_2 to V_1	$T = T_2 = $ const.
4:	Isochoric heating from T_2 to T_1	$V = V_1 = $ const.

Isochoric means with constant volume. Sketch the process in a P-V diagram. Determine the work and heat outputs for the individual steps. Calculate the efficiency

$$\eta_{\text{Stirling}} = \frac{-\Delta W}{\Delta Q_1}$$

for the case that the process is regarded as a *Stirling engine*. Here, ΔW is the sum of all work outputs, and ΔQ_1 is the heat supplied at the high temperature level T_1.

20 Chemical potential

The chemical potential μ is the generalized force that belongs to the external parameter N (number of particles). We present the thermodynamic relations for quasi-static processes with $dN \neq 0$. This includes in particular the Duhem-Gibbs relation.

Definition

In Part II, we discussed the microscopic foundations of thermodynamics. For introducing a new generalized force, the chemical potential, we briefly return to these microscopic foundations. The subsequent sections then return to thermodynamics.

The energy eigenvalues (6.5) for the ideal gas are of the form $E_r(V, N)$. This makes it clear that for a real gas, but also for other homogeneous systems such as solids or liquids, there are at least two external parameters $x = (V, N)$. So far, we have only considered processes with $dN = 0$ and suppressed the variable N. We drop this restriction now.

The external parameters are those variables $x = (x_1, ..., x_n)$, on which the Hamilton operator and thus the energies eigenvalues $E_r(x)$ depend. According to Chapter 8, the quasi-static change of x_i supplies the system with the work

$$\mathit{d}W_{\text{q.s.}} = \sum_{i=1}^{n} \overline{\frac{\partial E_r(x)}{\partial x_i}} \, dx_i = - \sum_{i=1}^{n} X_i \, dx_i \tag{20.1}$$

For $x_1 = V$ and $x_2 = N$, the generalized forces are

$$X_1 = P = - \overline{\frac{\partial E_r(V, N)}{\partial V}}, \qquad X_2 = -\mu = - \overline{\frac{\partial E_r(V, N)}{\partial N}} \tag{20.2}$$

The sign for μ is a convention. From (20.1) and (20.2) follows

$$\mathit{d}W_{\text{q.s.}} = -P \, dV + \mu \, dN \tag{20.3}$$

Together with $dE = \mathit{d}Q_{\text{q.s.}} + \mathit{d}W_{\text{q.s.}}$ (1st law) and $\mathit{d}Q_{\text{q.s.}} = T \, dS$ (2nd law) we obtain for a quasi-static process

$$\boxed{dE = T \, dS - P \, dV + \mu \, dN} \tag{20.4}$$

This is the exact differential of the thermodynamic potential energy $E(S,V,N)$. From it, all thermodynamic relations can be derived. In particular, we obtain for the chemical potential

$$\mu = -T \left(\frac{\partial S}{\partial N}\right)_{E,V} = \left(\frac{\partial E}{\partial N}\right)_{S,V} \tag{20.5}$$

The first expression is the standard definition of the generalized force as given in Chapter 10. According to the second expression the chemical potential is the energy required to add a particle to the thermally isolated system ($S = $ const.); the other external parameters (here V) shall be constant. Under these conditions, the chemical potential can be measured as the ratio dE/dN.

In practice, an added particle is taken away from another system; only the difference of chemical potentials is measurable. The zero point for μ is therefore arbitrary.

Thermodynamic potentials

The definitions (17.5) of the thermodynamic potentials remain unchanged, for example $F = E - TS$. However, because of (20.4), all differentials receive the additional term $\mu\, dN$:

$$\begin{aligned}
dE &= T\, dS - P\, dV + \mu\, dN \\
dF &= -S\, dT - P\, dV + \mu\, dN \\
dH &= T\, dS + V\, dP + \mu\, dN \\
dG &= -S\, dT + V\, dP + \mu\, dN
\end{aligned} \tag{20.6}$$

Thus N is an additional natural variable for all these potentials. For each potential there are now three Maxwell relations instead of one.

By a Legendre transformation with the term $-\mu N$ we could define a further potential for each of the potentials (20.6). However, we limit ourselves to the *grand-canonical potential*

$$J = E - TS - \mu N = F - \mu N \tag{20.7}$$

The designation "grand canonical" is chosen after the associated partition function which will be introduced in Part IV. From (20.6) and (20.7) follows

$$dJ = -S\, dT - P\, dV - N\, d\mu \tag{20.8}$$

This results in three Maxwell relations and the natural variables of J:

$$J = J(T, V, \mu) \qquad \text{(grand canonical potential)} \tag{20.9}$$

As mentioned in the footnote at the begin of Chapter 17, there are other designation for this potential, like "Landau potential", and Ω or Φ_{G} instead of J; we have chosen the letter J in continuation of E, F, H, ...

Exact differential

The chemical potential has a particularly simple relationship to the free enthalpy $G = G(T, P, N)$. As an extensive quantity, G is of the form (15.4),

$$G(T, P, N) = N\, g(T, P) \tag{20.10}$$

Then we get

$$\mu \overset{(20.6)}{=} \left(\frac{\partial G}{\partial N}\right)_{T,P} \overset{(20.10)}{=} g(T, P) = \frac{G}{N} \tag{20.11}$$

The function $g(T, P)$ introduced in (20.10) is therefore equal to the chemical potential μ:

$$G(T, P, N) = E - TS + PV = N\, \mu(T, P) \tag{20.12}$$

If we insert this into $J = E - TS - N\mu$, we get

$$J = -PV \tag{20.13}$$

This relationship complements the definition (20.7) of the grand canonical potential.
We write on the exact differential dG for $G = N\mu$ and compare this with dG from (20.6):

$$dG = N\, d\mu + \mu\, dN = -S\, dT + V\, dP + \mu\, dN \tag{20.14}$$

From this follows for the exact differential of the chemical potential:

$$\boxed{d\mu = -s\, dT + v\, dP \qquad \text{Duhem-Gibbs relation}} \tag{20.15}$$

Here $s = S/N$ and $v = V/N$. The natural and preferred variables of the chemical potential are T and P:

$$\mu = \mu(T, P) \tag{20.16}$$

Equilibrium condition

For given temperature T and the pressure P, the equilibrium follows from $G = N\mu = \text{minimal}$, (17.39). For a fixed number of particles N, this implies

$$\mu(T, P) = \text{minimal} \qquad \text{(equilibrium)} \tag{20.17}$$

Let the system consist of water molecules. For a given T and P the configuration or *phase* (water, water vapor or ice) is realized that has the smallest chemical potential.

Equilibrium for heat, volume and particle exchange

We consider two systems A and B that can exchange heat, volume and particles. A and B together form a closed system, so that

$$E = E_\text{A} + E_\text{B} = \text{const.}, \quad V = V_\text{A} + V_\text{B} = \text{const.}, \quad N = N_\text{A} + N_\text{B} = \text{const.} \tag{20.18}$$

The system state can be defined by the values for E_A, V_A and N_A. As a function of these variables, the entropy is maximum at equilibrium:

$$S(E_\text{A}, V_\text{A}, N_\text{A}) = S_\text{A}(E_\text{A}, V_\text{A}, N_\text{A}) + S_\text{B}(E - E_\text{A}, V - V_\text{A}, N - N_\text{A}) = \text{maximum} \tag{20.19}$$

One of the necessary conditions for this is

$$0 = \frac{\partial S}{\partial N_\text{A}} = \frac{\partial S_\text{A}}{\partial N_\text{A}} - \frac{\partial S_\text{B}}{\partial N_\text{B}} = -\frac{\mu_\text{A}}{T_\text{A}} + \frac{\mu_\text{B}}{T_\text{B}} \tag{20.20}$$

The analogous conditions $\partial S/\partial E_\text{A} = 0$ and $\partial S/\partial V_\text{A} = 0$ result in $T_\text{A} = T_\text{B}$ and $P_\text{A}/T_\text{A} = P_\text{B}/T_\text{B}$. Overall, this leads to

$$T_\text{A} = T_\text{B}, \quad P_\text{A} = P_\text{B}, \quad \mu_\text{A} = \mu_\text{B} \qquad \begin{array}{l}\text{(equilibrium for heat, volume}\\\text{and particle exchange)}\end{array} \tag{20.21}$$

An example of such a system are two *phases* of a substance, for example water (system A) and water vapor (system B). The condition $\mu_\text{A}(T, P) = \mu_\text{B}(T, P)$ are only fulfilled along a certain curve in the T-P plane. Along this *vapor pressure curve* there is a *phase equilibrium*, the phases water and water vapor coexist in equilibrium. This will be discussed in more detail in the next chapter.

Particle exchange without heat or volume exchange

We now consider two systems that can exchange particles but no heat or volume. We proceed as in Chapter 10 for volume exchange via an adiabatic wall. Analogous to the way from (10.24) to (10.26), the condition $S = \text{maximal}$ leads to

$$E_{\text{A}+\text{B}}(N_\text{A}) = E_\text{A}(S_\text{A}, V_\text{A}, N_\text{A}) + E_\text{B}(S_\text{B}, V_\text{B}, N - N_\text{A}) = \text{minimal} \tag{20.22}$$

Here S_A, S_B, V_A, V_B and N are fixed variables. From $\partial E_{\text{A}+\text{B}}/\partial N_\text{A} = 0$ follows

$$\mu_\text{A} = \mu_\text{B} \qquad \text{(particle exchange only)} \tag{20.23}$$

An example of such a system are two volumes (A and B) with helium II, which are connected by a "superleak" (Figure 38.6). The superleak only allows superfluid particles to pass through. The superleak particles have no entropy content and therefore do not transport heat. The consequences of the condition $\mu_\text{A} = \mu_\text{B}$ in this system will be discussed in Chapter 38.

Ideal gas

We specify the chemical potential for an ideal gas. To do this we use $PV = Nk_{\mathrm{B}}T$, the volume independence (16.7) of the energy, i.e. $E = E(T, N) = N\,e(T)$, and the entropy $S(T, V, N)$ of (16.32). From (20.12) follows

$$\mu = \frac{E}{N} - \frac{TS}{N} + \frac{PV}{N} = e(T) - T\left(\int^{T} dT' \,\frac{c_V(T')}{T'} + k_{\mathrm{B}} \ln \frac{V}{N} + \text{const.}\right) + k_{\mathrm{B}}T \tag{20.24}$$

The temperature dependencies were calculated in Part II for the monatomic gas, $e(T) = 3\,k_{\mathrm{B}}T/2$ and $c_V = 3\,k_{\mathrm{B}}/2$. For the diatomic gas they will be given in Chapter 27. Here we put them together in a not further specified function $f(T)$:

$$\frac{\mu}{k_{\mathrm{B}}T} = -f(T) - \ln \frac{V}{N} + \text{const.} = -f(T) + \ln \frac{P}{k_{\mathrm{B}}T} + \text{const.} \tag{20.25}$$

The first form yields $\mu(T, V/N)$, the second $\mu(T, P)$.

Exercises

20.1 *Maxwell relations for grand canonical potential*

Write down the Maxwell relations for $J(T, V, \mu) = F - \mu N$.

20.2 *Differential for energy per particle*

Show

$$de = T\,ds - P\,dv \tag{20.26}$$

where $e = E/N$, $s = S/N$ and $v = V/N$. The Duhem-Gibbs relations

$$d\mu = -s\,dT + v\,dP \qquad \text{and} \qquad \mu = \frac{G}{N}$$

are assumed to be known.

20.3 *Chemical potential for ideal gas*

Determine the chemical potential $\mu(T, P)$ for an ideal gas with the heat capacity $C_V(T, N)$. What results specifically for the monatomic gas with $C_V = 3\,N\,k_\mathrm{B}/2$?

20.4 *Derivation of the Duhem-Gibbs relation*

Derive the Duhem-Gibbs relation for a homogeneous system with

$$G = E - TS + PV = \mu N \tag{20.27}$$

by differentiating $S(\lambda E, \lambda V, \lambda N) = \lambda S(E, V, N)$ with respect to λ.

21 Exchange of particles

We discuss different applications of the equilibrium condition $\mu_A = \mu_B$ for particle exchange. One important application is the equilibrium between the phases of matter, for example between water and water vapor. We investigate the dependence of the transition temperature on the pressure (Clausius-Clapeyron equation) and on the concentration of a substance dissolved in the liquid phase (osmotic pressure, elevation of boiling point, depression of freezing point). Finally, we establish an equilibrium condition for chemical reactions.

Phase diagram

Depending on T and P, substances (made up by molecules of the same kind) can occur in different *phases*. Normally, at least the three phases solid, liquid and gaseous occur. The standard example is ice, water and water vapor. Water vapor is a gas consisting of H_2O molecules (not meant is visible vapor with small liquid droplets in air). The solid, crystalline phase may often divided into various phases with different crystal structures; this also applies to ice. Further examples of phase transitions are discussed in Part VI.

For a homogeneous system consisting of a certain kind of matter, there are two external parameters, V and N. The equilibrium states can be defined by E, V and N or by three other macroscopic variables. We choose the state variables T, P and N in the following. The variables T and P in themselves already determine the thermodynamic potential G per number of particles, $G(T, P, N)/N = \mu(T, P)$. Apart from the absolute size of the system, T and P therefore determine the thermodynamic state.

We consider a system in which two different phases A and B of a substance are present, for example water vapor and water. Particles can be exchanged between these two phases. For given T and P, the two phases are in equilibrium if

$$\mu_A(T, P) = \mu_B(T, P) \qquad \text{(phase equilibrium)} \qquad (21.1)$$

In the following, we associate μ_A with the gaseous phase and μ_B with the liquid phase; nevertheless, the formulas apply to two arbitrary phases. Since the internal structure of the phases differ from each other, the functions $\mu_A(T, P)$ and $\mu_B(T, P)$ are different. The equation (21.1) then defines in general a curve in the P-T diagram, Figure 21.1. For the assumed phase (gaseous and liquid) these curves are

$$
\begin{aligned}
P &= P_v(T) \qquad &\text{(vapor pressure curve)} \\
T &= T_b(P) \qquad &\text{(boiling temperature)}
\end{aligned}
\qquad (21.2)
$$

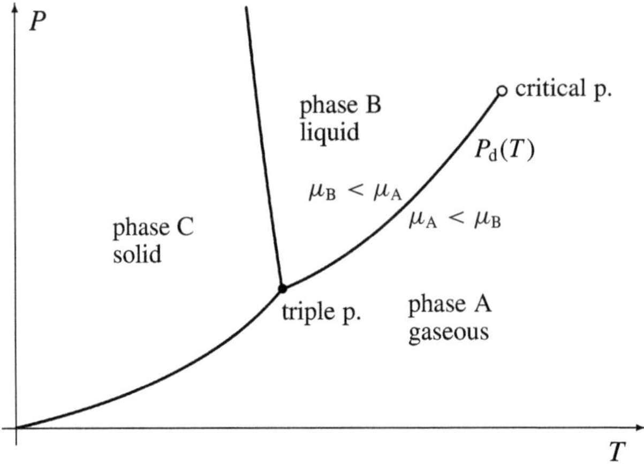

Figure 21.1   Schematic phase diagram of water. The vapor pressure curve $P_v(T)$ is given by the condition $\mu_A = \mu_B$; at the triple point $\mu_A = \mu_B = \mu_C$ holds. The equilibrium is given by $\mu =$ minimal. The phase with the smallest chemical potential will be realized.

We use $P_v(T) = P_{vapor}$ for the vapor pressure curve, and $T_b(T) = T_{boiling}$ for the boiling temperature. Next to the transition curve, $\mu_A \neq \mu_B$ applies. According to (20.17), $\mu(T, P)$ is minimal in equilibrium. Therefore, the phase with the smaller chemical potential will be realized.

These considerations do not show why there are different phases and what structure they have. However, they indicate that an equilibrium between two phases generally only exists along a curve in the P-T diagram. The P-T-diagram of a substance is then divided by curves into areas with different phases, Figure 21.1. The P-T-diagram with such a division is called *phase diagram* or *(thermodynamic) state diagram*. The transition across such a curve is a *phase transition*. The phase transition $A \rightarrow B$ is connected with the change $\mu_A \rightarrow \mu_B$. At the transition point, the chemical potential μ is continuous according to (21.1); the functions μ_A and μ_B are, however, distinctly different on both sides. Other variables such as $s_A \rightarrow s_B$ (with $s = -(\partial \mu / \partial T)_P$) but may be discontinuous at the transition point (Chapter 35).

The transition curve between the liquid and gaseous phase ends for all substances at a *critical point*. It is then possible that the systems changes from the gaseous phase (below vapor pressure curve) to the liquid phase (above the curve) on a path around the critical point. On such a path, $\mu(T, P)$ changes smoothly.

We now consider a substance with three phases. According to (21.1) there is a curve for the equilibrium $A \leftrightarrow B$, and another one for $B \leftrightarrow C$. In general, these two curves intersect at a point in the P-T diagram, Figure 21.1. This point is defined by

$$\mu_A(T, P) = \mu_B(T, P) = \mu_C(T, P) \qquad \text{(triple point)} \qquad (21.3)$$

The curve for the equilibrium $A \leftrightarrow C$ also passes through this *triple point*. At the triple point, all three phases are in equilibrium with each other. The triple point determines specific values for the temperature T_{triple} and the pressure P_{triple}. The triple point of water is used to define the Kelvin scale (Chapter 14).

Clausius-Clapeyron equation

The equilibrium condition (21.1) yields a relation for the equilibrium curve (21.2). We refer specifically to the vapor pressure curve, i.e. the gas–liquid transition (phases A and B). The vapor pressure curve $P = P_{\text{v}}(T)$ is given by

$$\mu_{\text{A}}(T, P_{\text{v}}(T)) = \mu_{\text{B}}(T, P_{\text{v}}(T)) \tag{21.4}$$

We calculate the total derivative with respect to the temperature:

$$\left(\frac{\partial \mu_{\text{A}}}{\partial T}\right)_P + \left(\frac{\partial \mu_{\text{A}}}{\partial P}\right)_T \frac{dP_{\text{v}}(T)}{dT} = \left(\frac{\partial \mu_{\text{B}}}{\partial T}\right)_P + \left(\frac{\partial \mu_{\text{B}}}{\partial P}\right)_T \frac{dP_{\text{v}}(T)}{dT} \tag{21.5}$$

We insert the partial derivatives of μ, which result from $d\mu = -s\, dT + v\, dP$:

$$(v_{\text{A}} - v_{\text{B}}) \frac{dP_{\text{v}}(T)}{dT} = s_{\text{A}} - s_{\text{B}} \tag{21.6}$$

The transition from B to A takes place at a certain point on the vapor pressure curve $P = P_{\text{v}}(T)$ in Figure 21.1; for example for water at normal pressure and at $T = 100\,^{\circ}\text{C}$. At this point, the phases are in equilibrium with each other; the process can be quasi-static. For constant pressure, (17.9) yields $\Delta s = \Delta h/T$, so that

$$s_{\text{A}} - s_{\text{B}} = \frac{h_{\text{A}} - h_{\text{B}}}{T} = \frac{q}{T} \tag{21.7}$$

The quantity $q = h_{\text{A}} - h_{\text{B}}$ is called *enthalpy of vaporization* (also called latent heat of vaporization). In such a transition, the pressure (and not the volume) is kept constant; therefore, the heat supplied for the conversion is correctly referred to as *enthalpy*. Since phase transitions are often induced by heat supply, the terms "heat of transformation" or "latent heat" are also common.

Referring to specific phase transformations, the respective terms can be used, like enthalpy of vaporization, condensation, melting (or fusion), freezing, sublimation and so on.

Equations (21.6) and (21.7) yield the Clausius-Clapeyron equation

$$\boxed{\frac{dP_{\text{v}}(T)}{dT} = \frac{q}{T\,(v_{\text{A}} - v_{\text{B}})}} \qquad \text{Clausius-Clapeyron equation} \tag{21.8}$$

This is a relation between the slope of the vapor pressure curve (the transition curve in the P-T diagram) and the associated entropy and volume change. The quantities q and v can optionally be related to the number of particles or moles. There are also phase transitions with $q = 0$; for them (21.8) implies $v_{\text{A}} = v_{\text{B}}$.

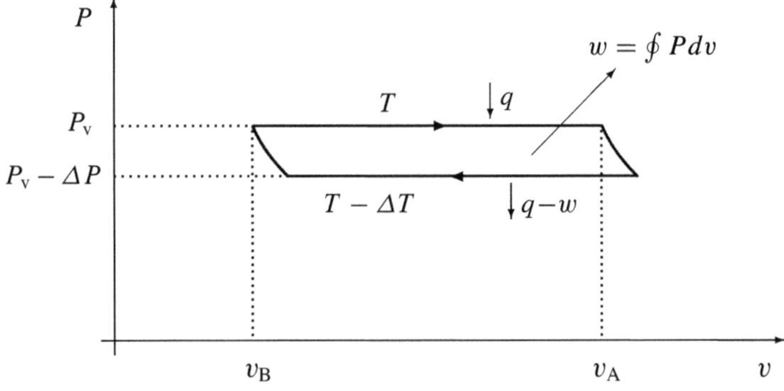

Figure 21.2 P-v diagram for a cycle process along the vapor pressure curve $P = P_v(T)$. For the quasi-static process, on obtains the ideal efficiency $\eta_{\text{ideal}} = \Delta T/T = w/q$ with $w \approx (v_A - v_B)\,\Delta P$. From this follows (21.8).

An alternative derivation of the Clausius-Clapeyron equation starts from the cyclic process shown in Figure 21.2: At a specific point P, T of the vapor pressure curve, one mole of liquid evaporates, and at the point $P - \Delta P$, $T - \Delta T$ of the curve, the gas condenses again; in the P-V diagram 21.2 these processes correspond to horizontal lines. By small expansion and compression sections, these lines are closed to form a cycle process. For the quasi-static process, the work performed is given by $w = \oint P\,dv \approx (v_A - v_B)\,\Delta P$. At the temperature T, the enthalpy of vaporization q is absorbed from a heat reservoir, at $T - \Delta T$ the heat $q - w$ is released. The cyclic process is a special Carnot process with the ideal efficiency

$$\eta_{\text{ideal}} = \frac{\Delta T}{T} = \frac{w}{q} = \frac{(v_A - v_B)\,\Delta P}{q} \tag{21.9}$$

From this follows the Clausius-Clapeyron equation (21.8).

At normal pressure $P \approx 1$ bar (at sea level), water boils at $T_b \approx 373\,\text{K}$ (or $100\,^\circ\text{C}$). We estimate the boiling temperature of water on a mountain with a height of $3\,\text{km}$ (like the Zugspitze in Germany). The enthalpy of vaporization of water is

$$q \approx 4 \cdot 10^4 \; \frac{\text{J}}{\text{mol}} \qquad \text{(water at } T_b = 373\,\text{K)} \tag{21.10}$$

We treat the water vapor as an ideal gas, i.e.

$$v_A = \frac{RT}{P} \tag{21.11}$$

The volume of water can be neglected compared to the volume of gas:

$$v_B \approx 18 \; \frac{\text{cm}^3}{\text{mol}} \ll v_A \approx 22 \cdot 10^3 \; \frac{\text{cm}^3}{\text{mol}} \tag{21.12}$$

Here we have used 1 mol $\widehat{=}$ 18 g, the density 1 g/cm^3 of water and the molar volume (14.32) for v_A. We insert (21.10)–(21.12) into (21.8):

$$\frac{T}{P}\frac{dP_v(T)}{dT} \approx \frac{q}{RT} \approx \frac{4\cdot 10^4}{8.3\cdot 373} \approx 13 \qquad (21.13)$$

According to the barometric formula (24.32), the air pressure changes at an altitude difference of $z = 3\,\text{km}$ by $\Delta P/P \approx \exp(-3/8) - 1 \approx -0.31$. We assume that the gradient dP_v/dT is approximately constant in this pressure change. Then, according to (21.13), the temperature changes along the vapor pressure curve by

$$\Delta T \approx \frac{\Delta P}{dP_v/dT} \approx \frac{\Delta P}{P}\,T\,\frac{RT}{q} \approx -0.31\cdot 373\,\text{K}\,\frac{1}{13} \approx -9\,\text{K} \qquad (21.14)$$

According to this, one can expect that water boils already at about $90\,^{\circ}\text{C}$ on the mountain. However, this estimate was based on some simplifications, in particular the barometric formula and temperature-independent values for dP_v/dT and q.

Humidity

Air usually contains a proportion of water vapor. The atmospheric pressure $P_0 \approx$ 1 bar is made up of the partial pressures of all constituents of the gas (see also Exercise 10.1):

$$P_0 = P(N_2) + P(O_2) + P(Ar) + P(CO_2) + \ldots + P(H_2O) \qquad (21.15)$$

The dots refer to other trace gases. The partial pressure of the water vapor at a given temperature is limited by the vapor pressure curve,

$$P(H_2O) \leq P_v(T) \qquad (21.16)$$

At $P(H_2O) = P_v(T)$, the water vapor is in equilibrium with water; here droplet formation sets in (fog, rain). This defines the measure for the relative air humidity P_H:

$$P_H = \frac{P(H_2O)}{P_v(T)} \qquad (21.17)$$

This quantity has the maximum value $P_H = 1 = 100\,\%$. In this case, we speak of 100 percent humidity.

It is now easy to calculate how much water is present in the air at a given humidity and temperature. For example, for a temperature $T = 20\,^{\circ}\text{C}$ one may read off from the phase diagram of water (see Figure 14.2):

$$P_v(20\,^{\circ}\text{C}) \approx 0.02\,\text{bar} \qquad (21.18)$$

At 100 percent air humidity, the partial pressure of the water vapor is therefore 2% of the total pressure $P_0 \approx 1$ bar. This corresponds to a density of

$$\varrho(100\,\%) \approx \frac{P_v(20\,^{\circ}\text{C})}{P_0}\,\frac{18\,\text{g}}{22.4\,\text{L}} \approx 17\,\frac{\text{g}}{\text{m}^3} \qquad (21.19)$$

For the density of pure water vapor, we have used the values (14.32) of an ideal gas, i.e. 22.4 liter for one mole (18 g). A air humidity of 30 % therefore means a contribution of about five grams of water (in the form of water vapor) per cubic meter. For comparison we note that one cubic meter of air has a mass of about 1,3 kg.

At temperatures below the triple point, the coexistence curve between the gaseous phase and the solid phase (see Figure 14.2) takes over the role of the vapor pressure curve. Also over an ice surface there is a finite partial pressure $P(H_2O)$, i.e. a finite humidity.

Osmotic pressure

In a solvent (e.g. water), N_c particles of another substance (such as salt) are dissolved. The particle density of the solvent is denoted by $n = N/V$, that of the dissolved substance by $n_c = N_c/V$. The solution has the concentration

$$c = \frac{N_c}{N} = \frac{n_c}{n} \tag{21.20}$$

We now consider an experiment in which the solution is separated from the pure solvent by a *semipermeable membrane*, Figure 21.3. Semipermeable means that the membrane is permeable to the molecules of the solvent but not for the molecules of the solute[1]. This creates a pressure difference on both sides of the semipermeable membrane, which is called *osmotic pressure*. We calculate this osmotic pressure.

Let Ω_{solvent} be the number of microstates of the solvent and Ω_c is that of the solute. We assume that the particles of the solvent and the solute are independent of each other. Then the total number of microstates of the solution is equal to the product $\Omega_{\text{solvent}} \Omega_c$. The solute is limited to the volume V of the solution. We further assume that the solute particles move independently in the available volume (i.e. without mutual interaction). Then Ω_c has the volume dependence known from the ideal gas

$$\Omega_c(V) \propto (V)^{N_c} \tag{21.21}$$

From the entropy

$$S_{\text{solution}} = k_B \ln(\Omega_{\text{solvent}} \Omega_c) = S_{\text{solvent}} + N_c k_B \ln V + \dots \tag{21.22}$$

we obtain the pressure P_{solution} of the solution:

$$\frac{P_{\text{solution}}}{T} = \left(\frac{\partial S_{\text{solution}}}{\partial V}\right)_E = \left(\frac{\partial S_{\text{solvent}}}{\partial V}\right)_E + \frac{N_c k_B}{V} = \frac{P_{\text{solvent}}}{T} + \frac{P_{\text{osm}}}{T} \tag{21.23}$$

The total pressure P_{solution} of the solution is the sum of the pressure P_{solvent} of the solvent and the pressure P_{osm} of the solute. This yields

$$\boxed{P_{\text{osm}} = n_c k_B T \qquad \text{van't Hoff's law}} \tag{21.24}$$

---

[1]The *solution* (salt in water) consists of the *solvent* (water), and of the dissolved substance called *solute* (salt).

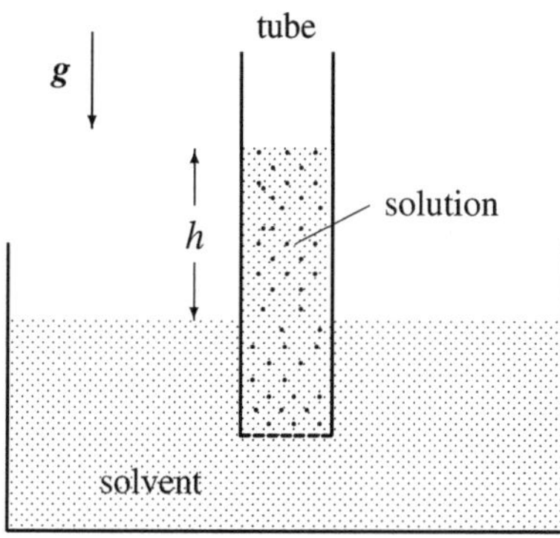

Figure 21.3 A tube is immersed in a vessel with water. The tube contains an aqueous salt solution. The tube is sealed at the bottom by a semi-permeable membrane through which only water molecules can pass. At equilibrium the liquid level in the tube is then higher than that of the water (in the gravity field). The difference in height h corresponds to the osmotic pressure P_{osm}.

Van't Hoff established this law as a phenomenological relationship.

On both sides of the semi-permeable membrane, the same pressure P_{solvent} must prevail for the solvent; otherwise solvent would flow through the membrane. This means that in Figure 21.3

$$\begin{aligned} \text{just above the membrane:} \quad P_1 &= P_{\text{solvent}} + P_{\text{osm}} \\ \text{just below the membrane:} \quad P_2 &= P_{\text{solvent}} \end{aligned} \qquad (21.25)$$

In the gravitational field, the pressure is proportional to the height difference. If the mass density ϱ of the solution and of the solvent are approximately equal, then the pressure difference $P_1 - P_2$ equals $\varrho g h$; here h is the height shown in Figure 21.3 and g is the gravitational acceleration. The osmotic pressure can then be read from the height difference:

$$P_{\text{osm}} = \varrho g h \qquad (21.26)$$

If the concentration in the solution is increased, the osmotic pressure increases according to (21.24). Then the solvent flows through the semi-permeable membrane until the new height h fulfills the condition (21.26). Effectively, the osmotic pressure thus acts in the direction of diluting the solution. The osmotic pressure enables plants to draw water upwards from the soil.

As the dissolved particles cannot move freely in the liquid, the ideal-gas approximation (21.21) does not seem plausible at first. However, remember that also molecules in a real gas cannot move freely; rather, they collide frequently with other molecules (the mean free path length is 10^{-5} cm in air at normal conditions). Nevertheless, the ideal gas law is a usable approximation for real gases. The reason is that the pressure only depends on how many particles per time are reflected at a wall and at what speed they hit the wall. This leads to $P \propto (N/V)$ and $P \propto T$; this pressure is approximately independent on the mean free path. In the solution, too, all particles participate in the general temperature movement; their kinetic energy is therefore on average equal to $3k_{\text{B}}T/2$. And the number of particles reflected at the membrane per time follows from their density and velocity, just as in the gas. Therefore, the ideal gas law is a usable approximation for the osmotic pressure.

Boiling point elevation and freezing point depression

At a given pressure, a liquid boils and freezes at certain temperatures. If the pressure is changed, these temperatures shift. Dissolving a substance in the liquid induces an effective change in pressure and therefore leads to a shift of the transition temperatures, namely to an elevation of the boiling point or a depression of the freezing point. It is assumed that the solute (the dissolved substance) is confined to the liquid phase; this is generally the case.

We assume that the pressure contribution of the solute is determined by van't Hoff's law. We consider the following pressures:

$$
\begin{aligned}
P_{\text{solution}} &= P && \text{pressure of the solution, total pressure} \\
P_{\text{osm}} &= n_c k_B T = \Delta P && \text{pressure of the dissolved substance} \\
P_{\text{solvent}} &= P - \Delta P && \text{pressure of the solvent}
\end{aligned}
\tag{21.27}
$$

The solution has the concentration c, (21.20). The chemical potential $\mu_c(T, P)$ of the solvent in the solution is is then equal to

$$
\mu_c(T, P) = \mu(T, P - \Delta P) \approx \mu(T, P) - \left(\frac{\partial \mu}{\partial P}\right)_T \Delta P = \mu(T, P) - c k_B T
\tag{21.28}
$$

This expansion presupposes $\Delta P \ll P$ or $c \ll 1$. In the last step, $(\partial \mu / \partial P)_T = v = V/N$ was used.

We refer to the liquid as phase B; the second phase A under consideration is gaseous or solid. For the pure solvent, the transition A $\leftrightarrow$ B occurs at the transition temperature $T_{tr}(P)$ where

$$
\mu_B(T_{tr}, P) = \mu_A(T_{tr}, P) \qquad (c = 0)
\tag{21.29}
$$

The transition (index tr) temperature T_{tr} might be specified as the boiling (index b) temperature T_b or the freezing (index f) temperature T_f.

Let now a substance be dissolved in the liquid phase B. Then the chemical potential $\mu_B(T, P)$ is given by $\mu_{c,B}(T, P)$ from (21.28). For $c \neq 0$, the condition (21.29) will be fulfilled at a different temperature $T_{tr} + \Delta T_{tr}$,

$$
\mu_{c,B}(T_{tr} + \Delta T_{tr}, P) = \mu_A(T_{tr} + \Delta T_{tr}, P) \qquad (c \neq 0)
\tag{21.30}
$$

This establishes a relationship between the concentration c and the temperature shift ΔT_{tr}. We expand both sides of (21.30) for small shifts using $d\mu = -s\,dT + v\,dP$:

$$
\begin{aligned}
\mu_{c,B}(T_{tr} + \Delta T_{tr}, P) &\overset{(21.28)}{\approx} \mu_B(T_{tr} + \Delta T_{tr}, P) - c k_B(T_{tr} + \Delta T_{tr}) \\
&\approx \mu_B(T_{tr}, P) - s_B \Delta T_{tr} - c k_B T_{tr} \\
\mu_A(T_{tr} + \Delta T_{tr}, P) &\approx \mu_A(T_{tr}, P) - s_A \Delta T_{tr}
\end{aligned}
$$

$$\tag{21.31}$$
$$\tag{21.32}$$

In (21.31), the term $c\,k_B\,\Delta T_{tr}$ was omitted because it is quadratic in the small quantities c and ΔT_{tr}. According to (21.30), the right-hand sides of (21.31) and (21.32) must be equal. Taking into account (21.29), this results in

$$\Delta T_{tr} = c\,\frac{k_B T_{tr}}{s_A - s_B} \tag{21.33}$$

The entropy difference between the two phases is characterized by the transformation enthalpy $q = h_A - h_B$ for the transition $B \to A$:

$$s_A - s_B = \frac{q}{T_{tr}} = \begin{cases} > 0 & \text{liquid} \to \text{gaseous} \\ < 0 & \text{liquid} \to \text{solid} \end{cases} \tag{21.34}$$

Here B denotes the liquid phase. From the last two equations follows

$$\boxed{\;\frac{\Delta T_{tr}}{T_{tr}} = c\,\frac{k_B T_{tr}}{q} \qquad \text{shift of the transition}\atop\text{temperature } T_{tr}\;} \tag{21.35}$$

If A is the gaseous phase, q and therefore ΔT_{tr} are positive; this results in a *boiling point elevation* $\Delta T_b > 0$. If A is the solid phase, q and therefore ΔT_{tr} are negative; this results in a *freezing point depression* $\Delta T_f < 0$. We use the index tr for transition, and more specifically b for boiling and f for freezing.

In winter, someone spreads salt on a wet paving resulting in a 10% solution. The enthalpy of transformation q for the transition from water to ice is $q \approx -6000\,\text{J/mol}$, and the transition takes place at the freezing temperature $T_f \approx 273$ K. We use these values in (21.35):

$$\Delta T_f = c\,\frac{R T_f}{q}\,T_f \approx 0.1\,\frac{8.3 \cdot 273}{-6000}\,273\text{ K} \approx -10\text{ K} \tag{21.36}$$

This means that the paving stays free of no ice for temperatures down to 10 degrees Celsius below zero.

Reaction equilibrium

We consider a mixture of m substances X_i that can be converted into each other according to the chemical reaction

$$\sum_{i=1}^{m} v_i\, X_i \rightleftharpoons 0 \tag{21.37}$$

As an example, we consider the synthesis of ammonium from nitrogen and hydrogen gas:

$$N_2 + 3\,H_2 \rightleftharpoons 2\,NH_3 \tag{21.38}$$

This reaction is of the form (21.37) with

$$X_1 = N_2, \quad X_2 = H_2, \quad X_3 = NH_3, \quad \nu_1 = 1, \quad \nu_2 = 3, \quad \nu_3 = -2 \quad (21.39)$$

We consider a closed system consisting of a homogeneous mixture of the substances X_i. We assume that this constitutes a *mixture of ideal gases*. As discussed in the last section on osmotic pressure, this is a usable approximation also if the substances X_i are dissolved in a solvent (not involved in the reaction). In this case, the model is referred to as an *ideal solution*.

The number of molecules of the substance kind X_i is equal to N_i. According to the ideal gas assumption, the chemical potentials of the individual gas kinds are independent of each other. This means that the free enthalpy (20.12) becomes

$$G(T, P, N_1,..., N_m) = \sum_{i=1}^{m} \mu_i(T, P)\, N_i \tag{21.40}$$

If now dk molecular reactions take place according to (21.37) or (21.38), then the number of X_i molecules changes by dN_i,

$$dN_i = \nu_i\, dk \qquad (i = 1,..., m) \tag{21.41}$$

The sign of $dk = \pm 1, \pm 2, \ldots$ depends on the direction of the reaction. For a given T and P, the equilibrium is determined by the minimum of G. At equilibrium G must therefore be minimal as a function of N_i. Because of (21.41), the N_i are functions of k. The necessary condition for a minimum is therefore

$$\frac{dG}{dk} = \sum_{i=1}^{m} \frac{\partial G(T, P, N_1,...., N_m)}{\partial N_i}\, \nu_i = 0 \tag{21.42}$$

From the last two equations follows

$$\sum_{i=1}^{m} \mu_i\, \nu_i = 0 \tag{21.43}$$

To evaluate this condition, we start from (20.25) for $\mu(T, V/N_i)$ and take into account $N_i/V = c_i\,(N/V) = c_i\,(P/k_B T)$:

$$\frac{\mu_i}{k_B T} = -f_i(T) - \ln \frac{V}{N_i} + \text{const.} = -f_i(T) + \ln c_i + \ln \frac{P}{k_B T} + \text{const.} \tag{21.44}$$

We multiply both sides with ν_i and sum over i. In equilibrium, the left-hand side then disappears because of (21.43). Thus we obtain

$$0 = -\sum_i \nu_i\, f_i(T) + \sum_i \nu_i\, \ln c_i + \sum_i \nu_i\, \ln P - \sum_i \nu_i\, \ln(k_B T) + \text{const.}$$

$$= \ln \prod_{i=1}^{m} (c_i)^{\nu_i} + \ln P^{\Sigma \nu_i} - \ln K(T) \tag{21.45}$$

with a function $K(T)$ which is not further specified here. From this we obtain for the product of the concentrations

$$\boxed{\prod_{i=1}^{m} \left(c_i\right)^{\nu_i} = \frac{K(T)}{P^{\sum \nu_i}} \qquad \text{law of mass action}} \qquad (21.46)$$

For the reaction (21.38), the law of mass action gives

$$\frac{c_{N_2} \cdot (c_{H_2})^3}{(c_{NH_3})^2} = \frac{K(T)}{P^2} \qquad (21.47)$$

An increase in pressure shifts the chemical equilibrium in favor of c_{NH_3} and thus favors the ammonium synthesis.

Another example is the formation of ions in water:

$$H_3O^+ + OH^- \rightleftharpoons 2\,H_2O, \quad \text{i.e. } \nu_1 = 1, \ \nu_2 = 1 \text{ and } \nu_3 = -2 \qquad (21.48)$$

In this case, the pressure dependence in the law of mass action cancels:

$$\frac{c_{H_3O^+} \cdot c_{OH^-}}{(c_{H_2O})^2} = K(T) \qquad (21.49)$$

At room temperature ($T \approx 300\,\text{K}$), $K(T) \approx 10^{-14}$. This means $c_{H_2O} \approx 1$ and $c_{H_3O^+} = c_{OH^-} \approx 10^{-7}$. The concentration of the H_3O^+ ions is expressed by the pH value:

$$\text{pH} = -\log c_{H_3O^+} \approx 7 \qquad (21.50)$$

Neutral water (pure water without dissolved substances) has a pH value of 7.

Exercises

21.1   Freezing point depression during ice skating

The pressure of an ice skate on the ice creates a freezing point depression. Is this effect sufficient to create a water film on which the skate glides?

The ice skater has a mass of 80 kg, and his skates each rest on a length of 10 cm and a width of 4 mm respectively. Use this to calculate the freezing point depression that results from the Clausius-Clapeyron equation.

21.2   *Vapor pressure curve from Clausius-Clapeyron equation*

Determine the vapor pressure curve from the Clausius-Clapeyron equation using the following assumptions: $v_A - v_B \approx v_A \approx RT/P$ and $q \approx$ const.

21.3   *Expansion coefficient along the vapor pressure curve*

Determine the thermal expansion coefficient

$$\alpha_v = \frac{1}{v_v}\left(\frac{\partial v_v}{\partial T}\right)_{\text{coex}}$$

for a gas that coexists with its liquid phase. Here $v_v(T, P)$ is the molar volume of the gas, which also called vapor (index v) in this context. The vapor is treated as an ideal gas. The molar volume of the liquid can be neglected compared to that of the gas.

21.4   *Coexistence curve for two gaseous phases*

A substance has two gaseous phases A and B, which satisfy the equations of state

$$P v_A = R_A T \qquad \text{and} \qquad P v_B = R_B T$$

Here, R_A and R_B are constants and v is the volume of a mole. For the specific heats at constant pressure, $c_{P,A}(T) = c_{P,B}(T) = c_P(T)$ holds. Calculate the coexistence curve $P_{\text{coex}}(T)$ at which the two phases are in equilibrium. Show that the transition enthalpy q is constant.

21.5   *Boiling of a salt solution*

For what salt concentration does water boil at $100\,^\circ$C on a 1 km high mountain?

21.6   *Dissolved substance in both phases*

If the solute is restricted to the liquid phase B then a concentration c yields the shift

$$\frac{\Delta T_{\text{tr}}}{T_{\text{tr}}} = c\,\frac{k_B T_{\text{tr}}}{q}$$

of the transition temperature T_{tr}. Here $q = T_{\text{tr}}(s_A - s_B)$ is the transformation enthalpy for the phase transition B $\to$ A. Generalize the expression for ΔT_{tr} to the case that there is a solute in both, phase A and phase B, with the concentrations are c_A and c_B, respectively.

IV Statistical ensembles

22 Partition functions

In this Part IV we introduce the canonical and the grand canonical ensemble and the corresponding partition functions (Chapter 22). In Chapter 23, we assign a thermodynamic potential to each of the partition functions. Classical systems are then analyzed with the help of the canonical partition function (Chapter 24). In Chapter 25, the ideal gas is used to demonstrate the practical equivalence of the various ensembles for macroscopic systems.

Canonical ensemble

The microcanonical partition function $\Omega(E, x)$ determines the equilibrium state for a given energy. In this chapter we introduce the canonical partition function $Z(T, x)$ which determines the equilibrium state for a given temperature. This is supplemented by the grand canonical partition function Y, for which the temperature and the chemical potential are given.

The statistical description of an equilibrium state is based on an ensemble of many systems of the same kind in which the microstates r occur with the probabilities P_r. As a fundamental postulate, we have introduced the assumption that all $\Omega(E, x)$ accessible microstates r of a closed system are equally probable:

$$P_r(E, x) = \begin{cases} \dfrac{1}{\Omega(E, x)} & \text{for } E - \delta E \le E_r(x) \le E \\[2mm] 0 & \text{otherwise} \end{cases} \tag{22.1}$$

The ensemble with these P_r is called *microcanonical*. The condition $\sum P_r = 1$ defines the microcanonical partition function Ω:

$$\Omega(E, x) = \sum_{r:\, E - \delta E \,\le\, E_r(x) \,\le\, E} 1 \tag{22.2}$$

By (22.1), the P_r of the closed system are given as a function of the energy E and the external parameters x (such as V and N). We now ask what the P_r look like if

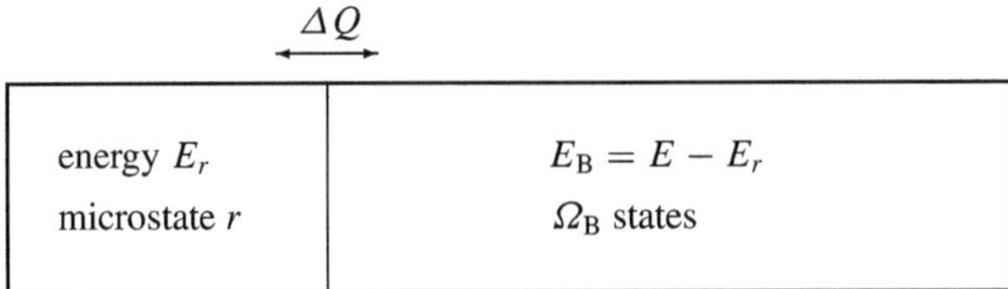

Figure 22.1   The closed system consists of the small left subsystem and the macroscopic residual system B. The two systems can exchange heat. With what probability P_r is the small subsystem in the microstate r with the energy E_r?

we specify the temperature T instead of the energy E:

$$P_r = \begin{cases} P_r(E, x) & \text{(microcanonical)} \\ P_r(T, x) & \text{(canonical)} \end{cases} \tag{22.3}$$

The statistical ensemble with the $P_r(T, x)$ is called *canonical*. In practice, the temperature is specified by contact with a heat bath; the system is therefore not closed. The microcanonical and canonical ensemble thus describe the equilibrium states of systems that are subject to different physical conditions.

To determine the probabilities $P_r(T, x)$ we proceed as follows: The system with the microstates r is brought into thermal contact with a much larger system (Figure 22.1). From the fundamental postulate for the overall system, the required $P_r(T, x)$ can then be derived. We do not introduce a new hypothesis for the equilibrium at a given temperature, but start again from the fundamental postulate.

The large system represents a heat bath for the small system; it specifies the temperature T. For the subsystem we assume

$$E_r \ll E \tag{22.4}$$

The total energy shall be much larger than the possible energy values E_r of the small system. This means that the large system must be macroscopic. The small system, on the other hand, can be macroscopic or microscopic. Examples for the small system in Figure 22.1 are

- A single air molecule in the lecture hall.
- A box with 1 liter of gas in the lecture hall.
- A beer bottle in a lake.

We determine the wanted probabilities $P_r(T, x)$. The $\Omega(E)$ accessible states of the overall system (E given) are equally probable. We now assume that the small system is in the microstate r. Then there are no longer $\Omega(E)$, but only $\Omega_B(E - E_r)$ possible states in the overall system. These $\Omega_B(E - E_r)$ states are equally probable. The probability to find one of the $\Omega_B(E - E_r)$ states among all $\Omega(E)$ states is given by

$$P_r = \frac{\Omega_B(E - E_r)}{\Omega(E)} \tag{22.5}$$

The external parameters are not displayed here. Because of (22.4), we can expand $\ln \Omega_\mathrm{B}(E - E_r)$ into powers of E_r:

$$\ln \Omega_\mathrm{B}(E - E_r) = \ln \Omega_\mathrm{B}(E) - \frac{\partial \ln \Omega_\mathrm{B}(E)}{\partial E}\, E_r + \ldots = \ln \Omega_\mathrm{B}(E) - \beta E_r + \ldots \quad (22.6)$$

According to (9.14, 9.15), $\partial \ln \Omega_\mathrm{B}/\partial E = 1/(k_\mathrm{B}T) = \beta$ determines the temperature T of system B. The heat bath B is sufficiently large, so that T is (practically) constant. The next terms in (22.6) –indicated by dots– are of the relative size $\mathcal{O}(E_r/E)$ and can be neglected. This yields

$$\Omega_\mathrm{B}(E - E_r) = \Omega_\mathrm{B}(E)\, \exp\left(-\beta E_r(x)\right) \quad (22.7)$$

The argument x stands for the external parameters of the small system. From (22.5) and (22.7) follows

$$\boxed{\; P_r(T, x) = \frac{1}{Z}\, \exp\left(-\frac{E_r(x)}{k_\mathrm{B}T}\right) \;} \quad (22.8)$$

The prefactor $\Omega_\mathrm{B}(E)/\Omega(E)$ was abbreviated by $1/Z$; it does not depend on r. The *Boltzmann factor* $\exp(-\beta E_r)$ determines the relative probabilities of the microstates r at a given temperature. From

$$\sum_r P_r = 1 \quad (22.9)$$

follows

$$\boxed{\; Z(T, x) = \sum_r \exp\left[-\beta E_r(x)\right] \;} \quad (22.10)$$

The quantity $Z(T, x)$ is called *canonical partition function*. The corresponding ensemble is called canonical ensemble or Gibbs ensemble.

With the probabilities P_r all relevant mean quantities can be calculated. In particular, the thermodynamic energy is given by

$$E(T, x) = \overline{E_r} = \sum_r P_r(T, x)\, E_r(x) \quad (22.11)$$

In the next chapter, the relationship with the thermodynamic observables will be determined.

We emphasize once more the essential difference between the microcanonical and the canonical ensemble:

- Microcanonical ensemble: A closed system in equilibrium is considered. The energy E is given. The probabilities $P_r(E, x)$ depend on the energy E and the external parameters x.

- Canonical ensemble: A system in equilibrium with a heat bath is considered. The temperature T is given. The probabilities $P_r(T, x)$ depend on the temperature T and the external parameters x.

To explain the differences and similarities, let us consider a system consisting of a single particle and, on the other hand, a system of macroscopically many particles:

1. A single particle:

   (a) The microcanonical ensemble describes the particle which is isolated from the environment; it could be placed in an otherwise empty, thermally insulated box. The energy eigenvalues of the particle are ε_r. In the microcanonical distribution implies

   $$w(\varepsilon_r) = \begin{cases} \text{const.} & \varepsilon - \delta\varepsilon \leq \varepsilon_r \leq \varepsilon \\ 0 & \text{otherwise} \end{cases} \qquad (22.12)$$

   This assumes $\delta\varepsilon \ll \varepsilon$. For the relative width of the distribution $w(\varepsilon)$ we obtain

   $$\frac{\Delta\varepsilon}{\overline{\varepsilon}} = \mathcal{O}\left(\frac{\delta\varepsilon}{\varepsilon}\right) \approx 0 \qquad (22.13)$$

   In the microcanonical ensemble, the energy is therefore close to the mean value $\overline{\varepsilon} = \overline{\varepsilon_r}$. All possible positions and momentum directions are equally probable.

   (b) The canonical ensemble describes the particle in a heat bath. This can be, for example, a selected air molecule in the lecture hall; the lecture hall specifies the temperature T. The probability distribution $w(\varepsilon_r)$ of the energy of this particle is given by the Boltzmann factor

   $$w(\varepsilon_r) \propto \exp(-\beta\varepsilon_r) \qquad (22.14)$$

   The complete expression for $w(\varepsilon_r)$ will be given in Chapter 24. The mean value $\overline{\varepsilon} = \overline{\varepsilon_r}$ and the width $\Delta\varepsilon$ of this distribution are both of size $k_B T$, so that

   $$\frac{\Delta\varepsilon}{\overline{\varepsilon}} = \mathcal{O}(1) \qquad (22.15)$$

   In contrast to (22.13), this distribution is fuzzy. The energy of a single particle may vary in a wide range; the size of this range is comparable to the energy of the particle itself. In the canonical ensemble, all possible positions and momenta (and not only all momentum directions) are equally probable.

   For a single particle, the two ensembles describe very different situations, (22.13) and (22.15). This is generally true for a system with few degrees of freedom.

2. Macroscopically many particles: We consider N independent particles (of the same kind) with the quantum numbers r_ν. The overall system then has the quantum numbers $r = (r_1, ..., r_N)$ and the energy

$$E_r = \sum_{\nu=1}^{N} \varepsilon_{r_\nu} \qquad (22.16)$$

(a) The microcanonical ensemble describes a system of N particles which is isolated from the environment. The probability for energy E_r is

$$W(E_r) = \begin{cases} \text{const.} & E - \delta E \le E_r \le E \\ 0 & \text{otherwise} \end{cases} \qquad (22.17)$$

The relative width of the distribution is

$$\frac{\Delta E}{\overline{E_r}} = \mathcal{O}\left(\frac{\delta E}{E}\right) \approx 0 \qquad (22.18)$$

All possible positions and momenta of the N particles are represented in the ensemble, but the total energy is fixed.

(b) The canonical ensemble describes N particles in a box that is placed in a heat bath. Specifically, this can be a gas box in the lecture hall. From (22.9) follows for the probability distribution of the energy E_r,

$$W(E_r) \propto \exp(-\beta E_r) = \exp\left[-\beta(\varepsilon_{r_1} + \ldots + \varepsilon_{r_N})\right] = \prod_{\nu=1}^{N} w(\varepsilon_{r_\nu}) \qquad (22.19)$$

Here $w(\varepsilon)$ is the smeared out distribution (22.14). According to the law of large numbers (4.14), however, the distribution $W(E_r)$ is a sharp distribution with

$$\frac{\Delta E}{\overline{E_r}} = \frac{1}{\sqrt{N}} \frac{\Delta \varepsilon}{\overline{\varepsilon}} = \frac{\mathcal{O}(1)}{\sqrt{N}} \approx 0 \qquad (22.20)$$

The Boltzmann factors in (22.19) describe the (large) fluctuations of the single-particle energies. However, the relative fluctuations of the total energy around the mean value are negligibly small. The energy E is therefore sharply defined, although only the temperature T is given.

The system has a well-defined energy for both ensembles, (22.18) and (22.20). This generally applies to macroscopic quantities, because (22.20) is based on the law of large numbers.

For microscopically small systems, the microcanonical and the canonical ensemble describe very different physical situations. For macroscopic systems, on the other hand, it makes practically no difference whether we specify the energy or the temperature. In Chapter 24, we will demonstrate this for the ideal gas by deriving the equations of state $PV = Nk_BT$ and $E = 3Nk_BT/2$ from $\Omega(E, V, N)$ as well as from $Z(T, V, N)$.

Grand canonical ensemble

We introduce the grand canonical ensemble. For this, as in the canonical ensemble, the temperature is specified instead of the energy. In addition, however, the chemical

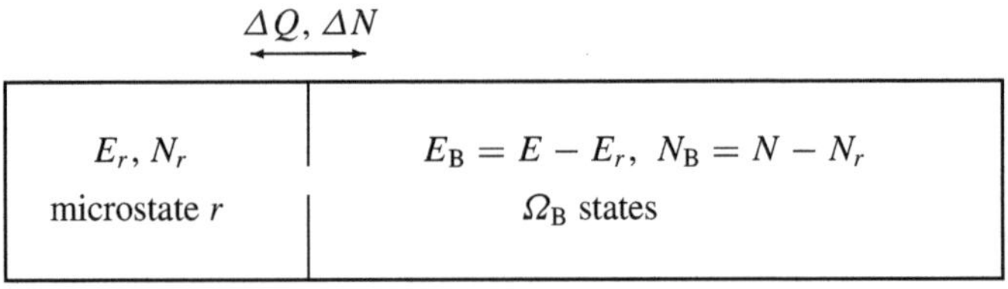

$$\xleftrightarrow{\quad \Delta Q,\ \Delta N \quad}$$

E_r, N_r	$E_B = E - E_r,\ N_B = N - N_r$
microstate r	Ω_B states

Figure 22.2   The closed system consists of the small left subsystem and the macroscopic system B. The two systems can exchange heat and particles. With what probability P_r is the small subsystem in the microstate r (with energy E_r and particle number N_r)?

potential is given instead of the number of particles; this is advantageous for a number of applications. The external parameters are then $x = (V, N)$. For a more general case, V might be replaced or supplemented by other external parameters.

We supplement (22.3):

$$P_r = \begin{cases} P_r(E, V, N) & \text{(microcanonical)} \\ P_r(T, V, N) & \text{(canonical)} \\ P_r(T, V, \mu) & \text{(grand canonical)} \end{cases} \qquad (22.21)$$

The new statistical ensemble is called *grand canonical*. In practice, the temperature and the chemical potential are fixed by contact with a heat and a particle reservoir.

We consider a small and a large system that can exchange both heat and particles (Figure 22.2). The microstate r of the small system

$$r = (E_r, N_r) \qquad \text{(microstate in the grand canonical ensemble)} \qquad (22.22)$$

determines not only the energy E_r but also the particle number N_r. We want to determine the probabilities for these $P_r(T, \mu)$ for these microstates under the condition of given temperature and chemical potential. According to the fundamental postulate, all $\Omega(E, N)$ states of the overall system are equally probable. If the small system is in the microstate r, then there are $\Omega_B(E - E_r, N - N_r)$ accessible states for the system B; this is also the number of states of the overall system. The probability to find one of the states $\Omega_B(E - E_r, N - N_r)$ among the total number $\Omega(E, N)$ states is

$$P_r = \frac{\Omega_B(E - E_r, N - N_r)}{\Omega(E, N)} \qquad (22.23)$$

Other external parameters were not displayed here. The large system shall be a heat and particle reservoir which specifies constant values for T and μ. We therefore require

$$E_r \ll E \qquad \text{and} \qquad N_r \ll N \qquad (22.24)$$

We expand $\ln \Omega_B(E - E_r, N - N_r)$ into powers of E_r and $-N_r$:

$$\ln \Omega_B(E - E_r, N - N_r) = \ln \Omega_B(E, N) - \frac{\partial \ln \Omega_B}{\partial E} E_r - \frac{\partial \ln \Omega_B}{\partial N} N_r + \dots \qquad (22.25)$$

The partial derivatives of $\ln \Omega_{\mathrm{B}}$ follow from (9.14), (9.15) and (20.5):

$$\beta = \frac{1}{k_{\mathrm{B}} T} = \frac{\partial \ln \Omega_{\mathrm{B}}(E, N)}{\partial E}, \qquad -\beta\mu = \frac{\partial \ln \Omega_{\mathrm{B}}(E, N)}{\partial N} \tag{22.26}$$

The next terms in (22.25) (indicated by dots) are of the relative size $\mathcal{O}(E_r/E)$ and $\mathcal{O}(N_r/N)$; they can be neglected. This gives us

$$\Omega_{\mathrm{B}}(E - E_r, N - N_r) = \Omega_{\mathrm{B}}(E, N) \, \exp\left(-\beta\left[E_r(V, N_r) - \mu N_r\right]\right) \tag{22.27}$$

Hereby (22.23) becomes

$$\boxed{P_r(T, V, \mu) = \frac{1}{Y} \, \exp\left(-\beta\left[E_r(V, N_r) - \mu N_r\right]\right)} \tag{22.28}$$

The prefactor $\Omega_{\mathrm{B}}(E, N)/\Omega(E/N)$ was abbreviated by $1/Y$; it does not depend on r. The temperature T and the chemical potential μ are given by the large system. Here the volume V is representative for all external parameters. The normalization $\sum P_r = 1$ determines Y:

$$\boxed{Y(T, V, \mu) = \sum_r \exp\left(-\beta\left[E_r(V, N_r) - \mu N_r\right]\right)} \tag{22.29}$$

The quantity $Y(T, V, \mu)$ is called *grand canonical partition function*. We take into account explicitly that r includes the specification of the particle number, (22.22). With possible particle exchange the number of particles N_r in the small system (Figure 22.2) can be arbitrary:

$$\begin{aligned} Y(T, V, \mu) &= \sum_{r'} \sum_{N_r=0}^{N} \exp\left(-\beta\left[E_{r'}(V, N_r) - \mu N_r\right]\right) \\ &= \sum_{N'=0}^{\infty} Z(T, V, N') \, \exp\left(\beta\mu N'\right) \end{aligned} \tag{22.30}$$

Here, r' runs over all quantum numbers to be specified for the microstate in addition to N_r. The sum over N_r runs over all possible particle numbers of the small system in Figure 22.2, i.e. from 0 to N. For the small system, only values with $N_r \ll N$ are relevant; only they contribute significantly to the partition function. Therefore, the upper summation limit can be set to infinity (instead of N). The summation index $N_r = N'$ can be named arbitrarily; however, the symbol N is reserved for the number of particles of the closed system. Equation (22.30) establishes the relation between the grand canonical and the canonical partition function.

With the probabilities P_r all relevant mean values can be calculated. In particular, these are the thermodynamic energy and particle number

$$E(T, V, \mu) \;=\; \overline{E_r} \;=\; \sum_r P_r(T, V, \mu)\, E_r(V, N_r) \qquad (22.31)$$

$$N(T, V, \mu) \;=\; \overline{N_r} \;=\; \sum_r P_r(T, V, \mu)\, N_r \qquad (22.32)$$

In the next chapter, the relationship with the other thermodynamic variables will be established.

The differences and similarities between the grand canonical and the other ensembles are analogous to the discussion above for the microcanonical and canonical ensemble. In particular, in the grand canonical ensemble, the fluctuations of the energy and the number of particle number are negligibly small if the system is macroscopic:

$$\frac{\Delta E}{\overline{E_r}} = \mathcal{O}\left(\frac{1}{\sqrt{\overline{N_r}}}\right), \qquad \frac{\Delta N}{\overline{N_r}} = \mathcal{O}\left(\frac{1}{\sqrt{\overline{N_r}}}\right) \qquad (22.33)$$

For macroscopic systems, it makes practically no difference whether (E, V, N), (T, V, N) or (T, V, μ) is given, or which of the three ensembles is used.

Exercises

22.1   Energy fluctuation in the canonical ensemble

Show that in the canonical ensemble the fluctuation ΔE of the energy is given by

$$\left(\Delta E\right)^2 = k_\mathrm{B} T^2\, \frac{\partial E(T, x)}{\partial T} \qquad (22.34)$$

where $E(T, x) = \overline{E_r}$. Show that the heat capacity $C_V = \partial E(T, V, N)/\partial T$ is positive. Justify $\Delta E / E = \mathcal{O}(N^{-1/2})$.

22.2   Particle number fluctuation in the grand canonical ensemble

Show that in the grand canonical ensemble the fluctuation ΔN of the number of particles is given by

$$\left(\Delta N\right)^2 = k_\mathrm{B} T \left(\frac{\partial N}{\partial \mu}\right)_{T,V} \qquad (22.35)$$

where $N(T, V, \mu) = \overline{N_r}$. Show that the isothermal compressibility

$$\kappa_T = -\frac{1}{V}\left(\frac{\partial V}{\partial P}\right)_{N,T} > 0 \qquad (22.36)$$

is positive. To do this, write out the differential dP for $P(T, V, N)$ in $N\,d\mu = V\,dP - S\,dT$. From this $(\partial N/\partial \mu)_{T,V}$ can be read off. Now use $P = P(T, V/N) = P(T, v)$. Justify $\Delta N / N = \mathcal{O}(N^{-1/2})$.

23 Associated potentials

The canonical and the grand canonical partition function are each closely connected to a specific thermodynamic potential. This potential determines all relevant thermodynamic relations.

We consider $x = (V, N)$ as external parameters; other parameters can be added if necessary. In the microcanonical ensemble the quantities E, V, N are given, in the canonical T, V, N and in the grand canonical T, V, μ. The probabilities P_r for the microstate r in the respective ensemble are

$$P_r(E, V, N) = \begin{cases} 1/\Omega & E - \delta E \leq E_r(V, N) \leq E \\ 0 & \text{otherwise} \end{cases} \tag{23.1}$$

$$P_r(T, V, N) = \frac{1}{Z} \exp\left(- \beta E_r(V, N)\right) \tag{23.2}$$

$$P_r(T, V, \mu) = \frac{1}{Y} \exp\left(- \beta\left[E_r(V, N_r) - \mu N_r\right]\right) \tag{23.3}$$

The respective partition functions follow from the normalization $\sum P_r = 1$,

$$\Omega(E, V, N) = \sum_{r: E - \delta E \leq E_r(V, N) \leq E} 1 \tag{23.4}$$

$$Z(T, V, N) = \sum_r \exp\left(- \beta E_r(V, N)\right) \tag{23.5}$$

$$Y(T, V, \mu) = \sum_r \exp\left(- \beta\left[E_r(V, N_r) - \mu N_r\right]\right) \tag{23.6}$$

All partition functions are determined by the eigenvalues $E_r(V, N)$, i.e. by the microscopic structure of the considered system. The connection with macroscopic thermodynamics is established by the following relations:

$$\boxed{\begin{aligned} S(E, V, N) &= k_\mathrm{B} \ln \Omega(E, V, N) \\ F(T, V, N) &= -k_\mathrm{B}T \ln Z(T, V, N) \\ J(T, V, \mu) &= -k_\mathrm{B}T \ln Y(T, V, \mu) \end{aligned}} \tag{23.7}$$

The first relation was used to define the entropy in Chapter 9. The other two relations will be derived in the following sections. From the exact differentials

$$dS = \frac{dE}{T} + \frac{P}{T}\,dV - \frac{\mu}{T}\,dN \tag{23.8}$$

$$dF = -S\,dT - P\,dV + \mu\,dN \tag{23.9}$$

$$dJ = -S\,dT - P\,dV - N\,d\mu \tag{23.10}$$

the partial derivatives of S, F and J can be read off. As discussed in Chapter 17, this allows for the determination of all thermodynamic relations, in particular for that of the thermal and of the caloric equation of state.

The derivation of the macroscopic properties of an equilibrium system from its microscopic structure, i.e. from the Hamilton operator or function $H(V, N)$, can now be represented by the following scheme:

$$H(V, N) \rightarrow E_r(V, N) \rightarrow \left\{ \begin{array}{ccc} \Omega(E, V, N) & \rightarrow & S(E, V, N) \\ Z(T, V, N) & \rightarrow & F(T, V, N) \\ Y(T, V, \mu) & \rightarrow & J(T, V, \mu) \end{array} \right. \begin{array}{c} \rightarrow \\ \rightarrow \\ \rightarrow \end{array} \left\{ \begin{array}{l} \text{all thermo-} \\ \text{dynamic} \\ \text{relations} \end{array} \right. \tag{23.11}$$

For the microcanonical ensemble, these steps have been explained for the ideal gas following (13.1). In Chapter 25 we will carry out the parallel steps for the canonical and grand canonical ensemble.

Derivation of $F = -k_{\mathrm{B}} T \ln Z$

We start from the exact differential

$$d \ln Z(\beta, V, N) = \frac{\partial \ln Z}{\partial \beta}\,d\beta + \frac{\partial \ln Z}{\partial V}\,dV + \frac{\partial \ln Z}{\partial N}\,dN \tag{23.12}$$

The argument T has been replaced by $\beta = (k_{\mathrm{B}} T)^{-1}$. We calculate the partial derivatives:

$$\frac{\partial \ln Z}{\partial \beta} = -\frac{1}{Z} \sum_r E_r \exp(-\beta E_r) = -\sum_r E_r\, P_r = -\overline{E_r} = -E \tag{23.13}$$

$$\frac{\partial \ln Z}{\partial V} = -\frac{\beta}{Z} \sum_r \frac{\partial E_r(V, N)}{\partial V} \exp(-\beta E_r) = -\beta\, \overline{\frac{\partial E_r}{\partial V}} = \beta P \tag{23.14}$$

$$\frac{\partial \ln Z}{\partial N} = -\frac{\beta}{Z} \sum_r \frac{\partial E_r(V, N)}{\partial N} \exp(-\beta E_r) = -\beta\, \overline{\frac{\partial E_r}{\partial N}} = -\beta\mu \tag{23.15}$$

Thus (23.12) becomes

$$d \ln Z(\beta, V, N) = -E\,d\beta + \beta P\,dV - \beta\mu\,dN \tag{23.16}$$

From this we get

$$d\left(\ln Z + \beta E\right) = \beta\left(dE + P\,dV - \mu\,dN\right) = \frac{1}{k_{\mathrm B}}\left(\frac{dE}{T} + \frac{P}{T}\,dV - \frac{\mu}{T}\,dN\right) \quad (23.17)$$

The comparison with (23.8) shows

$$k_{\mathrm B}\,d\left(\ln Z + \beta E\right) = dS \tag{23.18}$$

From this follows

$$S = k_{\mathrm B}\left(\ln Z + \beta E\right) + \text{const.} \tag{23.19}$$

If the constant vanishes, we obtain the desired relation

$$F(T, V, N) = E - TS = -k_{\mathrm B}T\,\ln Z(T, V, N) \tag{23.20}$$

We determine the constant in (23.19) from the limiting case $T \to 0$. W consider a quantum mechanical system with the energy eigenvalues $E_0 < E_1 \le E_2 \le \dots$. We write on the first terms of the partition function:

$$Z = \exp(-\beta E_0) + \exp(-\beta E_1) + \dots = \exp(-\beta E_0)\left[1 + \exp(-\beta \Delta) + \dots\right]$$
$$\tag{23.21}$$

For $k_{\mathrm B}T \ll \Delta = E_1 - E_0$ this yields

$$Z \xrightarrow{T \to 0} \exp(-\beta E_0), \qquad \ln Z \xrightarrow{T \to 0} -\beta E_0 \tag{23.22}$$

For $T \to 0$ the mean energy also approaches the lowest possible value E_0:

$$E = \overline{E_r} \approx E_0 + E_1\,\exp(-\beta\Delta) + \dots, \qquad E \xrightarrow{T \to 0} E_0 \tag{23.23}$$

By combining these results with the 3rd law, $S \to 0$ for $T \to 0$, we see that the constant vanishes:

$$\text{const.} = S - k_{\mathrm B}(\ln Z + \beta E) \xrightarrow{T \to 0} 0 - k_{\mathrm B}(-\beta E_0 + \beta E_0) = 0 \tag{23.24}$$

Derivation of $J = -k_{\mathrm B}T\,\ln Y$

We start from the exact differential

$$d\,\ln Y(\beta, V, \mu) = \frac{\partial \ln Y}{\partial \beta}\,d\beta + \frac{\partial \ln Y}{\partial V}\,dV + \frac{\partial \ln Y}{\partial \mu}\,d\mu \tag{23.25}$$

and calculate the partial derivatives:

$$\frac{\partial \ln Y}{\partial \beta} = -\frac{1}{Y}\sum_r (E_r - \mu N_r)\,\exp\left(-\beta[E_r - \mu N_r]\right)$$

$$= -\overline{E_r} + \mu\,\overline{N_r} = -E + \mu N \tag{23.26}$$

$$\frac{\partial \ln Y}{\partial V} = -\frac{\beta}{Y}\sum_r \frac{\partial E_r}{\partial V}\,\exp\left(-\beta[E_r - \mu N_r]\right) = -\beta\,\overline{\frac{\partial E_r}{\partial V}} = \beta P \tag{23.27}$$

$$\frac{\partial \ln Y}{\partial \mu} = \frac{\beta}{Y}\sum_r N_r\,\exp\left(-\beta[E_r - \mu N_r]\right) = \beta\,\overline{N_r} = \beta N \tag{23.28}$$

Thus (23.25) becomes

$$d \ln Y(\beta, V, \mu) = (-E + \mu N)\, d\beta + \beta P\, dV + \beta N\, d\mu \tag{23.29}$$

From this we get

$$d(\ln Y + \beta E - \beta \mu N) = \beta(dE + P\, dV - \mu\, dN) \tag{23.30}$$

According to (23.8), this is equal to dS/k_B. Therefore,

$$S = k_B(\ln Y + \beta E - \beta \mu N) \tag{23.31}$$

As in (23.19), the occurring constant vanishes. From (23.31) we obtain $k_B T \ln Y = TS - E + \mu N = -F + \mu N = -J$, i.e.

$$J = -k_B T \ln Y = -PV \tag{23.32}$$

Here we supplemented the basic result $J = -k_B T \ln Y$ by the relation $J = -PV$ known from (20.13).

Alternative derivation of the ensemble probabilities

This section (which may be skipped) presents an alternative of the various statistical ensembles: The entropy $S(P_1, P_2, ...)$ is written as a function of the probabilities P_r. The P_r then follow from the condition "$S =$ maximum" under the respective constraints.

We consider a statistical ensemble of M identical systems of which M_r systems are in the microstate r. For $M_r \gg 1$ this yields the probabilities $P_r = M_r/M$. In the following, we assume certain (initially unknown) values for M_r.

We sort the $M_1 + M_2 + M_3 + ... = \sum M_r$ microstates of the ensemble such so that the M_1 are put into a box labeled "state 1", the M_2 systems into a box labeled "state 2" and so on. There are obviously

$$\Gamma = \frac{M!}{M_1!\, M_2!\, M_3!\, ...} = \frac{M!}{\prod_r M_r!} \tag{23.33}$$

possibilities to distribute M systems into the boxes so that there are just M_r systems in the "state r" box. This can be compared with the number $N!/(N_1!\, N_2!)$ of possibilities to distribute N distinguishable particles into two partial volumes. Then Γ denotes the number of ensemble states.

Using $\ln n! \approx n \ln n - n$ we write the logarithm of Γ as a function of P_r:

$$\ln \Gamma = \ln M! - \sum_r \ln M_r! \approx M \ln M - M - \sum_r M_r \ln M_r + \sum_r M_r$$

$$= -\sum_r M_r(\ln M_r - \ln M) = -M \sum_r P_r \ln P_r \tag{23.34}$$

In Chapter 9 it was investigated how the energy $E = E_A + E_B$ is distributed between two subsystems. The number $\Omega(E_A)$ of microstates as a function of E_A has a prominent maximum. Almost all possible microstates lie close to this maximum. Therefore, the equilibrium was determined given by "$S = k_B \ln \Omega(E_A) =$ maximum".

Here we are faced with the analogous question of how the $M = \sum M_r$ systems of the ensemble are distributed among the microstates. The number $\Gamma(P_1, P_2,...)$ of the ensemble states has a prominent maximum as a function of the P_r (or M_r). Almost all possible ensemble states lie close to this maximum. Therefore, the equilibrium is determined by

$$S_{\text{ensemble}} = k_B \ln \Gamma(P_1, P_2,..) = \text{maximum} \qquad (23.35)$$

As in Chapter 9, we identify the logarithm of the number of possible states (multiplied by Boltzmann's constant) as the entropy. However, now we do not consider the Ω possible states *of one* macroscopic system, but the Γ possible states of an *ensemble* of M systems. Also this argumentation is based on the fundamental postulate: If all microstates are equally probable, then all possible ensemble states are equally probable.

From (23.35) and (23.34) we obtain for the entropy S *of one* system

$$S(P_1, P_2,...) = \frac{S_{\text{ensemble}}}{M} = -k_B \sum P_r \ln P_r \qquad (23.36)$$

This expression for the entropy can be applied to arbitrary macrostates $\{P_r\}$. In contrast to this, the definition of entropy in Chapter 9 was restricted to equilibrium states. Exercise 23.1 will show that (23.36) reduces to the known entropy expression when applied to equilibrium states. The form (23.36) for the entropy will be discussed again when we investigate approach to equilibrium in Chapter 41.

The equilibrium condition is $S(P_1, P_2,...) = $ maximum. The possible values of P_r are restricted by constraints which follow from the physical specifications:

$$\sum_r P_r = 1 \qquad \text{microcanonical, canonical, grand canonical} \qquad (23.37)$$

$$\sum_r P_r E_r = E \qquad \text{canonical, grand canonical} \qquad (23.38)$$

$$\sum_r P_r N_r = N \qquad \text{grand canonical} \qquad (23.39)$$

We evaluate now the condition of maximum entropy under various physical constraints:

1. *Closed system:* Only microstates with $E_r = E$ (more precisely $E - \delta E \leq E_r \leq E$) and $N_r = N$ are permitted. In addition to $\sum P_r = 1$, there is no further constraint. Thus (23.35) becomes

$$S(P_1, P_2, ...) - \lambda \sum_r P_r = \text{maximum} \qquad (23.40)$$

Here λ is a Lagrange multiplier which takes $\sum_r P_r = 1$ into account. The necessary condition for the maximum is

$$\frac{\partial}{\partial P_r}\left(S(P_1, P_2, \ldots) - \lambda \sum_{r'} P_{r'}\right) = 0 \qquad (23.41)$$

From this follows

$$P_r = \text{const.} \qquad (\text{microcanonical, } E_r = E) \qquad (23.42)$$

The constant depends on λ; it is fixed by the condition $\sum P_r = 1$.

2. *System in contact with a heat bath:* Various energy values E_r are permitted. However, the temperature determines mean value $E = \overline{E_r}$. The maximum of S must be calculated under the constraints (23.37) and (23.38):

$$S(P_1, P_2,\ldots) - \lambda_1 \sum_r P_r - \lambda_2 \sum_r E_r\, P_r = \text{maximum} \qquad (23.43)$$

The evaluation of $\partial S/\partial P_r = 0$ results in

$$P_r = \text{const.} \cdot \exp\left(-\lambda_2\, E_r/k_{\mathrm{B}}\right) \qquad (\text{canonical}) \qquad (23.44)$$

The Lagrange multiplier λ_2 ensures $E = \sum E_r\, P_r$; this yields $\lambda_2 = 1/T$. The constant in (23.44) depends on λ_1 and is fixed by $\sum P_r = 1$. The result is the canonical distribution. The statement (23.43) can then be written as $S - E/T = \text{maximal}$ or $F = E - TS = \text{minimal}$.

3. *System in contact with a heat and particle reservoir:* Various energy values E_r and particle numbers N_r are permitted. The mean values $E = \overline{E_r}$ and $N = \overline{N_r}$ are, however, determined by the temperature of the heat bath and by the chemical potential of the particle reservoir. The maximum of S has to be determined under the constraints (23.37) – (23.39):

$$S(P_1, P_2,\ldots) - \lambda_1 \sum_r P_r - \lambda_2 \sum_r E_r\, P_r - \lambda_3 \sum_r P_r\, N_r = \text{maximal} \qquad (23.45)$$

From $\partial S/\partial P_r = 0$ the P_r follows grand canonical distribution (Exercise 23.2). The constraints lead to $\lambda_2 = 1/T$ and $\lambda_3 = -\mu/T$. The result (23.45) can be also written as $J = E - TS - \mu N = \text{minimal}$.

Exercises

23.1   *Entropy for various macrostates*

The entropy of an arbitrary macrostate $\{P_r\}$ is given by

$$S = -k_B \sum P_r \ln P_r$$

Insert the known probabilities P_r of the (i) microcanonical, (ii) canonical and (iii) grand canonical ensemble; let the external parameters be V and N. Link the result to the statements

$$S = k_B \ln \Omega \,, \qquad F = -k_B T \ln Z \quad \text{and} \quad J = -k_B T \ln Y$$

23.2   *Maximum entropy under constraints*

An arbitrary macrostate $\{P_r\}$ has the entropy $S(P_r) = -k_B \sum P_r \ln P_r$. Determine the probabilities P_r of the grand canonical ensemble from the requirement

$$\frac{S(P_r)}{k_B} - \lambda_1 \sum_r P_r - \lambda_2 \sum_r E_r P_r - \lambda_3 \sum_r P_r N_r = \text{maximal} \qquad (23.46)$$

under the constraints

$$\sum_r P_r = 1 \,, \qquad \sum_r E_r P_r = E \,, \qquad \sum_r P_r N_r = N$$

What changes occur when the canonical ensemble is considered?

24 Classical systems

The canonical partition function is applied to classical systems. After a discussion of the basics, we derive Maxwell's velocity distribution, the barometric formula and the equipartition theorem.

Basics

A classical mechanical system is treated by introducing suitable generalized coordinates q_i and the Lagrange function

$$\mathcal{L} = \mathcal{L}(q_1,..., q_f, \dot{q}_1,..., \dot{q}_f) = \mathcal{L}(q, \dot{q}) \tag{24.1}$$

A possible explicit time dependency is not written on. The arguments are abbreviated by $q = (q_1,..., q_f)$ and $\dot{q} = (\dot{q}_1,..., \dot{q}_f)$, where f is the number of degrees of freedom.

In order to get from the Lagrange to the Hamilton function (Chapter 27 in [1]), we define the generalized momenta

$$p_i = p_i(q, \dot{q}) = \frac{\partial \mathcal{L}}{\partial \dot{q}_i} \qquad (i = 1,..., f) \tag{24.2}$$

These f relations are resolved for the velocities:

$$p_i = p_i(q, \dot{q}) \quad \longrightarrow \quad \dot{q}_k = \dot{q}_k(q, p) \tag{24.3}$$

The functions $\dot{q}_i(q, p)$ are inserted on the right-hand side of

$$H = H(q, p) = \sum_{i=1}^{f} \frac{\partial \mathcal{L}(q, \dot{q})}{\partial \dot{q}_i} \dot{q}_i - \mathcal{L}(q, \dot{q}) \tag{24.4}$$

The resulting Hamilton function $H(q, p)$ depends on the variables $q_1,..., q_f$, $p_1,..., p_f$ and possibly also on the time. The Hamilton function determines the canonical equations of motion, which are equivalent to the Lagrange equations. In statistical physics, we are not interested in the equations of motion themselves. What we need are the energy values of the microstates. The Hamilton function is the appropriate starting point for this.

The microstate of the classical system is defined by

$$r = (q, p) = (q_1,..., q_f, p_1,..., p_f) \tag{24.5}$$

Let the Hamilton function be equal to the energy, so that

$$E_r = H(q, p) = H(q_1,..., q_f, p_1,..., p_f) \tag{24.6}$$

As discussed in Chapter 5, a quantum mechanical state occupies a volume $(2\pi\hbar)^f$ in the $2f$-dimensional phase space. In the classical limit, such volumes are very small compared to the accessible phase space volume. The summation over the microstates (24.5) can therefore be replaced by integrals:

$$\sum_r \cdots = \frac{1}{(2\pi\hbar)^f} \int dq_1 \ldots \int dq_f \int dp_1 \ldots \int dp_f \cdots \tag{24.7}$$

The integration runs over all possible values of the coordinates and momenta. Since the prefactor is a borrowing from quantum mechanics this treatment could also be called semiclassical. The thermodynamic results do not depend on the prefactor.

The probability P_r for the state r in the canonical ensemble is

$$P_r = \frac{1}{Z} \exp(-\beta E_r) = \frac{1}{Z} \exp\left[-\beta H(q, p)\right] \tag{24.8}$$

The partition function $Z = \sum_r \exp(-\beta E_r)$ is then

$$\boxed{Z = \frac{1}{(2\pi\hbar)^f} \int dq_1 \ldots \int dq_f \int dp_1 \ldots \int dp_f \, \exp\left[-\beta H(q, p)\right]} \tag{24.9}$$

A classical quantity A of the system has a certain value $A_r = A(q, p)$ in the state r; an example of such a quantity is the energy $A_r = E_r = H(q, p)$. The equilibrium value of A is equal to the mean value $\overline{A} = \sum_r A_r P_r$,

$$\overline{A} = \frac{1}{(2\pi\hbar)^f} \int dq_1 \ldots \int dq_f \int dp_1 \ldots \int dp_f \, A(q, p) \, \frac{\exp\left[-\beta H(q, p)\right]}{Z} \tag{24.10}$$

From this follows the interpretation

$$\frac{\exp[-\beta H(q, p)]}{(2\pi\hbar)^f \, Z} \, d^f q \, d^f p = \begin{cases} \text{probability, to find the system} \\ \text{in the intervals } [q_i, q_i + dq_i] \\ \text{and } [p_j, p_j + dp_j] \end{cases} \tag{24.11}$$

The left-hand side without the volume elements $d^f q$ and $d^f p$ is the corresponding *probability density*.

One particle

We consider a single particle, i.e. the microstates

$$r = (\boldsymbol{r}, \boldsymbol{p}) = (x, y, z, p_x, p_y, p_z) \tag{24.12}$$

For the single particle, we denote the partition function and the Hamilton function by lower case letters:

$$z = \frac{1}{(2\pi\hbar)^3} \int d^3r \int d^3p \, \exp\left[-\beta h(\boldsymbol{r}, \boldsymbol{p})\right] \tag{24.13}$$

The mean value of a physical quantity a of the particle is

$$\bar{a} = \frac{1}{(2\pi\hbar)^3} \int d^3r \int d^3p \, a(\boldsymbol{r}, \boldsymbol{p}) \, \frac{\exp\left[-\beta h(\boldsymbol{r}, \boldsymbol{p})\right]}{z} \tag{24.14}$$

From this follows the interpretation

$$\frac{\exp\left[-\beta h(\boldsymbol{r}, \boldsymbol{p})\right]}{(2\pi\hbar)^3 \, z} \, d^3r \, d^3p = \begin{cases} \text{probability to find the} \\ \text{particle in the ranges} \\ d^3r \text{ at } \boldsymbol{r} \text{ and } d^3p \text{ at } \boldsymbol{p} \end{cases} \tag{24.15}$$

The left-hand side without the volume elements d^3r and d^3p is the corresponding probability density.

The particle shall be limited to a volume V, for example, a particle in a box. This limitation can be taken into account by appropriate integral limits in $\int d^3r$. Alternatively, one can also use the wall potential

$$U(\boldsymbol{r}) = \begin{cases} 0 & \boldsymbol{r} \in V \\ \infty & \boldsymbol{r} \notin V \end{cases} \tag{24.16}$$

Then the integrand in (24.13) and (24.14) contains the factor

$$\exp\left[-\beta U(\boldsymbol{r})\right] = \begin{cases} 1 & \boldsymbol{r} \in V \\ 0 & \boldsymbol{r} \notin V \end{cases} \tag{24.17}$$

which limits the spatial integration to V.

Maxwell's velocity distribution

We calculate the partition function z for a particle that moves freely in the volume V. The Hamilton function is

$$h(\boldsymbol{r}, \boldsymbol{p}) = \frac{\boldsymbol{p}^2}{2m} + U(\boldsymbol{r}) \tag{24.18}$$

For $U(\boldsymbol{r})$ from (24.16) we evaluate (24.13):

$$\begin{aligned} z &= \frac{1}{(2\pi\hbar)^3} \int d^3r \, \exp\left[-\beta U(\boldsymbol{r})\right] \int d^3p \, \exp(-\beta \, \boldsymbol{p}^2/2m) \\ &= \frac{V}{(2\pi\hbar)^3} \int d^3p \, \exp\left(-\frac{\boldsymbol{p}^2}{2mk_\mathrm{B}T}\right) = \frac{V(2mk_\mathrm{B}T)^{3/2}}{(2\pi\hbar)^3} \, 4\pi \int_0^\infty dx \, x^2 \exp\left(-x^2\right) \\ &= \frac{V(2\pi m k_\mathrm{B}T)^{3/2}}{(2\pi\hbar)^3} = \frac{V}{\lambda^3} \end{aligned} \tag{24.19}$$

The spatial integration results in the factor V due to (24.17). For the momentum integration we set $x^2 = p^2/(2m k_B T)$ with the dimensionless integration variable x; in addition, we use $d^3p = 4\pi p^2 dp$ The result can be expressed by the *thermal length* (also called de Broglie wave length):

$$\boxed{\lambda = \frac{2\pi\hbar}{\sqrt{2\pi m k_B T}} \qquad \text{thermal wavelength}} \tag{24.20}$$

This is the quantum mechanical wavelength of a particle with the kinetic energy $p^2/2m = \pi k_B T$.

We calculate the mean value of a quantity $a_r = a(p)$, which depends only on the momentum:

$$\overline{a} = \frac{V}{(2\pi\hbar)^3 z} \int d^3p \; a(p) \; \exp\left(-\frac{p^2}{2m k_B T}\right) \tag{24.21}$$

This results in the interpretation

$$\frac{V}{(2\pi\hbar)^3 z} \exp\left(-\frac{p^2}{2m k_B T}\right) d^3p = \left\{ \begin{array}{l} \text{probability that the} \\ \text{momentum lies} \\ \text{in the range } d^3p \text{ at } p \end{array} \right. \tag{24.22}$$

Since the probability density only depends on $p^2 = p^2$, we can replace d^3p with $4\pi p^2 dp$. Instead of the momentum $p = m\,v$ we may introduce the velocity v:

$$\frac{4\pi m^3 V}{(2\pi\hbar)^3 z} \exp\left(-\frac{m v^2}{2 k_B T}\right) v^2 dv = \left\{ \begin{array}{l} \text{probability that the} \\ \text{velocity } v = |v| \text{ lies} \\ \text{between } v \text{ and } v + dv \end{array} \right. \tag{24.23}$$

We write the left-hand side as $f(v)\,dv$ and use $z = V/\lambda^3$:

$$\boxed{f(v) = 4\pi \left(\frac{m}{2\pi k_B T}\right)^{3/2} v^2 \exp\left(-\frac{m v^2}{2 k_B T}\right) \qquad \text{Maxwell distribution}}$$

$$\tag{24.24}$$

This distribution $f(v)$ is called *Maxwell's velocity distribution*; it is sketched in Figure 24.1. As expected, this classical result is independent of $\hbar$. The Maxwell distribution is normalized:

$$\int_0^\infty dv\, f(v) = 1 \tag{24.25}$$

Thus, $f(v)\,dv$ is the probability of finding the velocity $v = |v|$ of the velocity in the interval $[v, v + dv]$. For small velocities, $f(v)$ initially increases because the phase space volume in an interval dv is proportional to v^2. For large values of v, however, $f(v)$ decreases exponentially due to the Boltzmann factor $\exp(-\varepsilon/k_B T)$.

The maximum of the Maxwell distribution follows from $df/dv = 0$ to

$$v_{\text{max}} = \sqrt{\frac{2 k_B T}{m}} \stackrel{\text{air}}{\approx} 400\,\frac{\text{m}}{\text{s}} \tag{24.26}$$

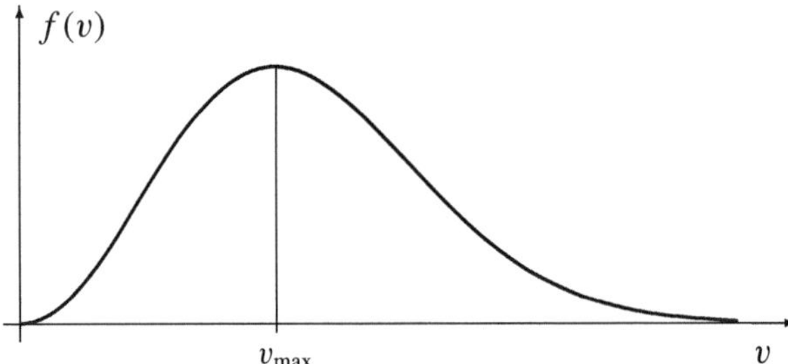

Figure 24.1 Maxwell's velocity distribution. For the air in the lecture hall the maximum of the distribution lies at about $v_{\mathrm{max}} \approx 400\,\mathrm{m/s}$.

For this estimate we used $m \approx 30\,\mathrm{GeV}/c^2$ (for O_2 or N_2 molecules), $c = 3\cdot 10^8\,\mathrm{m/s}$ and $k_{\mathrm{B}} T \approx \mathrm{eV}/40$ (room temperature).

If we consider a large number of particles, then the probability density $f(v)$ can also be interpreted as a particle density:

$$
F(v)\,dv = N\,f(v)\,dv =
\begin{cases}
\text{number of particles} \\
\text{with a velocity} \\
\text{between } v \text{ and } v + dv
\end{cases}
\tag{24.27}
$$

Barometric formula

For a classical particle in the gravitational field $u(\boldsymbol{r}) = m g z$ the Hamilton function is

$$
h(\boldsymbol{r},\,\boldsymbol{p}) = \frac{p^2}{2m} + m g z
\tag{24.28}
$$

We consider a molecule in the atmosphere. The coordinate z is the height above the ground. All other air molecules represent a heat bath for the selected molecule. For this heat bath we assume a position and time independent temperature T. According to (24.15) this implies

$$
\exp\left(-\frac{p^2/2m + m g z}{k_{\mathrm{B}} T}\right) d^3r\,d^3p \propto
\begin{cases}
\text{probability to find a particle} \\
\text{in } d^3r \text{ at } \boldsymbol{r} \text{ and in } d^3p \text{ at } \boldsymbol{p}
\end{cases}
\tag{24.29}
$$

We are now interested in the probability of finding the particle at a height between z and $z + dz$. We integrate over the x- and y-coordinate and over the momenta and obtain

$$
\exp\left(-\frac{m g z}{k_{\mathrm{B}} T}\right) dz \propto
\begin{cases}
\text{probability to find a particle at} \\
\text{a height between } z \text{ and } z + dz
\end{cases}
\tag{24.30}
$$

Since (24.30) applies to every single particle, the particle density $n = N/V$ is proportional to this probability density, i.e.

$$
n(z) = n(0)\,\exp\left(-\frac{m g z}{k_{\mathrm{B}} T}\right)
\tag{24.31}
$$

The proportionality constant $n(0)$ is the particle density at zero height. With the ideal gas law $n = N/V = P/k_B T$, this becomes the *barometric formula* for the pressure:

$$P(z) = P(0)\ \exp\left(-\frac{m g z}{k_B T}\right) \tag{24.32}$$

For the atmosphere this is only a crude approximation, because the assumption of a constant temperature (i.e. of an equilibrium against heat exchange) is not realistic. It is more realistic to assume an equilibrium with respect to volume exchange (Exercise 24.7).

The applications "Maxwell's velocity distribution" and "barometric formula" show an advantage of the canonical distribution as compared to the microcanonical one: For a microscopic system such as a single particle, the canonical ensemble often corresponds to the physical conditions. The microcanonical distribution, on the other hand, is not applicable for a singled out particle (because this particle does not represent a closed system).

Equipartition theorem

We now consider a general Hamilton function

$$H(p, q) = H(q_1,..., q_f, p_1,..., p_f) \tag{24.33}$$

and calculate the following mean value:

$$\overline{p_i \frac{\partial H}{\partial p_i}} = \frac{1}{Z\,(2\pi\hbar)^f} \int dq_1 ... \int dq_f \int dp_1 ... \int dp_f\ p_i \frac{\partial H}{\partial p_i}\ \exp\left[-\beta H\right]$$

$$= \frac{-k_B T}{Z\,(2\pi\hbar)^f} \int dq_1 ... \int dq_f \int dp_1 ... \int dp_f\ p_i \frac{\partial\ \exp\left[-\beta H(q, p)\right]}{\partial p_i} \tag{24.34}$$

Assuming

$$\left(p_i\ \exp\left[-\beta H(q, p)\right]\right)_{p_i = \pm\infty} = 0 \tag{24.35}$$

we can evaluate (24.34) by partial integration with respect to p_i,

$$\overline{p_i \frac{\partial H}{\partial p_i}} = \frac{k_B T}{Z\,(2\pi\hbar)^f} \int dq_1 ... \int dq_f \int dp_1 ... \int dp_f\ \exp\left[-\beta H(q, p)\right] = k_B T \tag{24.36}$$

The same consideration applies to the coordinate q_i if the boundary term vanishes, i.e.

$$\left(q_i\ \exp\left[-\beta H(q, p)\right]\right)_{\text{boundary of the } q_i \text{ integration}} = 0 \tag{24.37}$$

The integration limits depend on the meaning of the generalized coordinate.

Under the conditions (24.35) and (24.37), we obtain the mean values

$$\overline{p_i \frac{\partial H}{\partial p_i}} = k_B T, \qquad \overline{q_i \frac{\partial H}{\partial q_i}} = k_B T \tag{24.38}$$

This statement is called *equipartition theorem*. It is also known as the law of equipartition, equipartition of energy, or simply equipartition. We evaluate it under more specific assumptions. Let the Hamilton operator be of the form $H = \sum_i (p_i^2/2m + m\,\omega^2 q_j^2/2)$. The we obtain

$$\overline{p_i \frac{\partial H}{\partial p_i}} = \overline{\frac{p_i^2}{m}}\,, \qquad \left[p_i \exp(-\beta\,p_i^2/2m)\right]_{p_i=\pm\infty} = 0 \tag{24.39}$$

$$\overline{q_j \frac{\partial H}{\partial q_j}} = \overline{m\,\omega^2 q_j^2}\,, \qquad \left[q_j \exp(-\beta m\,\omega^2 q_j^2/2)\right]_{q_j=\pm\infty} = 0 \tag{24.40}$$

and

$$\overline{\frac{p_i^2}{2m}} = \frac{k_{\mathrm{B}}T}{2}\,, \qquad \overline{\frac{m\,\omega^2 q_j^2}{2}} = \frac{k_{\mathrm{B}}T}{2} \tag{24.41}$$

We formulate this result more generally as follows:

- *Each variable that occurs quadratically in the Hamilton function yields a contribution $k_{\mathrm{B}}T/2$ to the mean energy.*

It is assumed that the quadratic term in H is positive. The term "equipartition theorem" is used both, for (24.38) and for this more specific statement. In the following we discuss some applications of the equipartition theorem.

Monatomic ideal gas

The Hamiltonian operator of the monatomic ideal gas is given by

$$H = \sum_{\nu=1}^{N} \frac{\boldsymbol{p}_\nu^2}{2m} = \sum_{k=1}^{3N} \frac{p_k^2}{2m} \tag{24.42}$$

For this the equipartition theorem says

$$\overline{H} = \frac{3}{2}\,Nk_{\mathrm{B}}T = E(T) \tag{24.43}$$

This is the caloric equation of state of the monatomic ideal gas.

The generalized coordinates q_k of the system (24.42) are the $3N$ Cartesian coordinates. Since H is independent of them, the following applies

$$\overline{q_k \frac{\partial H}{\partial q_k}} = 0 \quad \text{because} \quad \frac{\partial H}{\partial q_k} = 0 \tag{24.44}$$

These coordinates therefore do not contribute to the energy $E = \overline{H}$. The right part of (24.38) does not apply in this case, because the precondition (24.37) is not fulfilled.

For conclusions drawn from the equipartition theorem, particular attention must be paid to the contribution of the wall potential. For occurring paradoxes and their resolution, we refer to an article by Thirring, Z.Physik 235 (1970) 339. We will restrict ourselves to unproblematic applications.

Diatomic ideal gas

We consider an ideal gas consisting of diatomic molecules. Here *ideal* means that the interaction between the molecules is neglected. The Hamiltonian function

$$H = \sum_{\nu=1}^{N} h(\nu) \tag{24.45}$$

is then a sum of the Hamilton functions $h(\nu)$ of the individual molecules. In $h(\nu)$, ν stands for the coordinates and momenta of the ν-th molecule. The possible degrees if freedom for a single molecule are:

1. Translations: The translation of the molecule is described by its center of mass coordinates x, y and z. The inertia parameter is the total mass M of the molecule.

2. Rotations: The orientation of the axis connecting the two atoms can be described by the familiar angles θ and ϕ. These angles are generalized coordinates that can describe the rotation around the two axes perpendicular to the connecting line. The associated moment of inertia is Θ.

3. Vibrations: Let the equilibrium distance between the two atoms be R_0. Displacements from the equilibrium position lead to vibrations, which are described by the coordinate $\xi(t) = R(t) - R_0$. The inertia parameter is the reduced mass m_r. For small displacements we expect harmonic oscillations.

We assume that the moment of inertia changes only slightly due the vibrations, i.e. $\Theta \approx$ const.; thereby we neglect a possible rotation-vibration coupling. Under these simplifications, the Lagrangian function for a single molecule reads:

$$\mathcal{L} = \frac{M}{2}\left(\dot{x}^2 + \dot{y}^2 + \dot{z}^2\right) + \frac{\Theta}{2}\left(\sin^2\theta\,\dot{\phi}^2 + \dot{\theta}^2\right) + \frac{m_r}{2}\left(\dot{\xi}^2 - \omega^2\xi^2\right) \tag{24.46}$$

This results in the Hamilton function

$$h = \frac{1}{2M}\left(p_x^2 + p_y^2 + p_z^2\right) + \frac{1}{2\Theta}\left(\frac{p_\phi^2}{\sin^2\theta} + p_\theta^2\right) + \left(\frac{p_\xi^2}{2m_r} + \frac{m_r\,\omega^2\xi^2}{2}\right) \tag{24.47}$$

We apply the equipartition theorem to the variables p_x, p_y, p_z, p_ϕ, p_θ, p_ξ and ξ, which all occur quadratically. The factor $1/\sin^2\theta$ in the term with p_ϕ does not interfere, because also this term is also equal to $p_\phi\,(\partial H/\partial p_\phi)/2$. Each of the quadratically occurring variables in h thus leads to an n average energy $k_B T/2$, i.e.

$$\overline{h} = \frac{7}{2}\,k_B T\,, \qquad \overline{H} = N\overline{h} = \frac{7}{2}\,Nk_B T = E(T) \tag{24.48}$$

From this follows the specific heat per particle

$$c_V = \frac{7}{2}\,k_B \qquad \text{(classical diatomic ideal gas)} \tag{24.49}$$

Real diatomic gases deviate more or less from this. One one hand, this is due to the interaction between the molecules. On the other hand, the rotational and vibrational energies are quantized. Therefore, they are only excited at sufficiently high temperatures (Chapter 27).

Brownian motion

Brownian particles are particles visible under a microscope that float on the surface of a liquid. They are so small that the statistical collisions (with the molecules of the liquid) lead to visible, irregular movements. This can be considered as a two-dimensional random walk. This *Brownian motion* constitutes a direct evidence that heat is the disordered motion of atoms and molecules.

A Brownian particle (mass m) can be described by a canonical ensemble with the temperature T. The surface area of the liquid shall be the x-y-plane. According to the law of equipartition, the Brownian particle then has the mean kinetic energy

$$\frac{m}{2}\,\overline{v^2} = \frac{m}{2}\left(\overline{v_x^2} + \overline{v_y^2}\right) = k_{\mathrm{B}}T \tag{24.50}$$

The first observations of Brownian motion were made by Leuwenhock in 1673 and Brown in 1828. Early attempts of interpretation explained the Brownian motion by animals, or the effect of light and temperature currents.

Exercises

24.1   Heat capacity in a two-level system

A system consists of N independent, distinguishable particles. There are two energy levels $\varepsilon_1 = 0$ and $\varepsilon_2 = \varepsilon > 0$ for each particle. Calculate the partition function $Z(T, N)$. At a given temperature, what is the average number of particles in the upper level? Sketch the specific heat of the system.

24.2   Heat capacity for N particles in the oscillator

The Hamilton operator

$$H = \sum_{v=1}^{N} \left[\frac{p_v^2}{2m} + \frac{m\omega^2}{2} r_v^2 \right]$$

describes N independent particles in a harmonic oscillator. The particles are in contact with a heat bath of temperature T. Calculate the partition function $Z(T, N)$ and the energy $E(T, N)$ of the system. How does the heat capacity $C(T, N)$ behave for low and high temperatures?

24.3   Velocity distribution for v_x

Determine the probability distribution for the x-component of the velocity of a free particle at a given temperature. Sketch this distribution and compare it with the Maxwell distribution. Calculate the mean value $\overline{v_x^2}$ and use this to determine $\overline{v^2}$.

24.4   Various mean values for Maxwell distribution

Show for the Maxwell distribution

$$\overline{v} = \sqrt{\frac{8}{\pi} \frac{k_B T}{m}} \qquad \text{and} \qquad \overline{v^2} = \frac{3 k_B T}{m} \tag{24.51}$$

Determine the numerical values for air at room temperature ($k_B T \approx \text{eV}/40$). Compare the results with the maximum v_{max} of the Maxwell distribution.

24.5   Distribution of the relative velocities

The velocities of the particles of a gas are isotropically distributed and satisfy the Maxwell distribution

$$f(v_i) = 4\pi \left(\alpha/\pi\right)^{3/2} v_i^2 \exp\left(-\alpha v_i^2\right) \qquad \text{with} \qquad \alpha = \frac{m}{2 k_B T} \tag{24.52}$$

For two selected particles ($i = 1, 2$) we define the center of mass velocity and the relative velocity by

$$V = \frac{v_1 + v_2}{2} \qquad \text{and} \qquad v = v_1 - v_2 \tag{24.53}$$

Calculate the Maxwell distribution $F(v)$ for the relative velocities. Compare $F(v)$ with $f(v)$.

24.6   Isotope separation

A box (volume V, temperature T) contains an ideal gas with two kinds of molecules, A and B. The molecules have different masses, $m_A > m_B$. Molecules can leave the container through porous walls. The individual pores are large compared to the dimensions of the molecules; their total area a is small compared to the area of the container walls. Calculate the concentration ratio $c_A(t)/c_B(t)$ of the molecules in the box as a function of time.

24.7   Convective equilibrium

Wind or convection means the exchange of volume elements. As a model of the atmosphere, an equilibrium is assumed with respect to the adiabatic quasi-static exchange of air. Under this assumption we obtain $dS = đQ_{\text{q.s.}}/T = 0$ and a position independent entropy density $s(\boldsymbol{r})$,

$$s(\boldsymbol{r}) = \text{const.}$$

In equilibrium, the pressure gradient $dP/dz = -\varrho g = -mg/v$ compensates for the gravity force. Here ϱ is the mass density of the air, m is the mass of an air molecule, $\boldsymbol{g} = -g\,\boldsymbol{e}_z$ the gravitational acceleration and $v = V/N$. For the air in the atmosphere, the ideal gas law $v = k_B T/P$ and $c_P \approx 7k_B/2$ shall be assumed.

Calculate the temperature distribution $T(z)$. What temperature drop results at an altitude of $1\,\text{km}$? Compare the pressure decrease for $\Delta z = 1\,\text{km}$ with that of the barometric formula.

24.8   Energy fluctuation in an ideal gas

In macroscopic systems, the microstates are so close together that mean values can be evaluated as integrals:

$$\overline{A} = \sum_r A_r\, P_r = \frac{1}{Z}\sum_r A_r \exp(-\beta E_r) = \int_0^\infty dE\, A(E)\, \omega(E)\, \exp(-\beta E)$$

Here $\omega(E)$ is the density of states. For an ideal gas, $\omega(E) \propto E^{3N/2}$ holds. Determine the fluctuation ΔE of the energy. To do this, expand the logarithm of $\omega(E)\exp(-\beta E)$ up to 2nd order around the maximum.

25 Monatomic ideal gas

For a macroscopic system, each of the statistical ensembles leads to the same thermodynamic relations. This is explicitly demonstrated for the ideal monatomic gas.

The fundamental postulate states that the equilibrium state of a closed system is represented by the microcanonical ensemble ($P_r = $ const.). This was the plausible starting point of our considerations in Part II; therefore, the microcanonical partition function Ω played the central role here. For special systems (Part V), the canonical or the grand canonical partition function Z or Y are usually easier to handle. Especially for the ideal monatomic gas, we evaluate all partition functions Ω, Z and Y and show explicitly that they all lead to the same thermodynamic results.

The Hamilton operator of the ideal monatomic gas was given in (6.2). We consider the corresponding Hamilton function:

$$H = \sum_{\nu=1}^{N} \left(\frac{\boldsymbol{p}_\nu^2}{2m} + U(\boldsymbol{r}_\nu) \right) = \sum_{\nu=1}^{N} h(\nu) \tag{25.1}$$

Here $U(\boldsymbol{r})$ is the wall potential (24.16) which restricts the N atoms (or molecules) to a volume V. In the argument of h, the letter ν stands for the position $\boldsymbol{r}_\nu$ and the momentum $\boldsymbol{p}_\nu$ of the ν-th particle.

Quantum mechanical effects can be neglected for all real gases, see $(6.8) - (6.10)$ and Chapter 30. We can therefore limit ourselves to the classical evaluation of the partition functions. The classical microstates of the ideal gas are given by

$$r = (\boldsymbol{r}_1,..., \boldsymbol{r}_N, \boldsymbol{p}_1,..., \boldsymbol{p}_N) = (x_1,..., x_{3N}, p_1,..., p_{3N}) \tag{25.2}$$

We denote the Cartesian coordinates by x_k and the corresponding momenta by p_k. The energy of the state r is

$$E_r = H(x, p) = \sum_{\nu=1}^{N} \frac{\boldsymbol{p}_\nu^2}{2m} = \sum_{k=1}^{3N} \frac{p_k^2}{2m} \tag{25.3}$$

For the sum over r we use (24.7). Then each spatial integration results in a factor V. For Z and Y this follows from (24.17), for Ω from the condition that we must sum over all accessible states. In addition, a factor $1/N!$ takes into account that the interchange of the positions and momenta of any two particles in (25.2) does not result in a new state. For the ideal gas of N indistinguishable particles, (24.7) therefore becomes

$$\sum_r \ldots = \frac{1}{N!} \frac{V^N}{(2\pi\hbar)^{3N}} \int dp_1 \ldots \int dp_{3N} \ldots \tag{25.4}$$

We obtain the partition functions Ω, Z and Y, by inserting (25.3) and (25.4) into the definitions (23.4) – (23.6).

We start with the microcanonical partition function:

$$\Omega(E, V, N) = \sum_{r\,:\,E-\delta E\,\leq\,E_r\,\leq\,E} 1 = \frac{1}{N!} \frac{V^N}{(2\pi\hbar)^{3N}} \underbrace{\int dp_1 \,\ldots\, \int dp_{3N}}_{E-\delta E\,\leq\,\Sigma_i\, p_i^2/2m\,\leq\,E} 1$$

$$= c^N \left(\frac{V}{N}\right)^N \left(\frac{E}{N}\right)^{3N/2} \tag{25.5}$$

This result was already obtained in (6.20).

We now calculate the canonical partition function. For a Hamilton function of the form $H = \sum_\nu h(\nu)$ we can write

$$Z(T, V, N) = \frac{1}{N!} \sum_{r_1} \ldots \sum_{r_N} \exp\left(-\beta\left[h(1) + \ldots + h(N)\right]\right)$$

$$= \frac{1}{N!} \prod_{\nu=1}^{N} \sum_{r_\nu} \exp\left[-\beta h(\nu)\right] = \frac{1}{N!} \left[z(T, V)\right]^N \tag{25.6}$$

Here $r_\nu = (\boldsymbol{r}_\nu, \boldsymbol{p}_\nu)$ stands for the position and momentum of the ν-th particle, and z denotes the partition function for a single particle. Equation (25.6) applies to any Hamilton function (or Hamilton operator) of the form $H = \sum_\nu h(\nu)$. The reduction from Z to z can be formulated somewhat more generally: Whenever subsystems are independent of each other, the partition function of the system can be reduced to the partition functions of the subsystems. The factor $1/N!$ must be taken into account for identical particles.

In (24.19), the classical partition function z for a single particle has already been calculated, $z = V/\lambda^3$. With this (25.6) yields

$$Z(T, V, N) = \frac{[z(T, V)]^N}{N!} = \frac{1}{N!} \frac{V^N}{\lambda^{3N}} \tag{25.7}$$

with the thermal wavelength

$$\lambda = \frac{2\pi\hbar}{\sqrt{2\pi m k_B T}} \tag{25.8}$$

For the grand canonical partition function we use (22.30):

$$Y(T, V, \mu) = \sum_{N=0}^{\infty} Z(T, V, N) \exp(\beta \mu N) = \sum_{N=0}^{\infty} \frac{1}{N!} \frac{V^N}{\lambda^{3N}} \exp(\beta \mu N) \quad (25.9)$$

We abbreviate $(V/\lambda^3) \exp(\beta \mu)$ with y and sum up the series:

$$Y(T, V, \mu) = \sum_{N=0}^{\infty} \frac{y^N}{N!} = \exp(y) = \exp(y) = \exp\left(\frac{V \exp(\beta \mu)}{\lambda^3}\right) \quad (25.10)$$

We compile the logarithms of all partition functions:

$$\ln \Omega(E, V, N) \;=\; N \ln\left(\frac{V}{N}\right) + \frac{3N}{2} \ln\left(\frac{E}{N}\right) + N \ln c \quad (25.11)$$

$$\ln Z(T, V, N) \;=\; N \ln\left(\frac{V}{N \lambda^3}\right) + N \quad (25.12)$$

$$\ln Y(T, V, \mu) \;=\; \frac{V}{\lambda^3} \exp(\beta \mu) \quad (25.13)$$

Each of these logarithms is associated with a thermodynamic potential or with the entropy $S(E, V, N)$; and this associated function contains the complete thermodynamic information (Chapter 17). We provide an overview of how the equations of state $P = P(T, V, N)$ and $E = E(T, V, N)$ are obtained from the associated quantity:

1. The microcanonical partition function yields the entropy $S(E, V, N) = k_{\mathrm{B}} \ln \Omega$. The partial derivative of S with respect to E leads to $T = T(E, V, N)$; this can be resolved to $E = E(T, V, N)$. The partial derivative with respect to V results in $P = P(E, V, N)$, in which $E = E(T, V, N)$ is inserted.

2. The canonical partition function yields the free energy $F(T, V, N) = -k_{\mathrm{B}} T \ln Z$. The partial derivatives of F with respect to T and V lead to $S = S(T, V, N)$ and $P = P(T, V, N)$. From this one also gets $E(T, V, N) = F(T, V, N) + T S(T, V, N)$.

3. The grand canonical partition function yields the thermodynamic potential $J(T, V, \mu) = -k_{\mathrm{B}} T \ln Y$. The partial derivatives of J with respect to T, V and μ lead to $S = S(T, V, \mu)$, $P = P(T, V, \mu)$ and $N = N(T, V, \mu)$. The last relationship is solved for $\mu = \mu(T, V, N)$ and inserted into the other two; this yields $S(T, V, N)$ and $P(T, V, N)$, and therefore also the energy $E(T, V, N) = J + T S + \mu N$.

In Table 25.1 the partition functions and the derivation of the caloric and thermal equations of state are shown at a glance. The desired relations can sometimes be

Table 25.1   Each statistical ensemble corresponds to certain physical conditions. From the energy values $E_r(V, N)$ of the microstates, the respective partition function is calculated. Each partition function corresponds to a thermodynamic potential or to $S(E, V, N)$. The bottom two lines show how the thermal and caloric equation of state are obtained. For the microcanonical ensemble, $E = E(T, V, N)$ is determined from $T(E, V, N)$ and inserted into $P(E, V, N)$. For the grand canonical ensemble, $\mu = \mu(T, V, N)$ is determined from $N(T, V, \mu) = -\partial J/\partial \mu$ and inserted into $E(T, V, \mu)$ and $P(T, V, \mu)$.

Ensemble	microcanonical	canonical	grand canonical
Physical condition	closed system	contact with heat bath	contact with heat and particle reservoir
Partition function	$\Omega(E, V, N)$	$Z(T, V, N)$	$Y(T, V, \mu)$
Associated thermodynamical potential	$S(E, V, N) = k_\mathrm{B} \ln \Omega$	$F(T, V, N) = -k_\mathrm{B}T \ln Z$	$J(T, V, \mu) = -k_\mathrm{B}T \ln Y$
Caloric equation of state $E(T, V, N)$	$\dfrac{1}{T(E, V, N)} = \dfrac{\partial S}{\partial E}$	$E(T, V, N) = -\dfrac{\partial \ln Z}{\partial \beta}$	$E(T, V, \mu) = -\dfrac{\partial \ln Y}{\partial \beta} + \mu N$
Thermal equation of state $P(T, V, N)$	$P(E, V, N) = T\,\dfrac{\partial S}{\partial V}$	$P(T, V, N) = -\dfrac{\partial F}{\partial V}$	$P(T, V, \mu) = -\dfrac{J}{V}$

obtained more directly, namely from the partial derivatives of $\ln \Omega$, $\ln Z$ and $\ln Y$; this is demonstrated below in (25.14) to (25.20).

The presented derivation of the caloric and thermal equation of state referred to arbitrary systems. For the ideal gas, we carry out this derivation once more for each partition function.

From the microcanonical partition function (25.11) we obtain

$$P \stackrel{(10.16)}{=} \frac{1}{\beta} \frac{\partial \ln \Omega}{\partial V} = \frac{N k_\mathrm{B} T}{V} \tag{25.14}$$

$$\frac{1}{k_\mathrm{B} T} \stackrel{(10.16)}{=} \frac{\partial \ln \Omega}{\partial E} = \frac{3}{2} \frac{N}{E} \tag{25.15}$$

From the canonical partition function (25.12) we obtain

$$P \stackrel{(23.14)}{=} \frac{1}{\beta} \frac{\partial \ln Z}{\partial V} = \frac{N k_\mathrm{B} T}{V} \tag{25.16}$$

$$E \stackrel{(23.13)}{=} -\frac{\partial \ln Z}{\partial \beta} = N \frac{\partial \ln \lambda^3}{\partial \beta} = \frac{3N}{2} \frac{\partial \ln \beta}{\partial \beta} = \frac{3}{2} N k_\mathrm{B} T \tag{25.17}$$

We have used $\lambda = \text{const.} \cdot \beta^{1/2}$.

From the grand canonical partition function (25.13) we get

$$N \stackrel{(23.28)}{=} \frac{1}{\beta} \frac{\partial \ln Y(\beta, V, \mu)}{\partial \mu} = \frac{V}{\lambda^3} \exp(\beta \mu) = \ln Y \tag{25.18}$$

$$P \stackrel{(23.27)}{=} \frac{1}{\beta} \frac{\partial \ln Y}{\partial V} = k_\mathrm{B} T \underbrace{\frac{\exp(\beta \mu)}{\lambda^3}}_{=N/V} = \frac{N k_\mathrm{B} T}{V} \tag{25.19}$$

$$E \stackrel{(23.26)}{=} -\frac{\partial \ln Y}{\partial \beta} + \mu N = -\frac{\partial}{\partial \beta} \left(\frac{V}{\lambda^3} \exp(\beta \mu) \right) + \mu N$$

$$= \frac{3}{2\beta} \underbrace{\frac{V}{\lambda^3} \exp(\beta \mu)}_{=N} - \mu \underbrace{\frac{V}{\lambda^3} \exp(\beta \mu)}_{=N} + \mu N = \frac{3}{2} N k_\mathrm{B} T \tag{25.20}$$

In the last two equations, μ was eliminated with the help of (25.18).

For the ideal monatomic gas, we have demonstrated that each partition function leads to the same thermal and caloric equations of state. In fact, this applies to any macroscopic system. The reasons for this were given in Chapter 22: If, for example, the temperature is specified the energy exhibits fluctuations around a mean values. For microscopic systems these fluctuations are, however, so small that they can be neglected. This applies accordingly to all other macroscopic quantities.

For deriving the thermodynamic relations, the various partition functions are equivalent. In the following Part V about special systems we use the partition function that is particularly easy to evaluate.

Exercises

25.1   *Gibbs paradox*

The canonical partition function of an ideal monatomic gas is

$$Z(T, V, N) = \frac{\left[z(T, V)\right]^{N}}{N!} = \frac{1}{N!}\frac{V^{N}}{\lambda^{3N}}, \qquad \lambda = \frac{2\pi\hbar}{\sqrt{2\pi m k_{B} T}} \qquad (25.21)$$

Use this to calculate the change ΔF of the free energy in the following process: A thermally insulated gas volume V is divided into two equal volumes by inserting a wall form the side. Alternatively, calculate ΔF from thermodynamic relations (to do this, consider the transferred quantities of work and heat).

Omit now the factor $1/N!$ in the expression for Z; this results in another expression F^{*} for the free energy. Determine the change ΔF^{*} in the process under consideration. The contradiction between this statistically calculated ΔF^{*} and the thermodynamically calculated ΔF is called *Gibbs paradox*. The contradiction was resolved by the introduction of the factor $1/N!$. This before was done before quantum mechanics could justify this factor (indistinguishability of particles).

V Special systems

26 Ideal spin system

In Part V we evaluate the canonical or the grand canonical partition function for a number of model systems. The relations to real systems and observable physical effects are discussed.

In this chapter we investigate an ideal system of N spin-1/2 particles. "Ideal" means that no interactions between the spins are taken into account; all spins adjust themselves independently of each other in an external magnetic field. This model explains the temperature dependence of paramagnetism.

We consider particles with mass m, charge q and spin s. Such particles have a magnetic moment $\boldsymbol{\mu} = \mu \boldsymbol{s}$ with $\mu = g q \hbar / 2mc$, where g is a numerical factor of the size 1. We specifically consider electrons with $q = -e$ and $g \approx 2$. For $g = 2$ this means

$$\boldsymbol{\mu} = -2\mu_{\mathrm{B}} \boldsymbol{s}, \qquad \mu_{\mathrm{B}} = \frac{e\hbar}{2m_{\mathrm{e}}c} \tag{26.1}$$

The quantity μ_{B} is the *Bohr magneton*. In an external magnetic field $\boldsymbol{B} = B\,\boldsymbol{e}_z$ the energy of a magnetic moment is equal to $\varepsilon = -\boldsymbol{\mu} \cdot \boldsymbol{B} = -\mu_z B$. Relative to the measurement direction (here the direction of $\boldsymbol{B}$) the projection of the spin can assume the values $s_z = \pm 1/2$ (the factor $\hbar$ is included in μ_{B}). Therefore, an electron has the possible energy values

$$\varepsilon = -\boldsymbol{\mu} \cdot \boldsymbol{B} = 2\mu_{\mathrm{B}} B\, s_z = \pm\mu_{\mathrm{B}} B \tag{26.2}$$

In the following we consider the spin degrees of freedom of a system of N electrons; other degrees of freedom are not taken into account. This spin system can serve as a model for the magnetic properties of a crystal of atoms where each atom has an unpaired electron with vanishing orbital angular momentum.

The microstate of the spin system of N particles is defined by

$$r = (s_{z,1}, s_{z,2}, \ldots, s_{z,N}), \qquad s_{z,\nu} = \pm 1/2 \tag{26.3}$$

The index $\nu = 1,\ldots, N$ counts the particles. For an ideal spin system, i.e. a system without interactions between the spins, the energy is the sum of the single particle energies (26.2):

$$E_r(B, N) = \sum_{\nu=1}^{N} \varepsilon_\nu = \sum_{\nu=1}^{N} 2\,\mu_{\mathrm{B}} B\, s_{z,\nu} \tag{26.4}$$

We evaluate the partition function $Z = \sum_r \exp(-\beta E_r(B, N))$:

$$Z(T, B, N) = \sum_{s_{z,1} = \pm 1/2} \cdots \sum_{s_{z,N} = \pm 1/2} \exp(-\beta \varepsilon_1) \cdot \ldots \cdot \exp(-\beta \varepsilon_N)$$

$$= \left(\sum_{s_z = \pm 1/2} \exp\left(- 2\beta \mu_{\mathrm{B}} B s_z \right) \right)^N = \left[z(T, B) \right]^N \qquad (26.5)$$

Here $z(T, B)$ is the partition function of a single spin,

$$z(T, B) = \sum_{s_z = \pm 1/2} \exp\left(- 2\beta \mu_{\mathrm{B}} B s_z \right) = 2 \cosh\left(\beta \mu_{\mathrm{B}} B \right) \qquad (26.6)$$

A factor $1/N!$ does not occur because the spins are localized at lattice sites and cannot readily be exchanged.

In the following, we use the indices $(+)$ and $(-)$ to denote the parallel and antiparallel setting of the magnetic moment relative to the field direction. The respective probabilities in the canonical ensemble are

$$P_\pm = \frac{\exp\left(\pm \beta \mu_{\mathrm{B}} B \right)}{2 \cosh\left(\beta \mu_{\mathrm{B}} B \right)} \qquad (26.7)$$

With parallel adjustment of the magnetic moment, the energy is lowered; therefore $P_+ \geq 1/2$. The mean magnetic moment $\overline{\mu}$ of a particle is

$$\overline{\mu} = -2 \mu_{\mathrm{B}} \overline{s_z} = \mu_{\mathrm{B}} \left(P_+ - P_- \right) \qquad (26.8)$$

Each external parameter x_i is associated with a generalized force X_i. The external parameters are the parameters on which the energy values E_r depend, i.e. B and N in (26.4). For the spin system, N is kept constant and will no longer be listed as an argument. We consider the generalized force belonging to B,

$$X_B \overset{(8.2)}{=} -\overline{\frac{\partial E_r}{\partial B}} = -\overline{\sum_{\nu=1}^{N} 2 \mu_{\mathrm{B}} s_{z,\nu}} = N \overline{\mu} = V M \qquad (26.9)$$

This is equal to the mean *magnetic moment* $N \overline{\mu}$ of the spin system. The fluctuations of the magnetic moment around the mean value are of the relative size $\mathcal{O}(N^{-1/2})$; therefore, $N \overline{\mu}$ is also simply referred to as the magnetic moment. Related the volume, this is the *magnetization*

$$M = \frac{N \overline{\mu}}{V} = n \overline{\mu} = \frac{\text{magnetic moment}}{\text{volume}} \qquad (26.10)$$

where $n = N/V$ is the particle density. We insert (26.8) with (26.7):

$$\boxed{M(T, B) = M_0 \tanh\left(\frac{\mu_{\mathrm{B}} B}{k_{\mathrm{B}} T} \right)} \qquad (26.11)$$

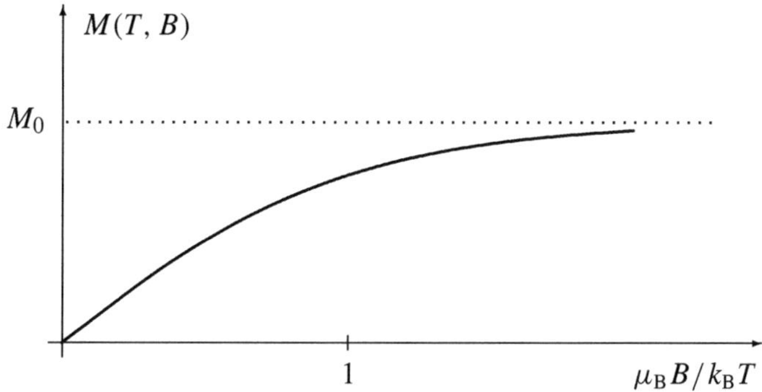

Figure 26.1 The magnetization $M = M_0 \tanh(y)$ is sketched as a function of $y = \mu_B B / k_B T$. This displays the dependence on the magnetic field B at fixed temperature, or the dependence on $1/T$ for a constant magnetic field.

Here $M_0 = n\mu_B$ is the maximum magnetization. Figure 26.1 shows the dependence of the magnetization $M = M_0 \tanh(y)$ on the dimensionless quantity $y = \beta\mu_B B$. We expand the tangent hyperbolicus for large and for small values of y:

$$
\tanh(y) = \begin{cases} y - y^3/3 \pm \ldots & (y \ll 1) \\[2ex] 1 - 2\exp(-2y) \pm \ldots & (y \gg 1) \end{cases}
\tag{26.12}
$$

The value $y = 1$ defines a temperature T_{mag}:

$$
k_B T_{\mathrm{mag}} = \mu_B B
\tag{26.13}
$$

For the magnetization, we obtain from (26.11, 26.12) the limiting cases

$$
M(T, B) = M_0 \cdot \begin{cases} \dfrac{\mu_B B}{k_B T} \pm \ldots & (T \gg T_{\mathrm{mag}}) \\[3ex] 1 - 2\exp\left(\dfrac{-2\mu_B B}{k_B T}\right) \pm \ldots & (T \ll T_{\mathrm{mag}}) \end{cases}
\tag{26.14}
$$

For high temperature or a weak magnetic field, the magnetization is proportional to the applied field. In these cases, the magnetic susceptibility is independent of the field:

$$
\chi_m = \frac{\partial M}{\partial B} = \frac{M_0 \mu_B}{k_B T} = \frac{\mathrm{const.}}{T} \qquad (\text{Curie's law, } T \gg T_{\mathrm{mag}})
\tag{26.15}
$$

This temperature dependence is referred to as *Curie's law*.

By averaging over (26.4) we obtain the energy

$$
E(T, B) = 2N\mu_B B \, \overline{s_z} = -N\overline{\mu}\, B = -V B M(T, B)
\tag{26.16}
$$

We assume that N and V are constant. For a given T and B, the equilibrium condition of the spin system reads

$$F(T, B) = E - TS = -VBM - TS = \text{minimal} \qquad (26.17)$$

The tendency towards the smallest possible free energy means that the term $-VBM$ acts to make M as large as possible (i.e. to align the spins), whereas the term $-TS$ acts to make S as large as possible (i.e. to occupy the spin states as evenly as possible). The strength of the first tendency increases with B, that of the second one with T. At low temperature the magnetic field (the order) wins, at high temperature the entropy (disorder) wins.

The ideal spin system is a simple model for paramagnetic behavior ($\chi_m > 0$) of matter. If the electrons in an atom are coupled to zero total spin, there is no resulting magnetic moment. Paramagnetism occurs in particular when there is an additional electron (in addition to the electrons coupled to zero). The spins in (26.3) are then the spins of these electrons; the index ν runs over the N atoms of the system. Regardless of the paramagnetic effect investigated here, an applied magnetic field induces currents in the atom that lead to a diamagnetic behavior ($\chi_m < 0$). If paramagnetism occurs, it masks the smaller (always present) diamagnetic effect.

In real systems, there are interactions between neighboring spins. We have neglected these interactions here. This is a valid approximation if $k_B T$ is large compared to the interaction energy. For specific materials, the interactions of neighboring spins can lead to a spontaneous magnetization, i.e. to ferromagnetism. Such a spontaneous magnetization sets in below a certain transition temperature (Chapter 36).

Achieving low temperatures

The following procedure is used to achieve very low temperatures (such as 10^{-3} K): At an initial stage there is a heat bath, for example with $T = 1$ K. The spin system is in contact with this heat bath. Then a magnetic field B_a is switched on isothermally. The field shall be so large that $M \approx M_0$. Now the system is thermally insulated. In the thermally insulated system, the external field is now switched off; after that there is only a weak internal residual field B_b. Immediately after switching off the field, the magnetization is still $M \approx M_0$. The high magnetization with a weak field means that the system now has a much lower temperature (see Exercise 26.1). The analogous procedure for the nuclear spins (instead of the electron spins considered here) may lead to *very* low temperatures (such as 10^{-6} K).

Negative (fictitious) temperature

The temperature introduced in Chapter 9 cannot be negative. From (26.7) and $T \geq 0$ follows for a state of equilibrium

$$\frac{P_+}{P_-} = \exp\left(2\beta\mu_B B\right) \geq 1 \qquad (26.18)$$

Experimentally, however, a state can be prepared in such a way that $P_+/P_- < 1$; for this, the energetically upper spin states must be more occupied than the lower ones. Such a state is a non-equilibrium state. This does not contradict the statement (26.18) which refers to equilibrium.

If for states with $P_+/P_- < 1$ one now uses $P_+/P_- = \exp(2\beta\mu_\mathrm{B}B)$ –although the preconditions for this are not fulfilled– then one formally obtains a *negative* (fictitious) temperature. If two spin systems with inverted occupation numbers are brought into contact with each other (but otherwise isolated), then the contact can lead to a quasi-equilibrium of these two systems with the negative (fictitious) temperatures $T_1 = T_2$. In this respect, the use of a negative temperature can be motivated.

In the context of this physics course, we stay with the temperature defined in Chapter 9. This (absolute) temperature cannot assume negative values.

Exercises

26.1 Adiabatic demagnetization

A system consists of N independent spin-1/2 particles (electrons) with the magnetic moment μ_B. The spins adjust themselves in a homogeneous magnetic field of strength B. Start from the partition function $Z(T, B)$ and from the differential

$$dF = -S\,dT - VM\,dB \tag{26.19}$$

of the free energy $F(T, B) = -k_B T \ln Z$ (with $N = \text{const.}$). From this, calculate the entropy S and the magnetization M. What results for $T \to 0$ and for $T \to \infty$?

Initially, let the temperature of the system be equal to T_a and the field equal to B_a. In the thermally insulated system, the field is now switched off slowly; due to the internal interaction, a weak field B_b remains. What temperature T_b is then reached? Sketch the entropy as a function of temperature for two different values $(B_a$ and $B_b)$ of the magnetic field.

26.2 Specific heat and susceptibility in the ideal spin system

A system consists of N independent spin $1/2$ particles (electrons) with the magnetic moment μ_B. It is located in an external magnetic field B. The system has the partition function

$$Z(T, B) = \left(2 \cosh x\right)^N \qquad \text{with} \quad x = \frac{\mu_B B}{k_B T} = \frac{T_{\text{mag}}}{T}$$

The number of particles N is constant. Calculate the specific heat $c_B(T, B) = (T/N)\,\partial S(T, B)/\partial T$ and the susceptibility $\chi_m = \partial M/\partial B$. Specialize the results for $x \gg 1$ and $x \ll 1$.

26.3 General ideal spin system

N independent spin particles are located in a homogeneous magnetic field $\mathbf{B} = B\,\mathbf{e}_z$ with $B > 0$. The spin $s > 0$ (here without the factor $\hbar$) can be a half integer or a whole number. There are the spin settings $s_z = s, s-1, s-2, \ldots, -s$. The magnetic moment is $\boldsymbol{\mu} = g\mu_0 s$, where g is the is the gyromagnetic factor, and $\mu_0 = q\hbar/(2mc)$. Calculate the canonical partition function (geometric series!), the free energy $F(T, B)$ and the magnetization $M(T, B)$. Determine the magnetization specifically for high and low temperatures.

27 Diatomic ideal gas

We calculate the heat capacity of a diatomic ideal gas. The relevant excitations are the translations, rotations and vibrations. The rotations and vibrations are treated quantum mechanically. The respective specific heats show characteristic temperature dependencies. The exchange symmetry leads to special effects (ortho- and parahydrogen).

Hamilton operator

Ideal means that the interactions between the gas molecules are neglected. The Hamilton operator is therefore a sum of independent Hamilton operators:

$$H = \sum_{\nu=1}^{N} h(\nu) \tag{27.1}$$

Here ν stands for all coordinates and moments of the ν-th molecule.

The diatomic ideal gas can be regarded as a model for air. Air consists mainly of N_2 (about 77%) and O_2 molecules (about 21%).

Under normal conditions, the mean free path in air is is 10^{-7} m, and the mean collision time is $2 \cdot 10^{-10}$ s. In this respect, the model (27.1) which neglects the interactions between the molecules is unrealistic. In the *statistical* treatment, however, this does not necessarily lead to incorrect results. For example, the pressure only depends on the mean density and the mean velocity of the gas molecules, but not whether the particles can transverse the gas volume without collisions. The energy depends primarily on the possible excitations (such as rotations, vibrations), but not how they are excited (e.g. by collisions). The ideal gas model is therefore a usable approximation for calculating the energy and the pressure. The neglected interactions are all the less important the more dilute the gas is (Chapter 28).

The Hamilton operator h consists of independent parts for the translation, vibration and rotation of a molecule:

$$h = h_{\text{trans}} + h_{\text{vib}} + h_{\text{rot}} \tag{27.2}$$

Excitations of the electrons are not considered here. The respective classical Hamilton function has already been discussed in (24.47).

The center of mass motion of the molecule (mass M) is described by

$$h_{\text{trans}} = \frac{\boldsymbol{p}_{\text{op}}^2}{2M} = -\frac{\hbar^2}{2M}\,\Delta \tag{27.3}$$

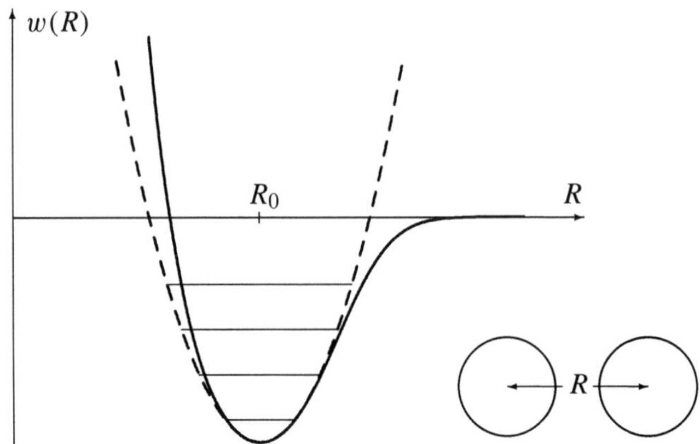

Figure 27.1   Sketch of the potential energy $w(R)$ of two atoms as a function of their distance R. (A more realistic atom-atom potential is shown in Figure 38.1). In the vicinity of the minimum at R_0, the potential can be approximated by an oscillator (dashed curve). The energy levels in the oscillator are indicated by horizontal lines. In addition to these oscillations (vibrations), there are thermally excited rotations.

Here $\boldsymbol{p}_{\mathrm{op}} = -\mathrm{i}\hbar\nabla$ is the operator of the center of mass momentum.

We now come to the internal degrees of freedom of the molecule. Figure 27.1 sketches the potential between the two atoms in the molecule (for example H_2 or O_2). The distance between the atomic nuclei is denoted by $R = R_0 + \xi$; where R_0 is the equilibrium distance. In the vicinity of the minimum we approximate the potential by a harmonic oscillator. Then the vibrations are described by the Hamilton operator

$$h_{\mathrm{vib}} = -\frac{\hbar^2}{2m_{\mathrm{r}}}\frac{d^2}{d\xi^2} + \frac{m_{\mathrm{r}}}{2}\,\omega^2\xi^2 \tag{27.4}$$

Here, m_{r} is the reduced mass of the two atoms and $m_{\mathrm{r}}\omega^2$ is equal to the second derivative of the potential $w(R)$ at R_0. The approximation by an oscillator deteriorates with increasing excitation energy; at the dissociation limit (R axis in Figure 27.1) it fails completely.

The mass of the molecule is essentially concentrated in the atomic nuclei. Therefore, the moment of inertia for a rotation around an axis perpendicular to the connecting line is given by

$$\Theta = m_{\mathrm{r}} R^2 \approx m_{\mathrm{r}} R_0^2 = \text{const.} \tag{27.5}$$

The approximation $R = R_0 + \xi \approx R_0$ is that of a *rigid rotator*; it neglects the coupling between rotations and vibrations. The Hamilton operator for the rigid rotator reads

$$h_{\mathrm{rot}} = \frac{\ell_{\mathrm{op}}^2}{2\Theta} = -\frac{\hbar^2}{2\Theta}\left(\frac{1}{\sin\theta}\frac{\partial}{\partial\theta}\sin\theta\frac{\partial}{\partial\theta} + \frac{1}{\sin^2\theta}\frac{\partial^2}{\partial\phi^2}\right) \tag{27.6}$$

Here ℓ_{op} is the angular momentum operator; the angles θ and ϕ define the orientation of the line connecting the atoms.

A rotation around the connecting line of the molecules does not occur because the overall wave function of the molecule is symmetric with respect to such a rotation. Then there is no quantum mechanical rotational state. A rotational state would be a superposition of wave functions rotated relative to each other; due to the symmetry this does not affect the wave function.

The eigenvalues of $h = h_{\mathrm{trans}} + h_{\mathrm{vib}} + h_{\mathrm{rot}}$ are

$$\varepsilon = \frac{\pi^2 \hbar^2}{2ML^2}\left(n_x^2 + n_y^2 + n_z^2\right) + \hbar\omega\left(n + \frac{1}{2}\right) + \frac{\hbar^2 l(l+1)}{2\Theta} \tag{27.7}$$

We assume a cubic volume $V = L^3$. The eigenfunctions of the translation are $\sin(\pi n_x x/L)$ for the x-direction, and correspondingly for the y- and z direction. The quantum numbers n_x, n_y and n_z can adopt the values $1, 2,....$ The eigenfunctions of the one-dimensional oscillator are of the form $H_n(\xi)\,\exp(-\alpha\xi^2)$, where the H_n are Hermitian polynomials; they are characterized by the quantum number $n = 0, 1, 2,....$ The eigenfunctions of ℓ_{op}^2 are the spherical harmonics $Y_{lm}(\theta, \phi)$; they depend on the quantum numbers $l = 0, 1, 2,...$ and $m = -l, -l+1,..., l$. The state of the ν-th molecule is thus defined by

$$r_\nu = \left(n_x^\nu, n_y^\nu, n_z^\nu, n_\nu, l_\nu, m_\nu\right) \tag{27.8}$$

We refrain from possible spin degrees of freedom for the time being. The microstate of the gas is then

$$r = (r_1, r_2, \ldots, r_N) \tag{27.9}$$

The energy of this state is

$$E_r = \sum_{\nu=1}^{N} \varepsilon_{r_\nu} = E_{\mathrm{trans}} + E_{\mathrm{vib}} + E_{\mathrm{rot}} \tag{27.10}$$

where ε is given by (27.7).

Quantum effects

The following quantum effects play a role:

1. Indistinguishability of the molecules: We take this effect into account by a factor $1/N!$ in the partition function:

$$\sum_r \cdots \;\Longrightarrow\; \frac{1}{N!}\sum_r \cdots = \left(\frac{1}{N!}\sum_{\mathrm{trans}}\right)\sum_{\mathrm{vib}}\sum_{\mathrm{rot}}\cdots \tag{27.11}$$

   This factor already occurs with the monatomic ideal gas (Chapters 6 and 25); it is independent of vibrations or rotations. We include this factor in the partition function of the translations. Otherwise, we evaluate this partition function classically.

The indistinguishability of the gas molecules implies a certain exchange symmetry of the many-body wave functions and therefore consequences in the counting of the states (Chapter 29). These effects are only partly taken into account by the factor $1/N!$. The additional effects of an exact quantum mechanical treatment are irrelevant for ordinary gases considered here; they will be investigated in Chapter 30.

2. Quantization of the energy levels: The average energy per degree of freedom is of the size $k_B T$. By $\Delta\varepsilon$ we denote the energy distances for the degrees of freedom under consideration. The quantization plays no role if

$$\Delta\varepsilon \ll k_B T \qquad \text{(classical limit)} \qquad (27.12)$$

The translations can almost always be treated classically for ordinary gases. For rotations and vibrations, however, (27.12) can be violated; these excitations must then be treated quantum mechanically.

3. Special symmetry effects: If the molecule consists of two identical atoms, the exchange symmetry of these atoms can restrict the possible rotational states. In the last section of this chapter, this effect is discussed for hydrogen molecules; it leads to the distinction between ortho- and parahydrogen.

Neighboring vibrational states have the energy distance $\Delta\varepsilon_{\text{vib}} = \hbar\omega$. The lowest excitation energy in the rotational spectrum is $\Delta\varepsilon_{\text{rot}} = \hbar^2/\Theta$. We introduce the corresponding temperatures:

$$\Delta\varepsilon_{\text{vib}} = \hbar\omega = k_B T_{\text{vib}} \qquad (27.13)$$

$$\Delta\varepsilon_{\text{rot}} = \frac{\hbar^2}{\Theta} = k_B T_{\text{rot}} \qquad (27.14)$$

For temperatures $T \lesssim T_{\text{vib}}$ or $T \lesssim T_{\text{rot}}$ these degrees of freedom must be treated quantum mechanically.

The energy gap between the first excited rotational state with $l = 1$ and the ground state with $l = 0$ is given by $\Delta\varepsilon_{\text{rot}}$. We estimate T_{rot} for a H_2 molecule. The reduced mass is $m_r \approx m_p/2$, where $m_p \approx 1\,\text{GeV}/c^2$ is the is the proton mass. A quantum mechanical treatment of the H_2 molecule results in the equilibrium distance $R_0 \approx 0.74\,\text{Å}$. With the moment of inertia (27.5) we obtain

$$\Delta\varepsilon_{\text{rot}} = \frac{\hbar^2}{\Theta} = \frac{2\hbar^2}{m_p R_0^2} \approx \frac{2}{0.74^2}\left(\frac{\hbar c}{\text{Å}}\right)^2 \frac{1}{\text{GeV}} \approx 170\,k_B\,\text{K} \qquad \text{(for } H_2\text{)} \quad (27.15)$$

For the numerical evaluation, $\hbar c/\text{Å} = (\hbar c/e^2)(e^2/\text{Å}) \approx 137 \cdot 14.4\,\text{eV}$ and $300\,k_B\,\text{K} \approx \text{eV}/40$ have been used. The estimated value is close to the actual value, $T_{\text{rot}} \approx 171\,\text{K}$. It is significantly higher than the temperature of about $20\,\text{K}$ at which hydrogen gas condenses under normal pressure.

For vibrations, we estimate

$$\Delta \varepsilon_{\text{vib}} \approx \frac{\hbar^2}{m_{\text{r}}\,\Delta \xi^2} \sim 4000\,k_{\text{B}}\,\text{K} \qquad \text{(for } H_2\text{)} \tag{27.16}$$

where $\Delta \xi$ stands for the width of the quantum mechanical ground state wave function (approximately the width of the lowest horizontal line in Figure 27.1). For a first numerical estimate, we have used $\Delta \xi \sim 0.2\,R_0$. The actual value is higher, $T_{\text{vib}} \approx 6244\,\text{K}$. The deviation is due to the crude approximation $\Delta \xi \sim 0.2\,R_0$.

Evaluation of the partition function

In the considered model, the various types of motion are independent of each other. The Hamilton operator is therefore a sum of independent Hamilton operators, (27.1) with (27.2). Therefore, the canonical partition function $Z = \sum_r \exp(-\beta E_r)/N!$ can be divided into three independent parts:

$$Z = \frac{1}{N!}\sum_{\text{trans}}\sum_{\text{vib}}\sum_{\text{rot}} \exp\left(-\beta(E_{\text{trans}} + E_{\text{vib}} + E_{\text{rot}})\right)$$

$$= \frac{1}{N!}\underbrace{\sum_{\text{trans}}\exp(-\beta E_{\text{trans}})}_{=\,Z_{\text{trans}}} \underbrace{\sum_{\text{vib}}\exp(-\beta E_{\text{vib}})}_{=\,Z_{\text{vib}}} \underbrace{\sum_{\text{rot}}\exp(-\beta E_{\text{rot}})}_{=\,Z_{\text{rot}}} \tag{27.17}$$

Because of

$$\ln Z(T, V, N) = \ln Z_{\text{trans}} + \ln Z_{\text{vib}} + \ln Z_{\text{rot}} \tag{27.18}$$

the contributions to pressure and energy can be calculated separately:

$$P(T, V, N) = k_{\text{B}}T\,\frac{\partial \ln Z(T, V, N)}{\partial V} = P_{\text{trans}} \tag{27.19}$$

$$E(T, V, N) = -\frac{\partial \ln Z(T, V, N)}{\partial \beta} = E_{\text{trans}} + E_{\text{vib}} + E_{\text{rot}} \tag{27.20}$$

The Hamilton operators h_{rot} and h_{vib} do not depend on the volume $V = L^3$; consequently this also applies to Z_{rot} and Z_{vib}. Therefore, these partition functions do not contribute to the pressure.

The energy $E_{\text{vib}}(n_1, ..., n_N)$ in (27.10) and (27.17) is a function of the quantum numbers; on the other hand $E_{\text{vib}}(T, N)$ in (27.20) is a function of T and N. The relationship between the two quantities is given by

$$E_{\text{vib}}(T, N) = \overline{E_{\text{vib}}(n_1, ..., n_N)} = \sum_{\nu=1}^{N} \hbar\omega\left(n_\nu + \frac{1}{2}\right) \tag{27.21}$$

As usual, we omit the bar for the thermodynamic quantity. This notation applies accordingly for E_{trans} and E_{rot}.

Translations

Apart from the factor $1/N!$, the translations can be treated classically. For ordinary gases the other quantum mechanical corrections are rather small (Chapter 30).

The partition function for translations is known from (25.7):

$$Z_{\text{trans}} = \frac{1}{N!}\,\frac{V^N}{\lambda^{3N}} \tag{27.22}$$

From this follows

$$P = P_{\text{trans}} = k_{\text{B}}T\,\frac{\partial \ln Z_{\text{trans}}}{\partial V} = \frac{Nk_{\text{B}}T}{V} \tag{27.23}$$

Since the rotations and vibrations do not contribute to the pressure, the thermal equation of state is independent of the type of molecules of the ideal gas.

For the caloric equation of state, the translations give the known contribution

$$E_{\text{trans}} = -\frac{\partial \ln Z_{\text{trans}}}{\partial \beta} = \frac{3}{2}\,Nk_{\text{B}}T \tag{27.24}$$

In the following we calculate the contributions $E_{\text{vib}}(T, N)$ and $E_{\text{rot}}(T, N)$.

Vibrations

We evaluate the partition function for the vibrations. The oscillator quantum number n_v of the v-th molecule can adopt the values $0, 1, 2, \ldots$:

$$Z_{\text{vib}}(T, N) = \sum_{n_1=0}^{\infty} \ldots \sum_{n_N=0}^{\infty} \exp\left[-\beta\hbar\omega\left(n_1 + 1/2 + \ldots + n_N + 1/2\right)\right]$$

$$= \left(\sum_{n=0}^{\infty} \exp\left[-\beta\hbar\omega(n + 1/2)\right]\right)^N = \left[z_{\text{vib}}(T)\right]^N \tag{27.25}$$

If the excitations of the individual molecules are independent of each other, the partition function is the product of the individual partition functions.

The partition function z_{vib} for the vibrations of an individual molecule is a geometric series that can be summed up:

$$z_{\text{vib}} = \sum_{n=0}^{\infty} \exp\left[-\beta\hbar\omega(n + 1/2)\right] = \frac{\exp(-\beta\hbar\omega/2)}{1 - \exp(-\beta\hbar\omega)} \tag{27.26}$$

From this follows

$$\ln z_{\text{vib}} = -\ln\left[1 - \exp(-\beta\hbar\omega)\right] - \beta\,\frac{\hbar\omega}{2} \tag{27.27}$$

The average energy of the vibrations is

$$
E_{\text{vib}}(T, N) = -N \frac{\partial \ln z_{\text{vib}}}{\partial \beta} = N\hbar\omega \left(\frac{1}{\exp(\beta\hbar\omega) - 1} + \frac{1}{2} \right)
$$

$$
= N\hbar\omega \left(\frac{1}{\exp(T_{\text{vib}}/T) - 1} + \frac{1}{2} \right) \tag{27.28}
$$

In the last expression, the temperature $T_{\text{vib}} = \hbar\omega/k_B$ was used. The vibrational component of the heat capacity $C_{\text{vib}} = \partial E_{\text{vib}}/\partial T$ is then

$$
\boxed{ C_{\text{vib}}(T, N) = Nk_B \frac{T_{\text{vib}}^2}{T^2} \frac{\exp(T_{\text{vib}}/T)}{\left[\exp(T_{\text{vib}}/T) - 1 \right]^2} } \tag{27.29}
$$

The specific heat $c_{\text{vib}}(T) = C_{\text{vib}}/N$ is sketched in Figure 27.2. Since the partition function of the vibrations does not depend on the volume, the term C_{vib} contributes equally to C_V and C_P; an index V or P is superfluous.

According to (27.21), the energy is of the form

$$
E_{\text{vib}}(T, N) = \overline{\sum_{\nu=1}^{N} \hbar\omega \left(n_\nu + \frac{1}{2} \right)} = N\hbar\omega \left(\overline{n} + \frac{1}{2} \right) \tag{27.30}
$$

The comparison with (27.28) gives the mean oscillator quantum number:

$$
\overline{n} = \frac{1}{\exp(\beta\hbar\omega) - 1} \approx \begin{cases} \exp\left(-\dfrac{\hbar\omega}{k_B T} \right) & \text{for } k_B T \ll \hbar\omega \\[2ex] \dfrac{k_B T}{\hbar\omega} & \text{for } k_B T \gg \hbar\omega \end{cases} \tag{27.31}
$$

For low temperatures, the excitation probability of an oscillation is exponentially small; this means that also c_{vib} is exponentially small. For large temperatures $\overline{n}$ adjusts itself such that the mean energy of the oscillator is $k_B T$. Then $c_{\text{vib}} = k_B$ holds; this limit case also follows from the equipartition theorem.

Rotations

The partition function for the rotations is reduced to the partition functions of the individual molecules:

$$
Z_{\text{rot}}(T, N) = \left[z_{\text{rot}}(T) \right]^N \tag{27.32}
$$

According to (27.8), the rotational state of a molecule is defined by the quantum numbers l and m; the respective energy is the last term in (27.7). This gives us

$$
z_{\text{rot}} = \sum_{l=0}^{\infty} \sum_{m=-l}^{l} \exp\left(-\frac{\hbar^2 l(l+1)}{2\Theta k_B T} \right) = \sum_{l=0}^{\infty} (2l+1) \exp\left(-\frac{T_{\text{rot}}\, l(l+1)}{2T} \right)
$$

$$
\tag{27.33}
$$

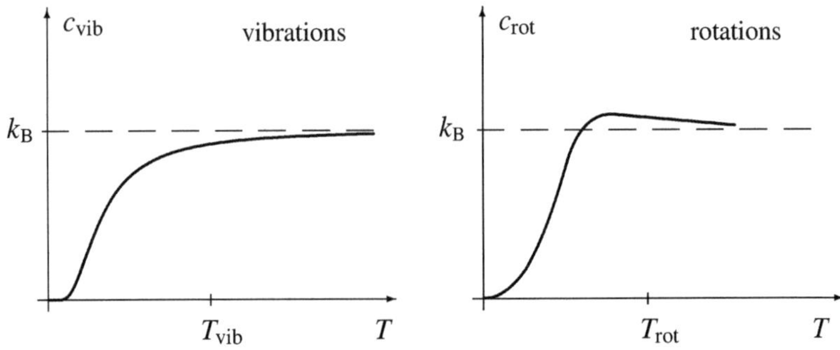

Figure 27.2 Temperature dependence of the vibrational and rotational component of the specific heat of a diatomic gas. In general, T_{vib} is significantly larger than T_{rot}. In the classical limit $c_{\text{vib}} = k_{\text{B}}$ and $c_{\text{rot}} = k_{\text{B}}$ hold.

The sum over m results in the factor $2l + 1$. The temperature $T_{\text{rot}} = \hbar^2/(\Theta k_{\text{B}})$ of (27.14) was used. We start by considering low temperatures, $T \ll T_{\text{rot}}$. Then only the first terms contribute:

$$z_{\text{rot}} = 1 + 3 \exp\left(-\frac{T_{\text{rot}}}{T}\right) + 5 \exp\left(-\frac{3\,T_{\text{rot}}}{T}\right) + \dots \qquad (T \ll T_{\text{rot}}) \quad (27.34)$$

From this follows for the energy $E = -\partial \ln Z_{\text{rot}}/\partial \beta$:

$$\begin{aligned}
E_{\text{rot}}(T, N) &= -N \frac{\partial}{\partial \beta} \ln\left(1 + 3\,\exp(-T_{\text{rot}}/T) + \dots\right) \\[2mm]
&= 3N k_{\text{B}} T_{\text{rot}} \exp\left(-\frac{T_{\text{rot}}}{T}\right) + \dots \qquad\qquad (27.35)
\end{aligned}$$

and the heat capacity $C_{\text{rot}} = \partial E_{\text{rot}}/\partial T$:

$$\boxed{\;C_{\text{rot}} \approx 3N k_{\text{B}} \frac{T_{\text{rot}}^2}{T^2} \exp\left(-\frac{T_{\text{rot}}}{T}\right) \qquad (T \ll T_{\text{rot}})\;} \quad (27.36)$$

For high temperatures, we replace the sum (27.33) by an integral and calculate corrections to it with Euler's summation formula:

$$\sum_{l=l_0}^{l_1} f(l) = \int_{l_0}^{l_1} dl\, f(l) + \frac{f(l_0) + f(l_1)}{2} - \frac{f'(l_0) - f'(l_1)}{12} + \frac{f'''(l_0) - f'''(l_1)}{720} \pm \dots$$

$$(27.37)$$

The derivation of Euler's summation formula may be found in standard maths textbooks. With

$$f(l) = (2l + 1) \exp\left(-\frac{T_{\text{rot}}\, l(l+1)}{2\,T}\right), \qquad l_0 = 0, \quad l_1 = \infty \qquad (27.38)$$

(27.37) becomes z_{rot}. We calculate the individual terms in (27.37). In the integral of the first term, we substitute $x^2 = T_{rot}\, l\,(l+1)/2T$ and $2x\,dx = T_{rot}\,(2l+1)\,dl/2T$:

$$\int_{l_0}^{l_1} dl\; f(l) = \int_0^\infty dl\;(2l+1)\,\exp\left(-\frac{l(l+1)\,T_{rot}}{2T}\right)$$

$$= \frac{4T}{T_{rot}}\int_0^\infty dx\; x\;\exp\left(-x^2\right) = \frac{2T}{T_{rot}} \tag{27.39}$$

The next terms yield

$$\frac{f(l_0) + f(l_1)}{2} = \frac{f(0)}{2} = \frac{1}{2} \tag{27.40}$$

$$-\frac{f'(l_0) - f'(l_1)}{12} = -\frac{1}{12}\frac{df(l)}{dl}\bigg|_{l=0} = -\frac{1}{6} + \frac{T_{rot}}{24\,T} \tag{27.41}$$

$$\frac{f'''(l_0) - f'''(l_1)}{720} = \frac{1}{720}\frac{d^3 f(l)}{dl^3}\bigg|_{l=0} = -\frac{T_{rot}}{120\,T} + \mathcal{O}\left(\frac{T_{rot}^2}{T^2}\right) \tag{27.42}$$

From this we get

$$z_{rot} = \frac{2T}{T_{rot}} + \frac{1}{3} + \frac{T_{rot}}{30\,T} + \mathcal{O}\left(\frac{T_{rot}^2}{T^2}\right) \qquad (T \gg T_{rot}) \tag{27.43}$$

and

$$\ln Z_{rot} = N \ln z_{rot} \approx N\,\ln\left(\frac{2T}{T_{rot}}\left(1 + \frac{T_{rot}}{6T} + \frac{T_{rot}^2}{60\,T^2}\right)\right)$$

$$\approx N \ln\left(\frac{2T}{T_{rot}}\right) + N\left(\frac{T_{rot}}{6T} + \frac{T_{rot}^2}{60\,T^2} - \frac{1}{2}\left(\frac{T_{rot}}{6T}\right)^2\right) \tag{27.44}$$

The logarithm of the second factor was expanded up to the order T_{rot}^2/T^2. From this we get the energy $E_{rot} = -\partial \ln Z_{rot}/\partial \beta$,

$$E_{rot}(T,N) = k_B T^2\,\frac{\partial \ln Z_{rot}}{\partial T} \approx N k_B T\left(1 - \frac{T_{rot}}{6T} - \frac{T_{rot}^2}{180\,T^2}\right) \tag{27.45}$$

and the heat capacity $C_{rot} = \partial E_{rot}/\partial T$,

$$\boxed{\;C_{rot} \approx N k_B\left(1 + \frac{T_{rot}^2}{180\,T^2}\right) \qquad (T \gg T_{rot})\;} \tag{27.46}$$

Since the partition function of the rotations does not depend on the volume the term C_{rot} contributes equally to C_V and C_P; an index V or P is superfluous.

   Figure 27.2 shows the specific heat $c_{rot}(T) = C_{rot}/N$ as a function of the temperature. For $T \gg T_{rot}$ we obtain $c_{rot} \approx k_B$; this limit also follows from the equipartition theorem. When approaching T_{rot} from above, c_{rot} initially increases slightly

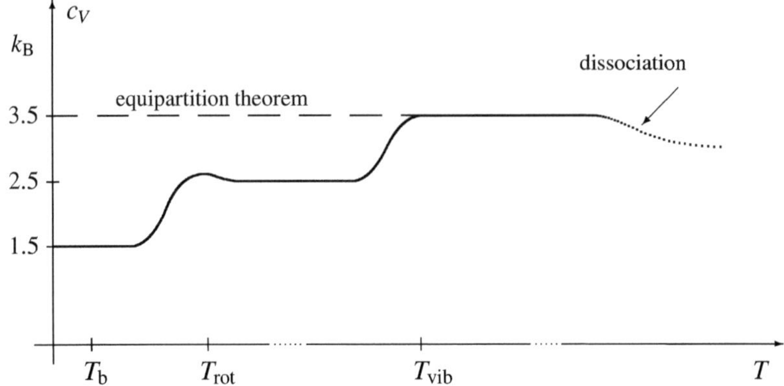

Figure 27.3 Temperature dependence of the specific heat c_V of a diatomic ideal gas. Real gases condense (at a given pressure) below the boiling temperature T_b to a liquid; this limits the applicability of the present model for low temperatures. At very high temperatures, the molecules dissociate to form atoms; the system could then be treated again as an ideal gas. The sketch is schematic only: For hydrogen gas (H_2 molecules), the relevant temperatures are $T_s \approx 20\,K$, $T_{rot} \approx 171\,K$ and $T_{vib} \approx 6244\,K$ (values from *Statistische Thermodynamik* by Findenegg and Hellweg, Springer Spektrum 2015).

above k_B. Then it decreases for $T < T_{rot}$ and finally disappears exponentially for $T \to 0$.

For the total specific heat c_V of a diatomic ideal gas, the result is sketched in Figure 27.3. The specific heat at constant pressure is $c_P = c_V + k_B$. For sufficiently high temperatures, the molecules dissociate into two atoms, each of which has the (translational) contribution $1.5\,k_B$ to c_V.

Ortho- and parahydrogen

The molecules of the diatomic gas consist of electrons, protons and neutrons. All of these particles are fermions; the wave function must be antisymmetric for each type of fermion. This symmetry requirement limits the possible values of the rotational angular momentum.

We deal with these effects using the example of hydrogen gas, which consists of H_2 molecules. This is the simplest diatomic molecule. For hydrogen gas, the temperature $T_{rot} \approx 171\,K$ of is significantly higher than the condensation temperature $T_s \approx 20\,K$ (for normal pressure).

In the ground state of the H_2 molecule, both electrons have the same spatial wave function. The distribution of this wave function across both atoms leads to a reduction in energy and thus to a bonding. The spins of the two electrons are coupled to $S_{el} = 0$. Thus their overall wave function is symmetric in position and antisymmetric in spin, i.e. in total antisymmetric. For the considered rotations, the structure of the electrons remains unchanged.

The wave function of the two protons (1 and 2) is of the form (see also Part VIII in [3]):

$$\Psi(1, 2) = \psi_0(R)\, Y_{lm}(\theta, \phi)\, |SS_z\rangle \tag{27.47}$$

We use the relative coordinate $\boldsymbol{R} = \boldsymbol{R}_1 - \boldsymbol{R}_2 := (R, \theta, \phi)$. The radial relative motion can be described by the oscillator ground state wave function $\psi_0(R) \propto \exp(-\gamma\,(R - R_0)^2)$ (lowest horizontal bar in Figure 27.1). We do not allow for vibrations; for $T \sim T_{\mathrm{rot}}$ they are hardly excited thermally. The spherical harmonics Y_{lm} describe the rotations. The two proton spins couple to the state $|SS_z\rangle$.

The exchange operator for the two protons is denoted by P_{12}. An exchange of the protons implies

$$\boldsymbol{R} \xrightarrow{P_{12}} -\boldsymbol{R}, \qquad \theta \xrightarrow{P_{12}} \pi - \theta, \qquad \phi \xrightarrow{P_{12}} \pi + \phi, \qquad |SS_z\rangle \xrightarrow{P_{12}} (-)^{S+1}|SS_z\rangle \tag{27.48}$$

Because of $Y_{lm}(\pi - \theta, \phi + \pi) = (-)^l\, Y_{lm}(\theta, \phi)$ we get in total

$$P_{12}\,\Psi(1, 2) = P_{12}\,\psi_0\, Y_{lm}\, |SS_z\rangle = (-)^{l+S+1}\,\Psi(1, 2) \overset{!}{=} -\Psi(1, 2) \tag{27.49}$$

Since the protons are fermions, $\Psi(1, 2)$ must be antisymmetric. So $l + S$ must be even. Therefore, only the following combinations of the quantum numbers l and S are allowed:

$$
\begin{aligned}
S &= 0: \quad & l &= 0, 2, 4, \ldots \quad & \text{(parahydrogen)} \\
S &= 1: \quad & l &= 1, 3, 5, \ldots \quad & \text{(orthohydrogen)}
\end{aligned}
\tag{27.50}
$$

This relationship has an impact on the contribution of the rotations to the specific heat:

1. For each odd l value, there are three spin states, $S_z = 0, \pm1$; for each even l-value there is only one spin state. Therefore, the odd l-values have a threefold weight in the partition function.

2. At temperatures $T \gg T_{\mathrm{rot}}$ the same number of even and odd l-values are accessible. Due to the triple degeneracy of the odd angular momentum states, the ratio of ortho- to parahydrogen then 3 to 1. For $T \ll T_{\mathrm{rot}}$, the ground state ($l = 0$) is predominantly occupied. At $T = 0$ the equilibrium consists of pure parahydrogen.

3. Special surprising effects result from the fact that the adjustment of the equilibrium ratio (ortho- to parahydrogen) takes a very long time; the relaxation times are of the order of a year. Therefore, the measured specific heat depends on the temperature at which the sample was previously stored. The system has a kind of long term memory.

The influence of the spin setting is based on the symmetry requirement for the wave function of the molecule. On the other hand, the spins make no significant direct contribution to the energy; the very small interaction between the magnetic dipoles of the protons is of the size $\delta E \approx 10^{-6}\, k_{\mathrm{B}}\, \mathrm{K}$.

We specify the discussed effects quantitatively. For this we write on the partition functions of the rotations of parahydrogen

$$z_{\mathrm{para}} = \sum_{l=0,2,4,\dots} (2l+1)\exp\left(-\frac{l(l+1)\,T_{\mathrm{rot}}}{2T}\right) = 1 + 5\exp\left(-\frac{3\,T_{\mathrm{rot}}}{T}\right) + \dots$$

$$(27.51)$$

and of orthohydrogen:

$$z_{\mathrm{ortho}} = \sum_{l=1,3,5,\dots} (2l+1)\exp\left(-\frac{l(l+1)\,T_{\mathrm{rot}}}{2T}\right) = 3\exp\left(-\frac{T_{\mathrm{rot}}}{T}\right) + \dots \quad (27.52)$$

After a sufficiently long time, equilibrium is reached and the partition function for the rotations becomes

$$z_{\mathrm{rot}} = \sum_{l=0}^{\infty} \sum_{m=-l}^{l} \sum_{S_z=-S}^{S} \exp\left(-\frac{l(l+1)\,T_{\mathrm{rot}}}{2T}\right) = 3\,z_{\mathrm{ortho}} + z_{\mathrm{para}} \quad (27.53)$$

The partition function runs over all quantum numbers of the microstate, i.e. also over S_z.

We determine the ratio η of ortho- to parastates in the equilibrium. To do this, we form the sum of the probabilities for orthostates and divide them by the sum of the probabilities for parastates, i.e.

$$\eta(T) = \frac{3\,z_{\mathrm{ortho}}(T)}{z_{\mathrm{para}}(T)} \approx \begin{cases} 3 & \text{for } T \gg T_{\mathrm{rot}} \\ 9\exp(-T_{\mathrm{rot}}/T) & \text{for } T \ll T_{\mathrm{rot}} \end{cases} \quad (27.54)$$

For high temperatures we have used $z_{\mathrm{ortho}} \approx z_{\mathrm{para}}$, and for low temperatures the expansions (27.51) and (27.52).

For a sample that has been stored at the temperature T_0 for a sufficiently long time, the specific heat is

$$c_{\mathrm{rot}}(T;T_0) = \frac{\eta(T_0)}{1+\eta(T_0)}\,c_{\mathrm{ortho}}(T) + \frac{1}{1+\eta(T_0)}\,c_{\mathrm{para}}(T) \quad (27.55)$$

Exercises

27.1 *Oscillators for high and low temperatures*

For N independent oscillators the heat capacity reads:

$$C_{\text{vib}}(T, N) = N k_{\text{B}} \frac{T_{\text{vib}}^2}{T^2} \frac{\exp(T_{\text{vib}}/T)}{\left[\exp(T_{\text{vib}}/T) - 1\right]^2}$$

Determine the leading temperature dependent terms for low ($T \ll T_{\text{vib}}$) and high ($T \gg T_{\text{vib}}$) temperatures.

27.2 *Anharmonic corrections for vibration*

The energies of the vibrational states of a diatomic molecule are

$$\varepsilon_n = \hbar \omega \left[\left(n + \frac{1}{2}\right) - \delta \left(n + \frac{1}{2}\right)^2\right]$$

The following calculations shall be carried out up to the first order in the small quantity δ. Determine the partition function z_{vib} for a single molecule. From this, calculate the vibrational energy E_{vib} for N independent molecules. Determine the leading contributions to the heat capacity for low and high temperatures.

27.3 *Equilibrium ratio H_2, D_2 and HD*

Three different hydrogen like gases are considered, namely the gases consisting of the molecules H_2, D_2 or HD. Taking into account the exchange symmetry, determine the partition function for rotations. What ratio $\eta(T_0)$ is obtained for even and odd l values if the samples are stored at a high temperature $T_0 \gg T_{\text{rot}}$ for a sufficiently long time? Calculate the specific heat c_{rot} of these samples for low temperatures ($T \ll T_{\text{rot}}$).

Notes: D denotes deuterium, i.e. a hydrogen atom with a deuteron (proton plus neutron) as a nucleus. The deuteron has the spin 1; in the D_2 molecule, these spins can couple to $S = 0$, 1 or 2.

27.4 *Law of mass action*

The chemical reaction equilibrium

$$2\text{O} \rightleftharpoons \text{O}_2$$

between monatomic and diatomic oxygen in a gas mixture is considered. The pressure P and the temperature T are given. The atom-atom potential is expanded around the minimum at R_0, yielding $V(R) \approx -\varepsilon + \mu\omega^2 (R - R_0)^2/2$. The temperature shall be so high that the rotations and vibrations of the O_2 molecules (moment of inertia $\Theta = \mu R_0^2$) are excited:

$$\frac{\hbar^2}{\Theta} \ll \hbar\omega \ll k_\mathrm{B} T \ll \varepsilon \tag{27.56}$$

On the other hand, the temperature so small ($T \ll \varepsilon$) that the approximate expression for $V(R)$ is usable. Treat the system as an ideal gas mixture. Determine the free energies of the two gases in the mixture. Use the high temperature approximation for the vibrations and rotations (in the O_2 molecule there are only rotational states with even l). From this, calculate the chemical potentials as a function of the temperature T, the pressure P and the respective concentration $c_i = N_i/N$. From the equilibrium condition for the chemical potentials, derive the *law of mass action*

$$\frac{c_1^2}{c_2} = \frac{K(T)}{P} \tag{27.57}$$

Discuss the temperature dependence of the function $K(T)$.

28 Dilute classical gas

We investigate the influence of the interaction between the atoms in a dilute gas on the caloric and thermal equation of state. This leads to a justification of the van der Waals equation.

We consider a monatomic gas with the Hamilton function

$$H = \sum_{\nu=1}^{N} \frac{p_\nu^2}{2m} + \sum_{\nu=2}^{N} \sum_{\nu'=1}^{\nu-1} w(|r_\nu - r_{\nu'}|) \tag{28.1}$$

The potential w shall depend only on the distance $|r_\nu - r_{\nu'}|$ of the respective atoms. The potential could have the form shown in Figure 38.1. The attractive part of w should be so weak that it does not come to molecular bonding. This applies to noble gases, or if the strength of the attractive part is small compared to $k_B T$.

For (28.1) we evaluate the classical partition function for a dilute gas. As a result we obtain a connection between the interaction in (28.1) and the corrections to the ideal gas law. The classical treatment of the translational motion is no significant restriction for an ordinary gas. We refer to a gas as *dilute* if the volume per atom is large compared to the volume of the atom itself.

We first consider the partitions function $Z(1) = Z(T, V, 1)$ and $Z(2) = Z(T, V, 2)$ for one and two particles in the volume V. For one particle, the result is known from (25.7),

$$Z(1) = \frac{1}{(2\pi\hbar)^3} \int_V d^3r \int d^3p \, \exp\left(-\frac{p^2}{2m k_B T}\right) = \frac{V}{\lambda^3} \tag{28.2}$$

Here $\lambda = 2\pi\hbar/\sqrt{2\pi m k_B T}$. For two particles, we take into account the interaction $w(r_{12}) = w(|r_1 - r_2|)$,

$$\begin{aligned} Z(2) &= \frac{1/2!}{(2\pi\hbar)^6} \int_V d^3r_1 \int_V d^3r_2 \int d^3p_1 \int d^3p_2 \, \exp\left(-\frac{\frac{p_1^2 + p_2^2}{2m} + w(r_{12})}{k_B T}\right) \\ &= \frac{1}{2!} \frac{1}{\lambda^6} \int_V d^3r_1 \int_V d^3r_2 \, \exp\left[-\beta w(r_{12})\right] \end{aligned} \tag{28.3}$$

For the treatment of the dilute gas, we start from the following consideration: In the ideal gas, the partition function of N particles can be reduced to that of one particle; namely $Z_{\mathrm{id}}(N) = Z(1)^N/N!$ according to (25.7). If we now only consider the interaction between just two atoms, it should be possible to reduce $Z(N)$ to $Z(2)$ and $Z(1)$. In the next order, three-particle effects had to be taken into account by $Z(3)$; they arise, for example, through the influence of a third atom on the interaction of the first two. In sufficiently dilute gas these effects will be small because the probability that three atoms are close together is low.

We develop a formalism for the quantitative application of this idea. For this purpose, we start from the relation (22.30) between the grand canonical and the canonical partition function:

$$Y(T, V, \mu) = \sum_{N=0}^{\infty} Z(T, V, N)\, \exp(\beta\mu N) = \sum_{N=0}^{\infty} Z(N)\, \exp(\beta\mu N) \qquad (28.4)$$

We use the abbreviation $Z(N) = Z(T, V, N)$. The first terms of this sum are

$$Y(T, V, \mu) = 1 + Z(1)\, \exp(\beta\mu) + Z(2)\, \exp(2\beta\mu) + \dots \qquad (28.5)$$

For the ideal gas (i.e. for $w = 0$) we use (25.18),

$$\exp(\beta\mu) = \frac{\lambda^3}{v} \ll 1 \qquad (w = 0) \qquad (28.6)$$

The last inequality applies specifically to a dilute gas with $v \gg \lambda^3$. One could now think that (28.5) is a usable expansion into powers of $\exp(\beta\mu) \ll 1$. However, the coefficients increase exponentially; for the ideal gas $Z_{\mathrm{id}}(N) = Z(1)^N/N!$ holds, (25.7). This is, however, quite different for the analogous expansion of $\ln Y$:

$$\ln Y = Z(1)\, \exp(\beta\mu) + \left[Z(2) - \frac{Z(1)^2}{2} \right] \exp(2\beta\mu) + \dots \qquad (28.7)$$

For this we write (28.5) as $Y = 1 + x$ and use the expansion $\ln Y \approx x - x^2/2$. Because $Z_{\mathrm{id}}(N) = Z(1)^N/N!$, the bracket expression vanishes for $w = 0$; this also applies to the following terms. The higher coefficients therefore disappear for $w \to 0$; that implies that they are small for weak interactions. Therefore (28.7) is a suitable starting point for the quantitative treatment of the dilute gas.

The first term on the right-hand side in (28.7) describes the ideal gas. The second takes term into account the interaction between just two atoms. The third term with $Z(3)$ takes into account the interaction between three atoms, as far as they go beyond the interaction between just two atoms.

From (23.32) and (23.28) follow

$$P = P(T, V, \mu) = \frac{k_{\mathrm{B}}T}{V}\, \ln Y(T, V, \mu) \qquad (28.8)$$

$$N = N(T, V, \mu) = \frac{1}{\beta}\, \frac{\partial \ln Y(T, V, \mu)}{\partial \mu} \qquad (28.9)$$

By eliminating μ from these two equations we obtain the thermal equation of state $P = P(T, V, N)$.

We introduce the abbreviations

$$Z_1 = Z(1) = \frac{V}{\lambda^3} \quad \text{and} \quad Z_2 = Z(2) - \frac{Z(1)^2}{2} \tag{28.10}$$

Then (28.8) and (28.9) with (28.7) become

$$\frac{PV}{k_{\mathrm{B}}T} = \ln Y = Z_1 \exp(\beta\mu) + Z_2 \exp(2\beta\mu) + \ldots \tag{28.11}$$

$$N = \frac{1}{\beta} \frac{\partial \ln Y}{\partial \mu} = Z_1 \exp(\beta\mu) + 2Z_2 \exp(2\beta\mu) + \ldots \tag{28.12}$$

From these two equations we get

$$\ln Y = N - Z_2 \exp(2\beta\mu) + \ldots \tag{28.13}$$

$$\exp(\beta\mu) = \frac{N}{Z_1} - \frac{2Z_2}{Z_1} \exp(2\beta\mu) + \ldots \tag{28.14}$$

If only the first terms on the right-hand sides are taken into account, we get the ideal gas result (28.7). The second terms are the leading corrections; in the following we neglect all higher terms. The lowest order of (28.14)

$$\exp(\beta\mu) \approx \frac{N}{Z_1} = \frac{N}{V/\lambda^3} = n\lambda^3 \quad \text{(lowest order)} \tag{28.15}$$

is now used the small correction term in (28.13):

$$\frac{PV}{k_{\mathrm{B}}T} = \ln Y \approx N - Z_2 \left(\frac{N}{Z_1}\right)^2 = N\left(1 - \frac{VZ_2}{Z_1^2} n\right) = N\left(1 + n\,B(T)\right) \tag{28.16}$$

Because of (28.15), (28.7) is effectively an expansion into powers of the particle density $n = N/V$. If further powers of $\exp(\beta\mu)$ are considered in (28.7) and in the following equations, one gets an expansion of the form

$$\frac{PV}{Nk_{\mathrm{B}}T} = 1 + n\,B(T) + n^2\,B_2(T) + \ldots \tag{28.17}$$

This expansion is called *virial expansion*. Equation (28.16) determines the first *virial coefficient $B(T)$*:

$$B(T) = -\frac{VZ_2}{Z_1^2} = -\frac{V}{Z(1)^2}\left(Z(2) - \frac{Z(1)^2}{2}\right) \tag{28.18}$$

$$= -\frac{1}{2V}\left(\int_V d^3r_1 \int_V d^3r_2 \, \exp\left[-\beta\,w(r_{12})\right] - \int_V d^3r_1 \int_V d^3r_2 \, 1\right)$$

We have used (28.2) and (28.3), and expressed V^2 by spatial integrals. The transformation to relative and center of mass coordinates

$$\boldsymbol{R} = \frac{\boldsymbol{r}_1 + \boldsymbol{r}_2}{2}, \quad \boldsymbol{r} = \boldsymbol{r}_2 - \boldsymbol{r}_1, \quad \int_V d^3 r_1 \int_V d^3 r_2 \ldots = \int_V d^3 R \int d^3 r \ldots \quad (28.19)$$

results in

$$B(T) = -\frac{1}{2V} \int_V d^3 R \int d^3 r \left(\exp\left[-\beta\, w(r) \right] - 1 \right) \quad (28.20)$$

The integrand $(\exp[-\beta w(r)] - 1)$ is non zero only in a range of a few ångström around $r = 0$. Therefore, the $\boldsymbol{r}$-integration can be extended to the whole space.

Since w only depends on $r = |\boldsymbol{r}|$, we can use $d^3 r = 4\pi r^2 dr$. The $\boldsymbol{R}$ integration results in the factor V. This gives us

$$\boxed{B(T) = -2\pi \int_0^\infty dr\; r^2 \left(\exp\left[-\beta\, w(r) \right] - 1 \right)} \quad (28.21)$$

In this central result, there is the *macroscopically measurable* viral coefficient $B(T)$ on the left. And on the right there is the *microscopic* interaction $w(r)$ between two atoms. This means that from the measurement of $B(T)$ conclusions can be drawn about $w(r)$.

An alternative method for determining $w(r)$ would be a scattering experiment. For crossed gas jets, for example, the cross section for elastic scattering may be measured as a function of the angle. In the Born approximation (Chapter 45 in [3]) the cross section yields the Fourier transform of $w(r)$. However, the scattering experiment is much more elaborate than the measurement of $B(T)$, which is carried out via (28.16).

Van der Waals equation

The determination of $w(r)$ from the measured $B(T)$ is done practically as follows: An approach is made for the potential, which depends on some parameters. This is used to calculate $B(T)$. Then one adjusts the parameters so that the calculated $B(T)$ comes as close as possible to the measured $B(T)$. We do this in the following for a potential that is characterized by two parameters.

The size d of the atoms amount to a few ångströms. For distances $r < d$, the potential $w(r)$ becomes repulsive. If one tries to push two atoms closer together the electrons have to move to higher states due to the Pauli principle. This requires an energy of a few electron volts per electron; because of eV $\approx 10^4\, k_\text{B} \text{K}$, this repulsion is practically infinitely strong (compared to $k_\text{B} T$). In the range $r > d$, the potential $w(r)$ becomes negative; for neutral atoms, induced electric dipoles create an attraction. The attractive part has a range of a few ångströms. This makes the

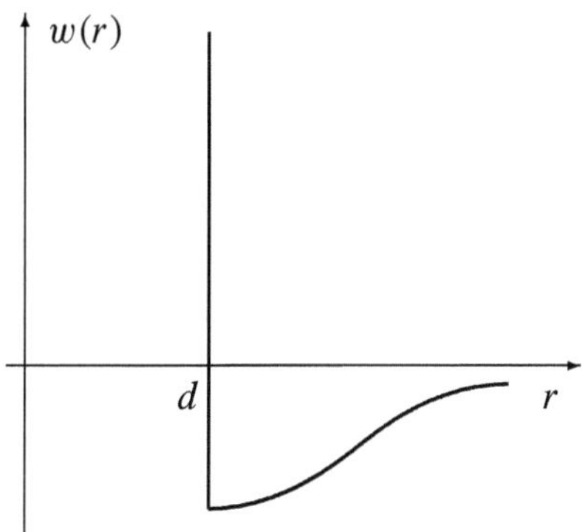

Figure 28.1 Sketch of the atom-atom potential used for evaluating (28.21). The potential has a hard core, i.e. $w = \infty$ for $r < d$. The strength of the attractive part $(w < 0)$ shall be small compared to $k_B T$.

following approach for $w(r)$ plausible:

$$
w(r) \begin{cases} = \infty & r < d \\ < 0 & \text{for} \quad r \gtrsim d \\ \approx 0 & r \gg d \end{cases}
\tag{28.22}
$$

Such a potential is sketched in Figure 28.1 Possible other potential approaches are (28.39) and (28.40).

Phenomenologically, the attractiveness of the potential is noticeable by the fact that all gases become liquids at sufficiently low temperatures. This phase transition occurs when $k_B T$ is comparable to the strength of the attractive potential. For the dilute gas, we require therefore

$$
\frac{|w(r)|}{k_B T} \ll 1 \quad \text{for } r > d
\tag{28.23}
$$

Then we may use

$$
\exp\left(-\beta\, w(r)\right) - 1 = \begin{cases} -1 & (r < d) \\ -\beta\, w(r) + \mathcal{O}(\beta^2 w^2) & (r > d) \end{cases}
\tag{28.24}
$$

From this we get

$$
B(T) \approx 2\pi \int_0^d dr\, r^2 + 2\pi\beta \int_d^\infty dr\, r^2\, w(r) = b - \frac{a}{k_B T}
\tag{28.25}
$$

with the positive constants

$$
b = \frac{2\pi d^3}{3}, \qquad a = -2\pi \int_d^\infty dr\, r^2\, w(r)
\tag{28.26}
$$

With this $B(T)$, Equation (28.16) becomes

$$
P = \frac{k_B T}{v}\left(1 + n\, B(T)\right) = \frac{k_B T}{v}\left(1 + \frac{b}{v}\right) - \frac{a}{v^2}
\tag{28.27}
$$

where $v = V/N$. We insert

$$1 + \frac{b}{v} \approx \frac{1}{1 - b/v} + \mathcal{O}\left(\frac{b^2}{v^2}\right) \tag{28.28}$$

into (28.27) and, as in (28.16), omit the terms of order $b^2/v^2 = n^2 b^2$. This yields the *van der Waals equation*:

$$\boxed{\; P + \frac{a}{v^2} = \frac{k_B T}{v - b} \qquad \text{van der Waals equation} \;} \tag{28.29}$$

The corrections to the ideal gas law be understood as follows: The finite size of the atoms reduces the available volume $v = V/N$ per particle by b. The attractive part of the interaction tends to hold the particles together and reduces the pressure (measurable at the vessel walls) by $-a/v^2$.

Equations (28.27) and (28.29) are equivalent for $v \gg b$. If one wants to use the equation of state outside the range of our derivative, then (28.29) is the appropriate form, because the van der Waals equation results in the sensible behavior $P \to \infty$ for $v \to b$. This corresponds to the behavior $P \to \infty$ for $v \to 0$ of the ideal gas.

Validity range

A necessary condition for the termination of the expansion (28.17) after the first correction term is

$$n\, B(T) \ll 1 \tag{28.30}$$

With (28.25) this becomes

$$\frac{b}{v} \ll 1 \quad \text{and} \quad \frac{a}{v\, k_B T} = \frac{\mathcal{O}\big(\langle |w_a| \rangle\, b\big)}{v\, k_B T} \ll 1 \tag{28.31}$$

The integral a in (28.26) can be calculated by the product of a mean value $\langle |w_a| \rangle$ of the attractive potential and the respective volume; the volume of the attractive region is of the same size as b. The second condition in (28.31) then follows from the first and from (28.23). This leaves the following two conditions

$$v \gg b \qquad \text{and} \qquad k_B T \gg \langle |w_a(r)| \rangle \tag{28.32}$$

The first condition means that the volume $v = V/N$ available per atom is large compared to the intrinsic volume $(4\pi/3)\,(d/2)^3 = b/4$ of an atom. If the second condition would be violated, the gas atoms would tend to clump together to form a liquid. Experimentally (28.32) can be fulfilled by low pressure and high temperature.

Since we used the expansion (28.7) into powers of $\exp(\beta\mu)$, one might think that it is sufficient if $\exp(\beta\mu) \ll 1$ holds. Because of (28.15) this condition is equivalent to $v \gg \lambda^3$. The requirement $v \gg \lambda^3$ is, however, much weaker than

$v \gg b$; it is practically always fulfilled, even if the system approaches the liquid phase. The condition $\exp(\beta\mu) \ll 1$ alone is therefore insufficient.

The van der Waals equation allows the description of real gases. In fact, it can be used beyond the specified range of validity. It is then understood as a phenomenological equation with empirical parameters a and b. Equation (28.26) can be regarded as an estimate for these parameters. As a phenomenological equation, the van der Waals equation allows a description of the of the phase transition gaseous–liquid (Chapter 37).

Energy

We determine the caloric equation of state of the dilute classical gas:

$$E = -\frac{\partial \ln Y}{\partial \beta} + \mu N = -\frac{\partial}{\partial \beta}\left(Z_1 \exp(\beta\mu) + Z_2 \exp(2\beta\mu) \right) + \mu N$$

$$= -\mu\left(Z_1 \exp(\beta\mu) + 2 Z_2 \exp(2\beta\mu) \right) - \frac{\partial Z_1}{\partial \beta} \exp(\beta\mu) \qquad (28.33)$$

$$-\frac{\partial Z_2}{\partial \beta} \exp(2\beta\mu) + \mu N = -\frac{\partial Z_1}{\partial \beta} \exp(\beta\mu) - \frac{\partial Z_2}{\partial \beta} \exp(2\beta\mu)$$

Here (28.11) and (28.12) were used. From

$$Z_1 = \frac{V}{\lambda^3} \propto \beta^{-3/2}, \qquad Z_2 \overset{(28.18)}{=} -\frac{Z_1^2}{V} B(T) = -\frac{V}{\lambda^6} B(T) \qquad (28.34)$$

follow the partial derivatives follow

$$\frac{\partial Z_1}{\partial \beta} = -\frac{3}{2\beta} Z_1, \qquad \frac{\partial Z_2}{\partial \beta} = -\frac{3}{\beta} Z_2 - \frac{V}{\lambda^6}\frac{\partial B(T)}{\partial \beta} \qquad (28.35)$$

We insert this into (28.33):

$$E = \frac{3}{2\beta} \underbrace{\left(Z_1 \exp(\beta\mu) + 2 Z_2 \exp(2\beta\mu) \right)}_{=\,N} + \underbrace{\exp(2\beta\mu)}_{\approx\, \lambda^6/v^2} \frac{V}{\lambda^6}\underbrace{\frac{\partial B(T)}{\partial \beta}}_{\approx\, -a} \qquad (28.36)$$

In the first term we used (28.12), in the second one (28.15) and (28.25). This gives us

$$E = E(T, V, N) = N\left(\frac{3}{2} k_{\mathrm{B}} T - \frac{a}{v} \right) \qquad (28.37)$$

According to (28.31), $a/v \sim \langle |w_{\mathrm{a}}(r)| \rangle\, (b/v)$. This is the strength of the attractive interaction $\langle |w_{\mathrm{a}}| \rangle$, multiplied by the probability $\mathcal{O}(b/v)$ that two particles are in the interaction area. The repulsive part of the potential $w(r)$ does not contribute to the energy, because the probability of being in this range is zero.

Exercises

28.1   *Van der Waals equation based on number of moles*

Consider the van der Waals equation

$$P + \frac{a}{v^2} = \frac{k_B T}{v - b} \qquad \text{with} \quad v = V/N \qquad (28.38)$$

Derive the analogous equation in which the volume $v' = V/\nu$ is related to the number of moles (instead of the number of particles).

28.2   *Virial coefficients from potential*

Given is the atom-atom potential

$$w(r) = \begin{cases} \infty & r \leq \sigma \\ -\varepsilon\left[1 - r^3/(8\sigma^3)\right] & \sigma < r < 2\sigma \\ 0 & r \geq 2\sigma \end{cases} \qquad (28.39)$$

Sketch the potential and calculate the virial coefficient $B(T)$ up to the first order in $1/(k_B T)$. Determine the parameters a and b of the van der Waals equation (28.38).

28.3   *Virial coefficients for Lennard-Jones potential*

A common and realistic approach for the atom-atom potential is the Lennard-Jones potential:

$$w(r) = 4\varepsilon\left(\frac{\sigma^{12}}{r^{12}} - \frac{\sigma^6}{r^6}\right) = \varepsilon\left(\frac{r_0^{12}}{r^{12}} - 2\frac{r_0^6}{r^6}\right) \qquad (28.40)$$

The two given forms are equivalent. The $1/r^6$ power of the attractive part corresponds to an induced dipole-dipole interaction. The $1/r^{12}$ potency is a phenomenological approach for the strong repulsion at smaller distances. Realistic parameters for ^{4}He atoms are $\varepsilon = 10.2\,k_B\text{K}$ and $r_0 = 2.87\,\text{Å}$.

Sketch the potential. Where are the zero and the minimum of the potential? Calculate the virial coefficient $B(T)$ using $|\beta w| \ll 1$ in the attractive region and $\exp(-\beta w) \approx 0$ in the range $r \leq \sigma$. Specify the parameters a and b for the van der Waals equation (28.38).

29 Ideal quantum gas

We present the quantum mechanical evaluation of the partition function for an ideal gas. The gas might consist of particles (such as atoms or electrons) or of quasi-particles (such as phonons or photons).

Basics

We first discuss the exchange symmetry of a wave function for N particles (for example atoms). From this symmetry follows how the states of an ideal gas are to be counted.

For an *ideal* gas consisting of N particles, the Hamilton operator H is a sum of single particle Hamilton operators h,

$$H = \sum_{\nu=1}^{N} h(\nu) \tag{29.1}$$

The argument ν stands for all coordinates (position, spin) and momentum operators of the ν-th particle. For non-relativistic particles (mass m) we assume a single particle Hamilton operator of the form

$$h(\nu) = -\frac{\hbar^2}{2m}\, \Delta_\nu + U(\boldsymbol{r}_\nu) \tag{29.2}$$

Here U is the wall potential (24.16), which limits the motion to the available volume V. In a cubic volume $V = L^3$, only the discrete momenta are allowed:

$$\boldsymbol{p} := (p_1,\, p_2,\, p_3)\,, \qquad p_i = \frac{\pi \hbar}{L}\, n_i = \Delta p\, n_i \qquad (n_i = 1, 2, ...) \tag{29.3}$$

For a complete definition of a single particle state, the spin must also be specified (except for spin less particles). The spin state of a particle with spin s is fixed by the spin component in a given direction; usually one takes the z-component s_z. A single particle state is then defined by the momentum quantum numbers $\boldsymbol{p}$ (or n_1, n_2, n_3) and the spin quantum number s_z. We abbreviate these quantum numbers by a,

$$a = (\boldsymbol{p}, s_z) \qquad (s_z = -s, -s+1, ..., s) \tag{29.4}$$

The single particle wave function is a product

$$\psi_a = \varphi_{\boldsymbol{p}}(\boldsymbol{r})\, \chi_{s_z} \tag{29.5}$$

of the spatial wave function φ_p and the spin function χ_{s_z}. In the cubic box, the spatial wave function is proportional to $\sin(p_1 x_1/\hbar)\,\sin(p_2 x_2/\hbar)\,\sin(p_3 x_3/\hbar)$. The spin part could be given in the form $|\theta_s, \phi_s\rangle$ or in the matrix representation (Chapter 37 in [3]). The wave function (29.5) is a solution of the single particle Hamilton operator:

$$h(v)\,\psi_a(v) = \varepsilon_a\,\psi_a(v) \tag{29.6}$$

In $\psi_a(v)$, the argument v stands for the position and spin coordinates of the v-th particle. For $h(v)$ from (29.2), the single particle energies depend only on the momentum $\varepsilon_a = p^2/2m$. In an external magnetic field B, on the other hand, the energy might be of the form $\varepsilon_a = p^2/2m - 2\mu_0 B\, s_z$, (26.2).

The essential quantum mechanical effect for an ideal quantum gas is the exchange symmetry of the many-body states. (The quantization of the momentum plays no role; because Δp becomes arbitrarily small for a macroscopic volume). We start with $N = 2$ particles and their eigenfunctions ψ_a and ψ_b. Any linear combination of the form

$$\Psi(1, 2) = \alpha\,\psi_a(1)\,\psi_b(2) + \beta\,\psi_a(2)\,\psi_b(1) \tag{29.7}$$

an eigenfunction of H:

$$H(1, 2)\,\Psi(1, 2) = \Big(h(1) + h(2)\Big)\,\Psi(1, 2) = (\varepsilon_a + \varepsilon_b)\,\Psi(1, 2) \tag{29.8}$$

The question now arises which of the linear combinations (29.7) should be taken. First we state that the Hamilton operator H is symmetric under the exchange of the two particles:

$$\big[H, P_{12}\big] = 0 \tag{29.9}$$

The permutation operator P_{12} is defined by

$$P_{12}\,\Psi(1, 2) = \Psi(2, 1) \tag{29.10}$$

for arbitrary wave functions $\Psi(1, 2)$. Because of (29.9) one can find simultaneous eigenfunctions of H and P_{12} (Chapter 17 in [3]). The eigenvalue equation of P_{12} is

$$P_{12}\,\Psi_\lambda(1, 2) = \lambda\,\Psi_\lambda(1, 2) \tag{29.11}$$

If we apply P_{12} to both sides once more and take into account $P_{12}^2 = 1$ (follows from the definition), then one finds $\lambda^2 = 1$ or $\lambda = \pm 1$. The respective eigenfunctions are called *antisymmetric* for $\lambda = -1$ and *symmetric* for $\lambda = 1$. We write on the eigenfunctions (29.7) of H such that they are simultaneous eigenfunctions of P_{12}:

$$\Psi_\pm(1, 2) = C\Big(\psi_a(1)\,\psi_b(2) \pm \psi_a(2)\,\psi_b(1)\Big) \tag{29.12}$$

Here C is a normalization constant. Numerous experiments show that, depending on the type of particle, only antisymmetric or symmetric wave functions occur in nature, namely:

- For particles with half-integer spin, the wave function is antisymmetric under the exchange of any two particles. These particles are called *fermions*.

- For particles with integer spin, the wave function is symmetric under the exchange of any two particles. These particles are called *bosons*.

Fermions include, in particular, electrons, neutrons and protons. Bound systems consisting of an odd number of fermions (e.g. ^{3}He atoms) are then also fermions, systems with an even number (e.g. ^{4}He atoms) are bosons. Bosons also include photons, which have spin 1.

The symmetry requirement applies to any number of particles. Therefore, only many-body wave functions $\Psi_{\pm}$ are permitted for which

$$P_{\nu\mu}\,\Psi_{\pm}(1,...,\nu,..,\mu,..,N) = \Psi_{\pm}(1,...,\mu,..,\nu,..,N) = \pm\Psi_{\pm}(1,...,\nu,...,\mu,..,N)$$

$$(29.13)$$

Here ν stands for the position and spin coordinate of the ν-th particle. Such a wave function $\Psi_{\pm}$ is also called *totally symmetric* or *totally antisymmetric*.

The antisymmetry of the wave function implies the *Pauli principle*. From (29.12) we obtain:

$$a = b \quad\Longrightarrow\quad \Psi_{-}(1, 2) \equiv 0 \qquad (29.14)$$

For the many-body wave function (29.13) this means: If the particles with the co-ordinates ν and μ are in the same single particle state, then $\Psi_{-}(..., \nu,..., \mu,...) = \Psi_{-}(..., \mu,..., \nu,...)$. On the other hand, the antisymmetry requires $\Psi_{-}(..., \nu,..,\mu,...) = -\Psi_{-}(..., \mu,..,\nu,...)$. This results in $\Psi_{-} \equiv 0$.

The Pauli principle is decisive for the structure of the atoms (periodic system) and for the atomic nucleus. The simplest model for these systems is the shell model. The shell model assumes that the particles move in a potential without mutual interaction. The Hamilton operator of the shell model is therefore also of the form (29.1). The shell model belongs to the class of ideal quantum gases (Chapter 47 in [3]). Statistical physics usually investigates systems with *very* many particles, like for example $N > 10^{10}$ particles. Therefore, the atomic and nuclear shell models are not dealt with in this book.

In contrast to fermions, any number of bosons can be in the same single particle state. This implies decisive differences for counting the possible states of the system, i.e. for the partition function. Therefore, Bose and Fermi systems lead to quite different partition functions.

From the symmetry of the wave function follows the *indistinguishability* of the particles. This means the following: In a system with the wave function (29.12), let a particle be in the state a, and another particle in state b. Obviously, in (29.12) it is not possible to say which particle is in which state. In this sense, the particles described by a symmetric or antisymmetric wave function are *indistinguishable*. This is more than similar or "of the same kind": That the particle are of the same kind is expressed by the symmetry (29.9) of the Hamilton operator; this symmetry also applies to the respective classical Hamilton function. It is the additional restriction to totally symmetric or antisymmetric wave functions, which makes particles of the same kind to indistinguishable particles.

Statistics

The quantum mechanical description of many-body systems implies a specific type of statistics. In this context, *statistics* means how to count the possibilities of distributing elementary particles into single particle states.

After the discretization (29.3) of the momenta, the single particle states can be brought into a countable order $(\boldsymbol{p}_1, s_{z,1})$, $(\boldsymbol{p}_2, s_{z,2})$, $(\boldsymbol{p}_3, s_{z,3})$, $\ldots$, starting with the lowest momenta. Indistinguishability means that we can only specify how many particles are in a certain single particle state, but not which particles. A microstate r of the quantum gas is therefore defined by specifying the number of particles in each single particle state:

$$r = \left(n_{\boldsymbol{p}_1}^{s_{z,1}}, n_{\boldsymbol{p}_2}^{s_{z,2}}, n_{\boldsymbol{p}_3}^{s_{z,3}}, \ldots \right) = \left\{ n_{\boldsymbol{p}}^{s_z} \right\} \tag{29.15}$$

The *occupation numbers* $n_{\boldsymbol{p}}^{s_z}$ may adopt the following values:

$$n_{\boldsymbol{p}}^{s_z} = \begin{cases} 0 \text{ or } 1 & \text{fermions} \\ 0, 1, 2, 3, \ldots & \text{bosons} \end{cases} \tag{29.16}$$

The energy of the microstate r is

$$E_r = \sum_{s_z,\, \boldsymbol{p}} \varepsilon_{\boldsymbol{p}}^{s_z}\, n_{\boldsymbol{p}}^{s_z} = \sum_{s_z,\, \boldsymbol{p}} \varepsilon_p\, n_{\boldsymbol{p}}^{s_z} \tag{29.17}$$

In the following assume $\varepsilon_{\boldsymbol{p}}^{s_z} = \varepsilon_p$; the single particle energy shall depend on the magnitude $(p = |\boldsymbol{p}|)$ of the momentum only.

For the microstate (29.15), the total number of particles is

$$N_r = \sum_{s_z,\, \boldsymbol{p}} n_{\boldsymbol{p}}^{s_z} \tag{29.18}$$

The microstate can also be defined by

$$r = \left(r', N_r \right) \tag{29.19}$$

where r' contains all specifications that are contained in r in addition to N_r. For some quantum gases the number of particles can be specified experimentally, for others not. An ordinary gas belongs to the first case, a gas of photons belongs to the second one.

If the number of particles N is fixed, there are two possibilities for the statistical treatment. On the one hand, one can restrict the treatment to microstates with $N_r = N$ and calculate the canonical partition function:

$$N_r = N, \quad E_{r'} = E_{r'}(V, N), \quad Z(T, V, N) = \sum_{r'} \exp\left(-\beta E_{r'}(V, N) \right) \tag{29.20}$$

On the other hand, the number of particles N_r can initially be left open using the grand canonical partition function:

$$Y(T, V, \mu) = \sum_r \exp\left(-\beta\left[E_r(V, N_r) - \mu N_r\right]\right), \qquad N = \overline{N_r} \qquad (29.21)$$

In this case, μ must be chosen so that the mean value of N_r is equal to N.

There are quantum gases in which the number of particles N is not predetermined. In a plasma, for example, the number of photons is not fixed; rather, it is a function of temperature. In this case, N does not occur as en external parameter in the Hamilton operator, and the energy eigenvalues are independent of N, i.e. $E_r(V, N) = E_r(V)$. This means that the generalized force belonging to this external parameter is zero:

$$\mu = \overline{\frac{\partial E_r(V)}{\partial N}} \equiv 0 \qquad \text{(particle number undetermined)} \qquad (29.22)$$

In this case, the canonical and grand canonical partition function coincide:

$$Z(T, V) = \sum_r \exp\left(-\beta E_r(V)\right) = Y(T, V, \mu = 0) \qquad (29.23)$$

We will treat all ideal quantum gases with the partition function Y (and use $\mu = 0$ if applicable).

By (29.15)–(29.18) the prerequisites for the calculation the partition function are given. The specific properties of the considered quantum gas are entered via the single particle energies ε_p and the spin. The spin determines which of the two possibilities of (29.16) has to be taken. The spin must also be taken into account when summing over the states.

Example

As an example, we consider a system with two single particle states that have the energies $\varepsilon_0 = 0$ and $\varepsilon_1 = \varepsilon$. In this system there are two particles. If they have spin, we only allow particles with aligned spin ($s_z = s$). The then possible states are depicted in Figure 29.1. This results in the following canonical partition functions:

$$
\begin{aligned}
\text{Fermi-Dirac statistics:} &\quad Z = \exp\left(-\beta\,\varepsilon\right) \\[2mm]
\text{Bose-Einstein statistics:} &\quad Z = 1 + \exp\left(-\beta\,\varepsilon\right) + \exp\left(-2\beta\,\varepsilon\right) \\[2mm]
\text{Maxwell-Boltzmann statistics:} &\quad Z = \frac{1 + 2\exp(-\beta\,\varepsilon) + \exp(-2\beta\,\varepsilon)}{2}
\end{aligned}
\qquad (29.24)
$$

Also the shorter designations such as the Bose, Fermi or Boltzmann statistics are common. The Boltzmann statistics is the previously used classical counting with the additional factor $1/N!$.

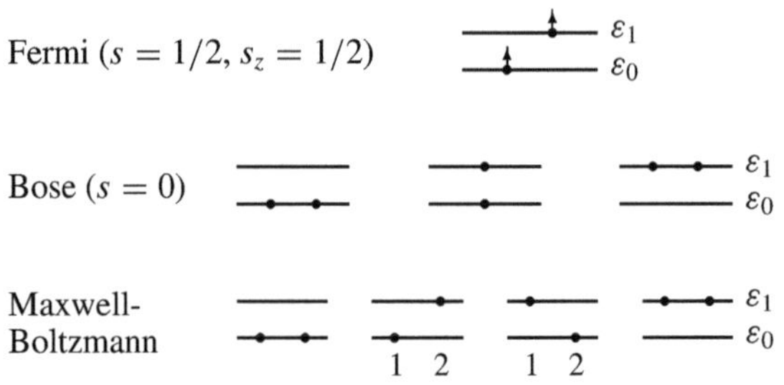

Fermi ($s = 1/2$, $s_z = 1/2$)

Bose ($s = 0$)

Maxwell-
Boltzmann

Figure 29.1 To illustrate quantum statistics, two particles in two energy levels are considered. In the case of Fermi statistics, the two spins shall be parallel, in the other cases the particles are assumed to be spinless.

This example shows that the way in which states are counted has an influence on the partition function and consequently on the thermodynamic properties of the system. It also makes it clear that the factor $1/N!$ in the classical statistics of the ideal gas (Chapters 6 and 25) does not fully take into account the indistinguishability of the particles.

Area of application

To evaluate the partition function, we must specify the spin and the single particle energies ε_p of the considered particles. The field of quantum gases includes all systems in which there are *excitations* which are characterized by a certain spin (i.e. fermions or bosons), and by a momentum $\boldsymbol{p}$ and an energy ε_p. This goes far beyond the range of ordinary gases.

In (29.2) we considered non-relativistic particles with the energy-momentum relationship $\varepsilon_p = p^2/2m$ The general relativistic energy-momentum relation is

$$\varepsilon_p = \sqrt{m^2 c^4 + c^2 p^2} \approx \begin{cases} m c^2 + \dfrac{p^2}{2m} & \text{(non-relativistic)} \\[2mm] c p & \text{(highly relativistic)} \end{cases} \tag{29.25}$$

In the non-relativistic case, the constant mc^2 is irrelevant for our applications and can therefore be omitted. The relativistic limit applies exactly for particles with zero mass, in particular for photons.

We list a few examples that may be treated as ideal gases consisting of particles with the energy ε_p. The following system are investigated in this way using (29.15)–(29.23):

- *Electrons in a metal*: In a metal, one electron per atom or lattice site can move approximately freely; its energy is $\varepsilon_p \approx p^2/2m$ with an effective mass m. The statistical treatment of the electron gas yields a contribution to the specific heat which is proportional to T.

- *Phonons in a crystal*: The vibrations of the lattice atoms around their rest position lead to oscillations and waves (including sound waves). For waves, the dispersion relation $\omega = \omega(k)$ describes the relationship between the frequency ω and the wave number k. The quantized excitations (with $\varepsilon_p = \hbar\omega$ and $p = \hbar k$) of the wave are called phonons. An oscillation mode can contain several excitation quanta; the phonons are therefore bosons. For low temperatures, the phonons yield a contribution to the specific heat which is proportional to T^3.

- *Photons in a radiation cavity*: Photons are the quanta of the electromagnetic field. They have spin 1 and are therefore bosons. Their energy is $\varepsilon_p = cp = 2\pi\hbar c/\lambda$. The statistical treatment as an ideal Bose gas leads to Planck's radiation law and to the Stefan-Boltzmann law.

- *Liquid helium*: Helium atoms occur as ^{4}He (bosons) or ^{3}He (fermions). As a rough approximation, also a helium liquid can be described as an ideal gas. The differences between the ideal Fermi and Bose gas are reflected by the behavior of the real ^{3}He and ^{4}He liquids.

- *Magnons in a ferromagnet*: Just like displacements of the lattice atoms from their center position lead to sound waves, the deflections of lattice spins from their aligned position result in spin waves. The quanta of spin waves are bosons called magnons. The magnons yield a contribution to the specific heat that is proportional to $T^{3/2}$.

In these applications, the energy ε_p may refer to a material particle, or to an excitation (quantized lattice oscillation, quantized electromagnetic oscillation).

According to Landau, we designate the particles or the quantized excitations as *quasiparticles*. The quasiparticles are the quantized elementary excitations of the system. An electron in a metal has different properties than a free electron; part of the interactions might be taken into account by an effective mass of the quasiparticle "electron in matter". A system consisting of many quasiparticles can often be treated approximately as an ideal gas.

Fermi statistics

We evaluate the grand canonical partition function

$$Y(T, V, \mu) = \sum_r \exp\left(-\beta\left(E_r - \mu N_r\right)\right) \tag{29.26}$$

for a system of fermions with spin 1/2. A microstate r is defined by (using arrows instead of $s_z = \pm 1/2$)

$$r = \left\{n_p^{s_z}\right\} = \left(n_{p_1}^\uparrow, n_{p_1}^\downarrow, n_{p_2}^\uparrow, n_{p_2}^\downarrow, n_{p_3}^\uparrow, \ldots\right) \tag{29.27}$$

We assume single particle energies of the form $\varepsilon_p^{s_z} = \varepsilon_p$. Then the partition function becomes

$$Y = \sum_{n_{p_1}^{\uparrow}=0}^{1} \sum_{n_{p_1}^{\downarrow}=0}^{1} \dots \exp\left(-\beta\left(\varepsilon_{p_1} - \mu\right) n_{p_1}^{\uparrow}\right) \exp\left(-\beta\left(\varepsilon_{p_1} - \mu\right) n_{p_1}^{\downarrow}\right) \cdots$$

$$= \left(1 + \exp\left[-\beta\left(\varepsilon_{p_1} - \mu\right)\right]\right)^2 \left(1 + \exp\left[-\beta\left(\varepsilon_{p_2} - \mu\right)\right]\right)^2 \cdots \tag{29.28}$$

or

$$Y(T, V, \mu) = \prod_p \left(1 + \exp\left[-\beta\left(\varepsilon_p - \mu\right)\right]\right)^2 \tag{29.29}$$

We now calculate the mean value of $n_p^{s_z}$. To do this, we choose a certain discrete momentum and spin value, p_i and $s_z = 1/2$. Compared to (29.28) an additional factor $n_{p_i}^{\uparrow}$ appears in the summand:

$$\overline{n_{p_i}^{\uparrow}} = \sum_r P_r\, n_{p_i}^{\uparrow} = \frac{1}{Y} \sum_r n_{p_i}^{\uparrow} \exp\left[-\beta\left(E_r - \mu N_r\right)\right]$$

$$= \frac{1}{Y} \cdot \ldots \cdot \sum_{n_{p_i}^{\uparrow}=0}^{1} n_{p_i}^{\uparrow} \exp\left(-\beta\left(\varepsilon_{p_i} - \mu\right) n_{p_i}^{\uparrow}\right) \cdots$$

$$= \frac{1}{Y} \cdot \ldots \cdot \exp\left[-\beta\left(\varepsilon_{p_i} - \mu\right)\right] \cdots = \frac{\exp\left[-\beta\left(\varepsilon_{p_i} - \mu\right)\right]}{1 + \exp\left[-\beta\left(\varepsilon_{p_i} - \mu\right)\right]}$$

$$= \frac{1}{\exp\left[\beta\left(\varepsilon_{p_i} - \mu\right)\right] + 1} \tag{29.30}$$

Because of $\varepsilon_p^{s_z} = \varepsilon_p$ the *mean occupation numbers* depend on $p = |\boldsymbol{p}|$ only:

$$\boxed{\overline{n_p^{s_z}} = \overline{n_p} = \frac{1}{\exp\left[\beta\left(\varepsilon_p - \mu\right)\right] + 1}} \tag{29.31}$$

If the single particle energies depend on the direction of the momentum or on the spin, then ε_p has to be replaced by $\varepsilon_p^{s_z}$.

To calculate the thermodynamic quantities, we start from

$$\ln Y(T, V, \mu) = 2 \sum_p \ln\left(1 + \exp\left[-\beta\left(\varepsilon_p - \mu\right)\right]\right) \tag{29.32}$$

The summation index is $\boldsymbol{p}$ and not p; according to (29.3), we must sum over all possible values of the components. From the partial derivatives of $\ln Y$ one obtains

$$N(T, V, \mu) = \frac{1}{\beta} \frac{\partial \ln Y}{\partial \mu} = 2 \sum_p \overline{n_p} \tag{29.33}$$

$$E(T, V, \mu) = -\frac{\partial \ln Y}{\partial \beta} + \mu N = 2 \sum_p \varepsilon_p\, \overline{n_p} \tag{29.34}$$

For systems with a given number of particles N, the chemical potential μ in (29.33) must be chosen such that $N(T, V, \mu)$ is equal to N. In this sense, μ can be understood as *normalization constant* of the mean occupation numbers (29.31). If one resolves $N = N(T, V, \mu)$ for $\mu = \mu(T, V, N)$ and substitutes this μ into $E = E(T, V, \mu)$, then the energy is obtained in the form $E = E(T, V, N)$. This is the caloric equation of state. The thermal equation of state $P = P(T, V, N)$ is obtained by inserting $\mu(T, V, N)$ into $P(T, V, \mu) = k_B T \ln Y / V$.

Bose statistics

For a system of bosons with zero spin, a microstate is given by

$$r = \{n_p\} = (n_{p_1}, n_{p_2}, n_{p_3}, \ldots) \tag{29.35}$$

We calculate the grand canonical partition function

$$
\begin{aligned}
Y &= \sum_r \exp\left[-\beta(E_r - \mu N_r)\right] \\
&= \sum_{n_{p_1}=0}^{\infty} \exp\left[-\beta(\varepsilon_{p_1} - \mu)\, n_{p_1}\right] \cdot \sum_{n_{p_2}=0}^{\infty} \exp\left[-\beta(\varepsilon_{p_2} - \mu)\, n_{p_2}\right] \cdots \\
&= \frac{1}{1 - \exp\left[-\beta(\varepsilon_{p_1} - \mu)\right]} \cdot \frac{1}{1 - \exp\left[-\beta(\varepsilon_{p_2} - \mu)\right]} \cdots \tag{29.36}
\end{aligned}
$$

This results in

$$Y(T, V, \mu) = \prod_p \frac{1}{1 - \exp\left[-\beta(\varepsilon_p - \mu)\right]} \tag{29.37}$$

Again, we determine the mean occupation number:

$$
\begin{aligned}
\overline{n_{p_i}} &= \frac{1}{Y} \sum_r n_{p_i} \exp\left[-\beta(E_r - \mu N_r)\right] \\
&= \frac{1}{Y} \sum_{n_{p_1}=0}^{\infty} \cdots \sum_{n_{p_i}=0}^{\infty} \cdots n_{p_i} \exp\left[-\beta(\varepsilon_{p_i} - \mu)\, n_{p_i}\right] \cdots \\
&= \frac{1}{Y} \sum_{n_{p_1}=0}^{\infty} \cdots \left(\frac{1}{\beta} \frac{\partial}{\partial \mu} \sum_{n_{p_i}=0}^{\infty} \exp\left[-\beta(\varepsilon_{p_i} - \mu)\, n_{p_i}\right] \right) \cdots \\
&= \frac{1}{Y} \sum_{n_{p_1}=0}^{\infty} \cdots \left(\frac{1}{\beta} \frac{\partial}{\partial \mu} \frac{1}{1 - \exp\left[-\beta(\varepsilon_{p_i} - \mu)\right]} \right) \cdots \\
&= \frac{\exp[-\beta(\varepsilon_{p_i} - \mu)]}{1 - \exp[-\beta(\varepsilon_{p_i} - \mu)]} = \frac{1}{\exp\left[\beta(\varepsilon_{p_i} - \mu)\right] - 1} \tag{29.38}
\end{aligned}
$$

If the single particle energies depend on the direction of the momentum or on the spin, then ε_p has to be replaced by $\varepsilon_p^{s_z}$.

The result (29.39) differs from that for fermions (29.31) by a sign in the denominator:

$$\boxed{\overline{n_p} = \overline{n_p} = \frac{1}{\exp\left[\beta(\varepsilon_p - \mu)\right] - 1}}$$

(29.39)

This sign leads to decisive differences between Fermi and Bose systems.

The energy and the number of particles may be obtained from

$$\ln Y(T, V, \mu) = -\sum_p \ln\left(1 - \exp\left[-\beta(\varepsilon_p - \mu)\right]\right)$$

(29.40)

or, more directly, by using (29.39):

$$E(T, V, \mu) = \sum_p \varepsilon_p \, \overline{n_p}, \qquad N(T, V, \mu) = \sum_p \overline{n_p}$$

(29.41)

For $s \neq 0$ the sum over s_z must be added. The chemical potential μ can again be considered as the normalization constant of the mean occupation numbers $\overline{n_p}$. The elimination of μ from $N(T, V, \mu)$, $E(T, V, \mu)$ and $P(T, V, \mu) = k_{\mathrm{B}}T \ln Y/V$ leads to the caloric and thermal equation of state.

Pressure

If the energy values are of the form $E_r = \sum \varepsilon_p n_p$ (ideal gas) and if ε_p is proportional to a power of p, then a simple relation between the energy the energy E and the pressure P can be established.

We start from the definition (8.8) of the pressure

$$P = -\overline{\frac{\partial E_r}{\partial V}} = -\sum_{s_z, \, p} \frac{\partial \varepsilon_p}{\partial V} \, \overline{n_p}$$

(29.42)

For a quasi-static change of V, the probabilities P_r and thus the $\overline{n_p}$ remain unchanged. The derivative with respect to V then acts only on ε_p.

In the quantum mechanical treatment, the particles (for example electrons or photons) are described by waves. The boundary conditions enforce discrete wave numbers, such as $k_i = i\,\pi/L$ for a wave in the interval $[0, L]$. From this follows the volume dependence $p_i = \hbar k_i \propto V^{-1/3}$ of the momenta. The statement $p_i \propto V^{-1/3}$ applies regardless of whether the particles are relativistic or not.

We consider two limiting cases of the energy-momentum relation:

$$\varepsilon_p = \begin{cases} p^2/2m & \propto \ V^{-2/3} \\[4pt] cp & \propto \ V^{-1/3} \end{cases}$$

(29.43)

The first case applies to a non-relativistic gas consisting of particles of mass m (such as a gas of atoms or electrons), the second case applies to photons, phonons or highly relativistic ($p \gg mc$) particles. The derivative with respect to V results in $(-2/3)\,\varepsilon_p/V$ in the first case, and to $(-1/3)\,\varepsilon_p/V$ in the second case. We insert this into (29.42):

$$P = \begin{cases} \dfrac{2}{3}\dfrac{E}{V} & \left(\varepsilon_p \propto p^2\right) \\[2ex] \dfrac{1}{3}\dfrac{E}{V} & \left(\varepsilon_p \propto p\right) \end{cases} \tag{29.44}$$

In these cases, the thermal equation of state $P = P(T, V, N)$ can be obtained directly from the energy $E(T, V, N)$.

The relationship (29.44) between E and P does not depend on the statistics, but only on the momentum dependence of the single particle energies. The prerequisite for this is that the energy is equal to the sum of the single particle energy (ideal gas). For a polyatomic ideal gas, the E has to be identified with translational part E_{trans} of the total energy.

Exercises

29.1   Quantum numbers in the infinite potential box

The Hamilton operator of a particle in an infinitely high box is

$$h = -\frac{\hbar^2}{2m}\,\Delta + U(r) \qquad \text{with} \qquad U(r) = \begin{cases} 0 & r \in V \\ \infty & r \notin V \end{cases}$$

Let the volume V of the box be cubic.

Specify the normalized eigenfunctions $\varphi_p(r)$. What values can p adopt?

29.2   Partition functions for three particles

Three particles are in two levels (with the energies ε_0 and ε_1). These are (i) classical, distinguishable particles, (ii) bosons with spin 0 or (iii) fermions with spin 1/2. Determine the respective canonical partition functions.

29.3   Fluctuation of the occupation numbers in a quantum gas

Derive

$$\left(\Delta n_j\right)^2 = -\,k_{\mathrm{B}}T\,\frac{\partial \overline{n_j}}{\partial \varepsilon_j}$$

for the fluctuation Δn_j of the occupation numbers n_j of an ideal quantum gas. Here, $j = (p, s_z)$ stands for the quantum numbers of a single particle state. Determine the relative fluctuation $(\Delta n_j)^2 / \overline{n_j}^2$ for a Fermi and a Bose gas.

30 Dilute quantum gas

For a dilute, ideal gas of N particles we calculate the leading quantum mechanical corrections to the classical results. The size of the quantum corrections is determined by the ratio of the thermal wavelength to the mean particle distance.

The single particle energies shall be of the form $\varepsilon_p = p^2/2m$. Under the condition

$$\exp(-\beta\mu) \gg 1 \tag{30.1}$$

the differences between the Fermi and Bose statistics can be neglected:

$$\overline{n_p} = \frac{1}{\exp\left[\beta(\varepsilon_p - \mu)\right] \pm 1} \approx \text{const.} \cdot \exp(-\beta\varepsilon_p) \tag{30.2}$$

Due to (30.1) and $\exp(\beta\varepsilon_p) \geq 1$, the first term in the denominator is much larger than 1. In the result, the term $\exp(\beta\mu)$ was written as "const." (independent of p). The result corresponds to the Maxwell distribution, i.e. the ideal classical gas.

If the quantum corrections are small, the classical result $\exp(\beta\mu) = \lambda^3/v = \lambda^3/(V/N)$, (25.18), holds approximately. With this, the condition (30.1) becomes

$$\lambda^3 \ll v \qquad \text{(dilute gas)} \tag{30.3}$$

The temperature determines the thermal wavelength λ. The condition (30.3) is then fulfilled for a sufficiently *dilute* gas. Under this condition, we determine the leading quantum mechanical corrections to the classical ideal gas.

For the Fermi gas with $s = 1/2$ we expand $\ln Y$ of (29.32) into powers of $\exp(\beta\mu)$:

$$\ln Y_{\text{Fermi}} = 2\sum_p \ln\left(1 + \exp\left[-\beta(\varepsilon_p - \mu)\right]\right) \tag{30.4}$$

$$= 2\sum_p \left(\exp\left[-\beta(\varepsilon_p - \mu)\right] - \frac{1}{2}\exp\left[-2\beta(\varepsilon_p - \mu)\right] + \ldots\right)$$

For the Bose gas with $s = 0$ we expand $\ln Y$ of (29.40) accordingly:

$$\ln Y_{\text{Bose}} = -\sum_p \ln\left(1 - \exp\left[-\beta(\varepsilon_p - \mu)\right]\right) \tag{30.5}$$

$$= \sum_p \left(\exp\left[-\beta(\varepsilon_p - \mu)\right] + \frac{1}{2}\exp\left[-2\beta(\varepsilon_p - \mu)\right] + \ldots\right)$$

In the form

$$\ln Y = (2s+1) \sum_p \left(\exp\left[-\beta\left(\varepsilon_p - \mu\right)\right] \mp \frac{1}{2} \exp\left[-2\beta\left(\varepsilon_p - \mu\right)\right] + \dots \right) \quad (30.6)$$

we can summarize both cases. The upper sign applies to the Fermi gas, the lower sign to the Bose gas. If the single particle energies do not depend on the spin, the sum over the spin states results in the factor $2s + 1$.

To evaluate the sum over the momenta, we consider a cubic volume $V = L^3$. The Cartesian momentum components can take the values $p_i = \Delta p\, n_i$ where $n_i = 1, 2, \dots$. The distances $\Delta p = \pi\hbar/L$ between neighboring momenta are very small compared to the mean momenta (for a macroscopic volume V):

$$\Delta p \ll \overline{p} \quad (30.7)$$

As the momentum values are close together, the sum over the momenta can be replaced by integrals:

$$
\begin{aligned}
\sum_p \dots &= \sum_{n_1=1}^{\infty} \sum_{n_2=1}^{\infty} \sum_{n_3=1}^{\infty} \dots = \int_0^{\infty} dn_1 \int_0^{\infty} dn_2 \int_0^{\infty} dn_3 \dots \\[2mm]
&= \frac{1}{(\Delta p)^3} \int_0^{\infty} dp_1 \int_0^{\infty} dp_2 \int_0^{\infty} dp_3 \dots \\[2mm]
&= \frac{L^3}{(2\pi\hbar)^3} \int_{-\infty}^{\infty} dp_1 \int_{-\infty}^{\infty} dp_2 \int_{-\infty}^{\infty} dp_3 \dots = \frac{V}{(2\pi\hbar)^3} \int d^3p \dots
\end{aligned}
\quad (30.8)
$$

It was assumed that the integrand is symmetric with respect to $p_i \leftrightarrow -p_i$. The quantization of the momentum values with $\Delta p = \pi\hbar/L$ determines the prefactor of the integral. As a result, the shape of the volume V does not matter. Replacing the sum with the integral is independent of the condition (30.1). This substitution is also used in the following chapters, too.

We first consider the leading term in (30.6), i.e. the 0th order. Here we replace the sum over the momenta by an integral as in (30.8):

$$\ln Y = (2s+1) \frac{V}{(2\pi\hbar)^3} \int d^3p \, \exp\left(-\frac{p^2}{2m k_{\mathrm{B}} T}\right) \exp(\beta\mu) \qquad \text{(0th order)}$$

$$(30.9)$$

The integration results in

$$\ln Y = (2s+1) \frac{V}{\lambda^3} \exp\left(\beta\mu\right) \qquad \text{(0th order)} \qquad (30.10)$$

In this zeroth order we obtain the ideal gas law

$$N = \frac{1}{\beta} \frac{\partial \ln Y}{\partial \mu} = (2s+1) \frac{V}{\lambda^3} \exp\left(\beta\mu\right) = \ln Y = \frac{PV}{k_{\mathrm{B}} T} \qquad \text{(0th order)} \quad (30.11)$$

From this also follows

$$\exp(\beta\mu) = \frac{1}{2s+1}\frac{\lambda^3}{v} \qquad \text{(0th order)} \tag{30.12}$$

This shows that (30.6) is a expansion into powers of λ^3/v. Such an expansion is possible for sufficiently small densities (v large) or high temperatures (λ small). In the limit $\lambda^3/v \to 0$, the chemical potential goes to minus infinity, $\mu \to -\infty$.

We now evaluate the correction term in (30.6), i.e. the 1st order. In the exponent there is an additional factor 2 compared to first term. This leads to a factor $2^{-3/2}$ when integrating with d^3p. In 1st order we thus obtain

$$\ln Y = (2s+1)\frac{V}{\lambda^3}\left(\exp(\beta\mu) \mp \frac{1}{2^{5/2}}\exp(2\beta\mu)\right) \tag{30.13}$$

Higher terms are neglected here and in the following. From (30.13) follows

$$N = \frac{1}{\beta}\frac{\partial \ln Y}{\partial \mu} = (2s+1)\frac{V}{\lambda^3}\left(\exp(\beta\mu) \mp \frac{1}{2^{3/2}}\exp(2\beta\mu)\right) \tag{30.14}$$

The last two equations yield

$$\ln Y = N \pm \frac{2s+1}{2^{5/2}}\frac{V}{\lambda^3}\exp(2\beta\mu) \tag{30.15}$$

The second term on the right-hand side is a 1st order term. In this small term, we can use the zeroth order for $\exp(\beta\mu)$, i.e. (30.12):

$$\ln Y = N \pm \frac{1}{2^{5/2}}\frac{1}{2s+1}\frac{\lambda^3}{V}N^2 \tag{30.16}$$

With $\ln Y = -J/k_BT = PV/k_BT$ we write this in the form

$$\boxed{\frac{PV}{Nk_BT} = 1 + \frac{B_{\mathrm{qm}}(T)}{v} \qquad \begin{array}{l}\text{dilute}\\ \text{quantum gas}\end{array}} \tag{30.17}$$

with the *quantum mechanical virial coefficient*

$$B_{\mathrm{qm}}(T) = \begin{cases} +\dfrac{\lambda^3}{2^{7/2}} & \text{Fermi gas } (s = 1/2) \\[2ex] -\dfrac{\lambda^3}{2^{5/2}} & \text{Bose gas } (s = 0) \end{cases} \tag{30.18}$$

With (29.44) we obtain for the energy

$$E = \frac{3}{2}PV = \frac{3}{2}Nk_BT\left(1 + \frac{B_{\mathrm{qm}}(T)}{v}\right) \tag{30.19}$$

We have thus derived the leading quantum mechanical correction to the ideal gas.

Discussion

Because of $\exp(\beta\mu) \sim \lambda^3/v$, (30.6) is a expansion into powers of λ^3/v. We discuss the condition $\lambda^3/v \ll 1$ for specific systems.

The thermal wavelength λ is the quantum mechanical wavelength of a particle with a kinetic energy of the size $k_\mathrm{B}T$:

$$\lambda = \frac{2\pi\hbar}{\sqrt{2\pi m k_\mathrm{B}T}} \quad \Longrightarrow \quad \frac{\hbar^2}{2m}\left(\frac{2\pi}{\lambda}\right)^2 = \pi k_\mathrm{B}T \tag{30.20}$$

Except for a numerical factor, $v^{1/3}$ equals the mean distance between two gas particles. The corrections to the ideal gas law are characterized by the ratio

$$\frac{\lambda}{v^{1/3}} \approx \frac{\text{thermal wavelength}}{\text{mean particle distance}} \tag{30.21}$$

For sufficiently low temperatures, $\lambda \propto 1/\sqrt{T}$ eventually becomes comparable to $v^{1/3}$; then the quantum corrections are no longer small. We consider the corresponding transition (index tr) temperature T_tr for which the quantum corrections become relevant:

$$\Delta\varepsilon_\mathrm{tr} = \frac{\hbar^2}{m\,v^{2/3}} = k_\mathrm{B}T_\mathrm{tr} \tag{30.22}$$

We evaluate this numerically for air. The air molecules (O_2 or N_2) have the mass $m \approx 30\ \mathrm{GeV}/c^2$ (one nucleon mass is approximately equal to $1\ \mathrm{GeV}/c^2$). For a dilute gas (v large), the temperature T_tr is small and the quantum correction play no role. In order the quantum corrections become relevant we must consider small v values. The volume per particle v must be larger than the intrinsic volume of the molecules. For $v^{1/3} = 4\ \text{Å}$ we obtain

$$\Delta\varepsilon_\mathrm{tr} \approx \frac{\hbar^2}{30\,\mathrm{GeV}/c^2 \cdot (4\,\text{Å})^2} \approx 10^{-5}\,\mathrm{eV} \approx 0.1\,k_\mathrm{B}\mathrm{K} \qquad \text{(for air)} \tag{30.23}$$

Here, $\hbar c/\text{Å} = (\hbar c/e^2)(e^2/\text{Å}) \approx 137 \cdot 14.4\ \mathrm{eV}$ was used.

The scale $\Delta\varepsilon_\mathrm{tr} = k_\mathrm{B}T_\mathrm{tr}$ for quantum effects is not a consequence of momentum quantization with $\Delta p = \pi\hbar/V^{1/3}$. This momentum quantization only leads to extremely small energy gaps:

$$\Delta\varepsilon' = \frac{(\Delta p)^2}{2m} \approx \frac{\hbar^2}{m\,V^{2/3}} = \frac{\Delta\varepsilon_\mathrm{tr}}{N^{2/3}} \ll \Delta\varepsilon_\mathrm{tr} \tag{30.24}$$

The quantization $\Delta\varepsilon'$ generally plays no role for a macroscopic system. The quantum corrections described by B_qm are rather based on the exchange symmetry of the particles, which leads to noticeable differences in the counting of the states (see Figure 29.1).

According to (30.23), quantum effects for the translation (plus exchange symmetry) become noticeable only at temperatures al low as $T_\mathrm{tr} \lesssim 0.1\ \mathrm{K}$. Ordinary gases become liquid and solid long before such low temperatures are reached. In the gaseous phase, the calculated quantum effects are therefore quite small corrections. The approximations leading to (30.17) are therefore valid.

The noble gases ^{3}He and ^{4}He constitute remarkable exceptions. They condense at about 5 K, but at normal pressure they will not become solid for $T \to 0$. Therefore, the degrees of freedom of the translation are retained, even if modifications are expected due to the interactions within the liquid. We repeat the estimation (30.23), using the smaller mass $m \approx 4\,\text{GeV}/c^2$ of the helium atoms and the value $v^{1/3} \approx 3.6\,\text{Å}$ for liquid helium. This yields

$$\Delta \varepsilon_{\text{tr}} = \frac{\hbar^2}{m\,v^{2/3}} \approx 1\,k_{\text{B}}\,\text{K} \qquad (\text{for }{}^4\text{He}) \tag{30.25}$$

In fact, in the vicinity of 2 K there are dramatic differences in the behavior of the real ^{3}He and ^{4}He liquids. The reason for this is the different exchange symmetry for bosons (^{4}He atoms) and fermions (^{3}He atoms). The real liquids show qualitative similarities with the corresponding ideal quantum gases (Chapters 31 and 32).

For real gases consisting of atoms or molecules, the quantum mechanical correction $\mathcal{O}(\lambda^3/v)$ is superimposed by interaction effects. For the dilute, interacting classical gas, we had obtained corrections of the size

$$B(T) \sim b \gg B_{\text{qm}}(T) \sim \lambda^3 \tag{30.26}$$

where $B(T)$ is the virial coefficient from (28.21) and b is four times the intrinsic volume of the atoms. Because of $\lambda^3 \ll b$, the quantum mechanical corrections are practically always negligibly small. In addition, quantum mechanical effects had to be taken into account when evaluating the partition function $Z(2)$ in (28.3): For the attractive potential to act, a gas particle must restrict its position to the range $\mathcal{O}(b)$. This requires a kinetic energy of the size $\hbar^2/(m\,b^{2/3})$; thereby the effect of the attractive interaction is reduced. This quantum mechanical correction is of the size $\hbar^2/(m\,b^{2/3}) = \mathcal{O}(\hbar^2)$; it is larger than $B_{\text{qm}} = \mathcal{O}(\hbar^3)$.

For the electron gas in the metal, the calculated quantum mechanical corrections are not sufficient. The smaller mass $m_{\text{e}} \approx 0.5\,\text{MeV}/c^2$ results in much larger (factor 10^4) transition temperature T_{tr}. The electron gas must therefore be treated fully quantum mechanically (Chapter 32).

31 Ideal Bose gas

We investigate the ideal Bose gas with a fixed number of particles. The ideal Bose gas is a remarkable model because it leads to an exactly calculable phase transition, the so-called Bose-Einstein condensation. The close relationship of this phase transition to the λ transition in the real ^{4}He liquid will be discussed in Chapter 38.

We consider non-relativistic bosons with spin 0 and mass m. The single particle states with momentum $\boldsymbol{p}$ have the energy

$$\varepsilon_p = \frac{p^2}{2m} \tag{31.1}$$

According to (29.39), the mean number of bosons in the respective single particle state is given by

$$\overline{n_p} = \frac{1}{\exp\left[\beta(\varepsilon_p - \mu)\right] - 1} \tag{31.2}$$

The thermodynamic energy and the particle number are

$$E(T, V, \mu) = \sum_p \varepsilon_p \,\overline{n_p}\,, \qquad N(T, V, \mu) = \sum_p \overline{n_p} \tag{31.3}$$

As in (30.8), the sum over the discrete momenta can be carried out as an integral:

$$\sum_p \ldots = \frac{V}{(2\pi\hbar)^3} \int d^3p \,\ldots \tag{31.4}$$

The *ideal Bose gas* is defined by (31.1)–(31.4). Eliminating μ from (31.3) yields the energy $E(T, V, N)$. From this follows he specific heat c_V and the pressure (29.44), and all other thermodynamic quantities.

For the mean occupation number (31.2) we note

$$\overline{n_p} \stackrel{\varepsilon_p - \mu \,\to\, 0}{\longrightarrow} \infty \tag{31.5}$$

For positive μ, the mean occupation number becomes singular for $\varepsilon_p \to \mu$, and the integral (31.4) over $\overline{n_p}$ would diverge. Therefore we must require

$$\mu \leq 0 \tag{31.6}$$

and $\varepsilon_p - \mu \geq 0$. For $\varepsilon_p - \mu > 0$ we expand $\overline{n_p}$ into powers of $\exp(-\beta(\varepsilon_p - \mu)) < 1$,

$$
N = \sum_p \overline{n_p} = \sum_p \frac{\exp(-\beta(\varepsilon_p - \mu))}{1 - \exp(-\beta(\varepsilon_p - \mu))} = \sum_p \sum_{l=1}^{\infty} \left(\exp[-\beta(\varepsilon_p - \mu)] \right)^l
$$

$$
= \frac{V}{(2\pi\hbar)^3} \sum_{l=1}^{\infty} \exp(\beta\mu l) \int d^3p \, \exp\left(-\frac{p^2 l}{2mk_{\mathrm{B}}T} \right) \tag{31.7}
$$

For $l = 1$ the integral is already known from (24.19) and (24.20),

$$
\frac{1}{(2\pi\hbar)^3} \int d^3p \, \exp\left(-\frac{p^2}{2mk_{\mathrm{B}}T} \right) = \frac{1}{\lambda^3} \quad \text{with} \quad \lambda = \frac{2\pi\hbar}{\sqrt{2\pi m k_{\mathrm{B}}T}} \tag{31.8}
$$

In the integral in (31.7) we substitute $p^2 l \to p'^2$ and $d^3p \to d^3p'/l^{3/2}$. Compared with (31.8), this results in an additional factor $1/l^{3/2}$, i.e.

$$
N(T, V, \mu) = \frac{V}{\lambda^3} \sum_{l=1}^{\infty} \frac{\exp(\beta\mu l)}{l^{3/2}} = \frac{V}{\lambda^3} g_{3/2}(z) \tag{31.9}
$$

In the last step, we have introduced the generalized Riemann zeta function

$$
g_\nu(z) \equiv \sum_{l=1}^{\infty} \frac{z^l}{l^\nu} \tag{31.10}
$$

with the argument

$$
z = \exp(\beta\mu) \tag{31.11}
$$

For given $v = V/N$ and T, Equation (31.9) determines the chemical potential $\mu = \mu(T, v)$. The temperature dependence of $\mu(T, v)$ is depicted in Figure 31.1. For $T \to \infty$ we get $\lambda^3 \to 0$, and therefore $g_{3/2}(z) \to 0$, i.e. $z \to 0$ and $\mu \to -\infty$. With decreasing temperature λ^3 and thus also $g_{3/2}(z)$ increases. Since $g_{3/2}(z)$ is a monotonic function of z, the variable z must then become larger. This increase is limited by $\mu \leq 0$ and $z \leq 1$. The maximum value of $g_{3/2}(z)$ is obtained for $\mu = 0$ or $z = 1$. This maximum is reached for at a finite temperature T_{cr},

$$
\frac{\lambda_{\mathrm{cr}}^3}{v} = g_{3/2}(1) = \zeta(3/2) \tag{31.12}
$$

The generalized Riemann zeta function $g_\nu(z)$ reduces to Riemann's zeta function $\zeta(\nu)$:

$$
g_\nu(1) = \zeta(\nu) \equiv \sum_{l=1}^{\infty} \frac{1}{l^\nu} \approx \begin{cases} \infty & (\nu = 1/2) \\ 2.6124 & (\nu = 3/2) \\ 1.3415 & (\nu = 5/2) \end{cases} \tag{31.13}
$$

The numerical values will be needed later. The critical or transition temperature T_{cr} associated with $\mu = 0$ follows from (31.12):

$$\boxed{k_B T_{cr} = \frac{2\pi}{[\zeta(3/2)]^{2/3}} \frac{\hbar^2}{m\, v^{2/3}} \qquad \text{transition temperature}} \qquad (31.14)$$

As we will see, this defines the transition (or critical, index cr) temperature T_{cr} of a phase transition. For $T \sim T_{cr}$ the quantum mechanical wavelength is equal to the mean particle distance. For the parameters of liquid ^{4}He we obtain

$$T_{cr} = 3.13 \,\text{K} \qquad \left(^4\text{He},\ v = 46\,\text{Å}^3\right) \qquad (31.15)$$

So far we have seen that μ increases with decreasing temperature, namely from $-\infty$ at $T \to \infty$ to $\mu = 0$ for $T \to T_{cr}$. For $\mu > 0$, (31.9) has no solution. This results in a restriction of the temperature range:

$$\mu \leq 0 \quad \overset{(31.9)}{\Longrightarrow} \quad \frac{\lambda^3}{v} \leq \zeta(3/2) \quad \text{or} \quad T \geq T_{cr} \qquad (31.16)$$

However, the existing particles must somehow be distributed over the available levels also for $T \leq T_{cr}$. This means that the equation $N = \sum \overline{n_p}$ *must* have a solution also for $T \leq T_{cr}$. In contradiction to this, we have derived the condition $T \leq T_{cr}$ in (31.16). This points to an error in the evaluation of the particle number condition (31.7).

For a given density N/V follows from equation (31.9)

$$\mu \quad \overset{T \to T_{cr}^+}{\longrightarrow} \quad 0^- \qquad (31.17)$$

The indices $+$ and $-$ indicate from which side the limit is approached. With $\mu \to 0^-$ the average occupation number of the lowest level with $\varepsilon_0 = 0$ becomes arbitrarily large:

$$N_0 = \overline{n_0} \quad \overset{\mu \to 0^-}{\longrightarrow} \quad \infty \qquad (31.18)$$

In the following, we denote the number of particles at the lowest level by N_0. Figure 31.2 sketches the occupation numbers for $T < T_{cr}$, as they result from (31.2) with $\mu = 0$ and the following discussion.

The contribution of the N_0 particles got lost when we replaced the sum by an integral in (31.7): For $\mu = 0$ and $p \to 0$ we get $\overline{n_p} \propto 1/p^2$. In the integration $d^3p = 4\pi p^2\, dp$ this yields a contribution proportional to dp only, i.e. a vanishing contribution in an infinitesimal environment of $p = 0$. As the contribution with $\overline{n_0}$ is not included in the integral, it must be taken into account separately in the step from the sum to the integral:

$$N = \overline{n_0} + \sum_{p \neq 0} \overline{n_p} = \overline{n_0} + \frac{V}{(2\pi\hbar)^3} \int d^3p\ \overline{n_p} = N_0 + \frac{V}{\lambda^3} g_{3/2}(z) \qquad (31.19)$$

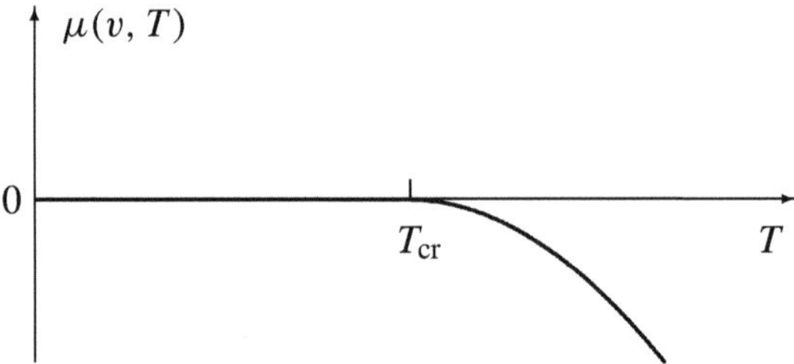

Figure 31.1 Chemical potential of the ideal Bose gas as a function of the temperature ($v = \text{const.}$).

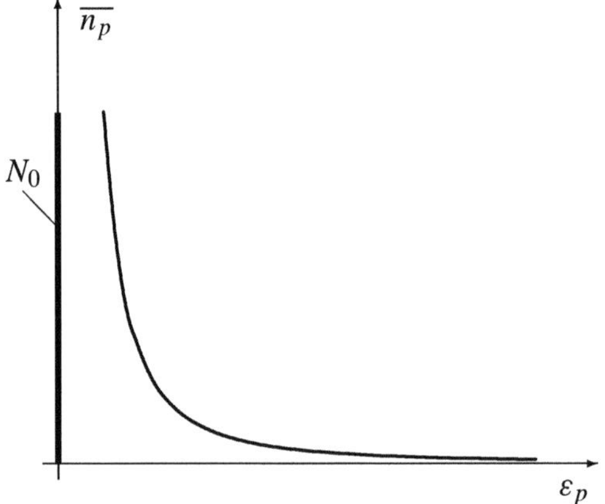

Figure 31.2 Occupation numbers of the ideal Bose gas for $T < T_{\mathrm{cr}}$. The thick bar represents the macroscopic occupation of the lowest single particle level for $T < T_{\mathrm{cr}}$, where $N_0 = \overline{n_0} = \mathcal{O}(N)$.

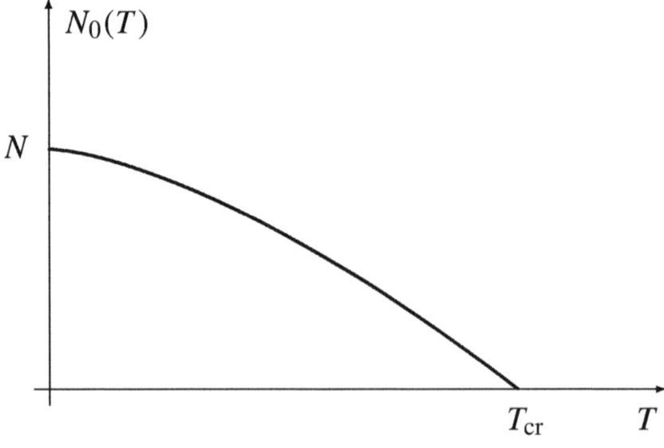

Figure 31.3 Number N_0 of the condensed particles in the ideal Bose gas as a function of T.

For $T > T_{cr}$, the contribution $\overline{n_0} = \mathcal{O}(1)$ can be omitted; and (31.9) remains valid for $T > T_{cr}$. For $N_0 \gg 1$ we get from (31.2)

$$1 \ll N_0 = \frac{1}{\exp\left[\beta(\varepsilon_0 - \mu)\right] - 1} \approx \frac{k_B T}{\varepsilon_0 - \mu} \overset{V \to \infty}{=} \frac{k_B T}{-\mu} \tag{31.20}$$

We do not take into account effects that are due to the finite size of the system. Therefore, we consider the limiting case

$$N \to \infty, \qquad V \to \infty, \qquad v = \frac{V}{N} = \text{const.} \tag{31.21}$$

In this limit, the energy of the lowest single particle level approaches zero, $\varepsilon_0 \approx \hbar^2/(m\,V^{2/3}) \to 0$. This also applies to the chemical potential below T_{cr}:

$$\mu = -\frac{k_B T}{N_0} = -\frac{k_B T}{\mathcal{O}(N)} \overset{N \to \infty}{=} 0^- \qquad (T \le T_{cr},\ N \to \infty) \tag{31.22}$$

The limiting case (31.21) is called *thermodynamic limit*. The finiteness of a concrete macroscopic system is generally irrelevant due to the large number of particles (about $N = 10^{24}$). Therefore, the theoretical treatment uses this thermodynamic limit.

We insert $\mu = 0$ into (31.19):

$$N = N_0 + \frac{V}{\lambda^3}\,\zeta(3/2) \qquad (T \le T_{cr}) \tag{31.23}$$

For the second term, we write

$$\frac{V}{\lambda^3}\,\zeta(3/2) = \frac{\lambda_c^3}{\lambda^3}\frac{V}{\lambda_{cr}^3}\,\zeta(3/2) = \left(\frac{T}{T_{cr}}\right)^{3/2}\frac{V}{\lambda_{cr}^3}\,\zeta(3/2) \overset{(31.12)}{=} N\left(\frac{T}{T_{cr}}\right)^{3/2} \tag{31.24}$$

This determines the temperature dependence of $N_0(T)$:

$$\frac{N_0}{N} = \begin{cases} 1 - \left(\dfrac{T}{T_{cr}}\right)^{3/2} & (T \le T_{cr}) \\[2ex] \mathcal{O}\left(\dfrac{1}{N}\right) \approx 0 & (T > T_{cr}) \end{cases} \tag{31.25}$$

This result is sketched in Figure 31.3. For $T < T_{cr}$ a finite fraction N_0/N of all atoms occupies the lowest state. These atoms are referred to as condensed particles or a as the condensate. For $T = 0$ we obtain $N_0 = N$, i.e. the quantum mechanically expected ground state.

The feature that in the ideal Bose gas exhibits a discrete transition temperature T_{cr}, and that for $T \le T_{cr}$ a finite fraction of all particles goes into the lowest state is referred to as *Bose-Einstein condensation*. This theoretical effect was discovered in 1924 by Einstein, who analyzed the ideal gas using the statistics introduced by

Bose. The designation "condensation" is due to the fact that, as in the gaseous-liquid transition, the condensed particles no longer contribute to the pressure.

The ideal Bose gas exhibits a phase transition, i.e. at a certain temperature T_{cr} the behavior of the system changes qualitatively. We will discuss phase transitions in Part VI in more detail and will also come back to the Bose-Einstein condensation.

We calculate the energy of the ideal Bose gas:

$$
\sum_p \varepsilon_p\, \overline{n_p} = \frac{V}{(2\pi\hbar)^3} \sum_{l=1}^{\infty} \exp(\beta\mu l) \int d^3 p\, \frac{p^2}{2m}\, \exp\left(-\frac{p^2 l}{2m k_B T}\right)
$$

$$
= \frac{V}{(2\pi\hbar)^3} \sum_{l=1}^{\infty} \exp(\beta\mu l)\left(-\frac{1}{\beta}\frac{\partial}{\partial l}\right) \int d^3 p\, \exp\left(-\frac{p^2 l}{2m k_B T}\right)
\tag{31.26}
$$

As in (31.7), the integral including the prefactor $1/(2\pi\hbar)^3$ is equal to $1/(\lambda^3\, l^{3/2})$. The partial derivative $\partial/\partial l$ then gives $-(3/2)/(\lambda^3\, l^{5/2})$, i.e.

$$
E(T, V, \mu) = \frac{3}{2}\, k_B T\, \frac{V}{\lambda^3} \sum_{l=1}^{\infty} \frac{\exp(\beta\mu l)}{l^{5/2}} = \frac{3}{2}\, k_B T\, \frac{V}{\lambda^3}\, g_{5/2}(z)
\tag{31.27}
$$

This also applies to $T < T_{cr}$, because the additional term $N_0\, \varepsilon_0$ from the lowest level does not contribute due to $\varepsilon_0 = 0$.

Caloric equation of state

We calculate the specific heat c_V. We start with the case $T \leq T_{cr}$ (i.e. with $\mu = 0$ and $z = 1$) and obtain for this

$$
\frac{c_V(T)}{k_B} = \frac{1}{N k_B}\left(\frac{\partial E}{\partial T}\right)_{V,N} = \frac{15}{4}\frac{v}{\lambda^3}\, g_{5/2}(1) = \frac{15}{4}\frac{\zeta(5/2)}{\zeta(3/2)}\left(\frac{T}{T_{cr}}\right)^{3/2} \qquad (T \leq T_{cr})
\tag{31.28}
$$

The differentiation resulted in a factor of $5/2$ due to $T/\lambda^3 \propto T^{5/2}$. In the last step, (31.24) was used. This result can be interpreted in the following way: Each of the $N(T/T_{cr})^{3/2}$ non-condensed particles makes a contribution of the size $\mathcal{O}(k_B)$ to the specific heat; the condensed particles, on the other hand, do not contribute.

We now come to the case $T \geq T_{cr}$, where μ and z are functions of T and v. From (31.27) we then obtain

$$
\frac{c_V(T)}{k_B} = \frac{1}{N k_B}\left(\frac{\partial E}{\partial T}\right)_{V,N} = \frac{15}{4}\frac{v}{\lambda^3}\, g_{5/2}(z) + \frac{3}{2}\frac{v}{\lambda^3}\, T\, g'_{5/2}(z)\left(\frac{\partial z}{\partial T}\right)_v
\tag{31.29}
$$

For $T > T_{cr}$, the chemical potential follows from the particle number condition (31.9):

$$
1 = \frac{v}{\lambda^3}\, g_{3/2}(z) \qquad (T > T_{cr})
\tag{31.30}
$$

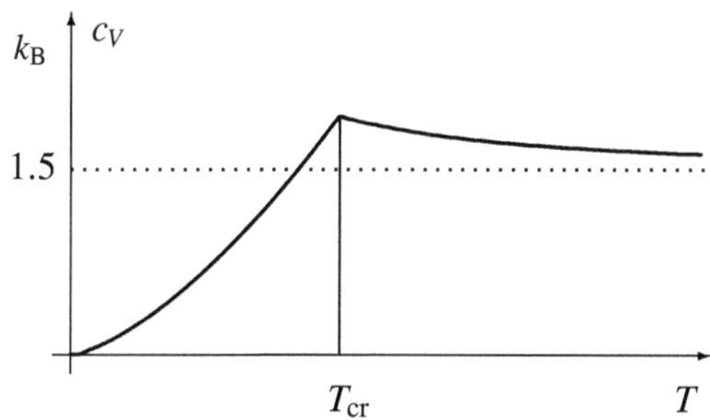

Figure 31.4 Specific heat of the ideal Bose gas as a function of temperature.

We differentiate this with respect to the temperature (for $v = $ const.) and cancel the factor v/λ^3:

$$0 = \frac{3}{2T}\, g_{3/2}(z) + g'_{3/2}(z)\left(\frac{\partial z}{\partial T}\right)_v \tag{31.31}$$

From this, we determine the partial derivative $(\partial z/\partial T)_v$. We insert this derivative into (31.29), whereby we also use (31.30),

$$\frac{c_V(T)}{k_B} = \frac{15}{4}\frac{g_{5/2}(z)}{g_{3/2}(z)} - \frac{9}{4}\frac{g'_{5/2}(z)}{g'_{3/2}(z)} = \frac{15}{4}\frac{g_{5/2}(z)}{g_{3/2}(z)} - \frac{9}{4}\frac{g_{3/2}(z)}{g_{1/2}(z)} \qquad (T \geq T_{\mathrm{cr}}) \tag{31.32}$$

In the last step, $z\, g'_\nu(z) = g_{\nu-1}(z)$ was used; this follows immediately from the definition (31.10).

For high temperatures $(\lambda^3/v \ll 1)$ we may use $z = \exp(\beta\mu) \ll 1$. With $g_\nu(z) \approx z + z^2/2^\nu$ we calculate (31.32) in first order in z:

$$\frac{c_V(T)}{k_B} \approx \frac{3}{2}\left(1 + \frac{z}{2^{7/2}}\right) \approx \frac{3}{2}\left[1 + \frac{\zeta(3/2)}{2^{7/2}}\left(\frac{T_c}{T}\right)^{3/2}\right] \qquad (T \gg T_{\mathrm{cr}}) \tag{31.33}$$

The leading contribution $c_V(T) = 3k_B/2$ is the classical limit. In the (small) correction term with $z/2^{7/2}$, we used in the leading order $z \approx \lambda^3/v$, and (31.24). The result (31.33) may also be obtained from (30.17).

From (31.28) and (31.32) (note $g_{1/2}(1) = \infty$) we get the same value for $T = T_{\mathrm{cr}}$:

$$\frac{c_V(T)}{k_B} = \frac{15}{4}\frac{\zeta(5/2)}{\zeta(3/2)} \approx 1.925 \qquad (T = T_{\mathrm{cr}}^{\pm}) \tag{31.34}$$

Just like $N_0(T)$, the specific heat $c_V(T)$ is continuous at $T = T_{\mathrm{cr}}$ but has a kink. The approximation (31.33) yields $c_V(T_{\mathrm{cr}}) \approx 1.846\, k_B$; it is therefore valid beyond the range $T \gg T_{\mathrm{cr}}$. From (31.28), (31.33) and (31.34) we obtain the behavior of the specific heat. It is shown in Figure 31.4.

In (31.15), the density of liquid ^{4}He leads to the transition temperature of 3.13 K. In fact, liquid ^{4}He exhibits a phase transition at $T_{\mathrm{cr}} = 2.17$ K, the so-called λ transition. Since liquid ^{3}He does not exhibit such a transition, this transition must be associated with the exchange symmetry (the interactions between the atoms are practically the same for ^{4}He and ^{3}He). Since the exchange symmetry and not the interaction is decisive for the λ–transition, it makes sense to consider the ideal Bose gas as the simplest model for ^{4}He, and to compare the Bose-Einstein condensation with the λ-transition (Chapter 38).

Thermal equation of state

From (29.44) we obtain for the thermal equation of state

$$P(T, V, N) = \frac{2}{3}\frac{E(T, V, N)}{V} \tag{31.35}$$

For $T > T_{\text{cr}}$, the energy $E(T, V, N)$ results from $E(T, V, \mu)$ and $N(T, V, \mu)$, (31.19) and (31.27), by eliminating μ. For $T < T_{\text{cr}}$ it is sufficient to insert $\mu = 0$ into $E(T, V, \mu)$.

We discuss some features of the thermal equation of state for $T < T_{\text{cr}}$. For $\mu = 0$, Equation (31.35) becomes

$$P(T) = \frac{2}{3}\frac{E(T, V, \mu = 0)}{V} \overset{(31.27)}{=} \frac{k_{\text{B}}T}{\lambda^3}\,\xi(5/2) \tag{31.36}$$

Surprisingly, this pressure does not depend on the volume V (or on V/N), but only on the temperature. Because of $(\partial P/\partial V)_T = 0$, the system offers no resistance against a reduction of its volume. Rather, a reduction in volume simply forces more particles into the condensate.

In the P-V diagram, $(\partial P/\partial V)_T = 0$ means a horizontal isotherm (for $T < T_{\text{cr}}$). This can be compared with the horizontal isotherms of the van der Waals gas (Figure 37.2). Along this horizontal line, the transition between two phases (liquid to gaseous in Figure 37.2, condensed to non-condensed here) takes place.

The statement $\partial P/\partial V = 0$ for $T < T_{\text{cr}}$ has to be modified in two respects. On the one hand, the condensed particles in a finite volume have a non-vanishing zero point energy. This means that they also make a (tiny) contribution to the energy and the pressure. This contribution disappears in the thermodynamic limit (31.21); however, every real system has a finite volume. On the other hand, in the real system (for example in liquid helium, which will be discussed in more detail in Chapter 38) there are interactions, which lead to a finite value of $\partial P/\partial V$.

Ideal Bose gas in an oscillator

We transfer some results to the case of an ideal Bose gas to an oscillator potential. The single particle energies in the spherical oscillator are $\varepsilon_{n_x n_y n_z} = \hbar\omega(n_x + n_y + n_z + 3/2)$. The particle number condition

$$N = \sum_{n_x=0}^{\infty}\sum_{n_y=0}^{\infty}\sum_{n_z=0}^{\infty}\frac{1}{\exp[\beta(\varepsilon_{n_x n_y n_z} - \mu)] - 1} \tag{31.37}$$

determines the chemical potential μ. The summand is the mean occupation number for the level with the quantum numbers n_x, n_y and n_z. The lowest level has the energy $\varepsilon_{000} = 3\hbar\omega/2$ (instead of $\varepsilon_0 \to 0$ for $V \to \infty$). The transition point at which it comes to a macroscopic occupation of the ground state results from

$$\mu \to \mu_{\mathrm{cr}} = \frac{3}{2}\,\hbar\omega \tag{31.38}$$

For $T \leq T_{\mathrm{cr}}$, the chemical potential is $\mu = \mu_{\mathrm{cr}}$ and $\varepsilon_{n_x n_y n_z} - \mu = \hbar\omega(n_x + n_y + n_z)$. For $k_{\mathrm{B}}T \gg \hbar\omega$, the sum in (31.37) can be evaluated as an integral. Thereby the occupation number N_0 of the oscillator ground state must be taken into account separately:

$$N = N_0 + \int_0^\infty dn_x \int_0^\infty dn_y \int_0^\infty dn_z \,\frac{1}{\exp[\beta\hbar\omega(n_x + n_y + n_z)] - 1} \qquad (T \leq T_{\mathrm{cr}}) \tag{31.39}$$

The integrand can be expressed by the geometric series $\sum_{l=1}^\infty \exp(-[\dots]l)$; the individual integrals can then be then be carried out elementarily. This results in

$$N = N_0 + \zeta(3)\left(\frac{k_{\mathrm{B}}T}{\hbar\omega}\right)^3 \qquad (T \leq T_{\mathrm{cr}}) \tag{31.40}$$

where $\zeta(3) = \sum_{l=1}^\infty l^{-3} \approx 1.202$. The critical temperature results from $N_0 \to 0$, i.e.

$$k_{\mathrm{B}}T_{\mathrm{cr}} = \hbar\omega\left(\frac{N}{\zeta(3)}\right)^{1/3} \approx 0.94\,\hbar\omega\,N^{1/3} \tag{31.41}$$

This turns (31.40) into

$$\frac{N_0}{N} = 1 - \left(\frac{T}{T_{\mathrm{cr}}}\right)^3 \qquad (T \leq T_{\mathrm{cr}}) \tag{31.42}$$

This can be compared with (31.25). The condensate consists of all atoms that are in the oscillator ground state.

The thermodynamic limit (31.21) requires a finite density, $V/N = \mathrm{const}$. The volume of the system now under consideration is proportional to b^3, where $b = \hbar/m\omega$ is the oscillator length. The thermodynamic limit is therefore given by

$$N \to \infty, \qquad \hbar\omega \to 0, \qquad N\,(\hbar\omega)^3 = \mathrm{const}. \tag{31.43}$$

With this, (31.41) yields a finite value for the transition temperature T_{cr}.

Real Bose gas in an atomic trap

Atoms with an even total number of fermions, such as a hydrogen atom (1 proton, 1 electron) or a lithium atom (7 nucleons, 3 electrons), have integer spin and are therefore bosons. We consider specifically atoms with an odd number of electrons.

Such atoms have a magnetic moment of the size of the Bohr magneton. One can now place some 10^3 to about 10^7 of these atoms together in a magnetic trap and bring them to low temperatures (below 10^{-6} K). Such an arrangement was first used in 1995 for ^{87}Rb- and ^{23}Na atoms to observe Bose-Einstein condensation[1]. Alternatively, optical atom traps are used in which the laser light induces electric dipole moment of the atoms.

The forces in an atom trap can be approximated by an oscillator potential. If we neglect the interactions between the atoms, we can use the results of the previous section. From (31.41) we obtain

$$T_{cr} \sim 3 \cdot 10^{-7}\,\text{K} \qquad \left(\text{for } \hbar\omega \approx 10^{-8}\,k_B\text{K and } N = 4 \cdot 10^4\right) \qquad (31.44)$$

as a typical value for the transition temperature. Because of the finite number of atoms, the thermodynamic limit (31.43) holds at most approximately. As a consequence, the transition is not completely sharp; a kink like the one at T_{cr} in Figure 31.4 is therefore smeared out.

The condensate consists of the atoms that are in the oscillator ground state. In momentum space, the condensate forms an excessive portion with small momenta (corresponding to the oscillator ground state), which stands out clearly from the broader momentum distribution of the other particles. If one switches off the magnetic field of the trap, the atoms dispers freely according to their the initial momenta; the density of this expanding cloud of atoms can be observed by light absorption.

In an infinite system, the condensate cannot be spatially separated from the other particles. This is different for Bose-Einstein condensation in an atom trap: Here, the spatial density distribution consists of a broad distribution and additionally an pronounced, narrow Gaussian distribution, which is proportional to the square of the oscillator ground state wave function (width $b = \hbar/m\omega \sim 10^{-6}$ m). This density distribution can be measured by light scattering. Such experiments demonstrate the Bose-Einstein condensation in space and in momentum space.

The density n of the gas in the magnetic trap is rather small:

$$n\,|a|^3 < 10^{-3} \qquad (31.45)$$

Here a is the scattering length of the s-wave; $a = (m/4\pi\hbar^2)\int d^3r\,V(r)$ with the atom-atom potential $V(r)$. Under the conditions considered (temperature $T \lesssim T_{cr}$ and normal pressure), the thermodynamic equilibrium state of the system is actually a solid body. Because of (31.45) collisions are so rare, however, that the gas state is (meta)stable for a sufficiently long time.

So far, we have treated the atoms in the trap as an ideal gas. Neglecting the interaction between the atoms assumes that their energy contribution E_{int} is small compared to the kinetic energy E_{kin}. For the ratio of these energies one obtains

$$\frac{E_{int}}{E_{kin}} = \frac{N\,|a|}{b} \qquad (31.46)$$

---

[1]For a brief introduction to the experiments, we refer to W. Petrich, Physikalische Blätter 52 (1996) 345, for a detailed overview of the *Theory of trapped Bose-condensed gases* by F. Dalfovo et al., see also www.arxiv.org/cond-mat/9806038.

Here a is the scattering length and $b = \hbar/m\omega$ is the oscillator length. The ratio (31.46) can be much larger than 1, even if the gas is very dilute (31.45). This leads to deviations from the ideal behavior. The transition temperature remains, however, approximately the same.

Exercises

31.1   Specific heat of the Bose gas for high temperatures

The specific heat of the ideal Bose gas reads

$$\frac{c_V(T)}{k_\mathrm{B}} = \frac{15}{4}\frac{g_{5/2}(z)}{g_{3/2}(z)} - \frac{9}{4}\frac{g_{3/2}(z)}{g_{1/2}(z)} \qquad (T \geq T_\mathrm{cr})$$

Here $g_\nu(z)$ is the generalized Riemann zeta function, $z = \exp(\beta\mu)$, and T_cr is the transition temperature. Determine the two leading terms of the specific heat for $T \gg T_\mathrm{cr}$.

31.2   Bose gas in the oscillator

The condensate of an ideal Bose gas in the harmonic oscillator consists of the N_0 particles in the ground state. If a condensate is present, then the particle number condition reads

$$N = N_0 + \int_0^\infty dn_x \int_0^\infty dn_y \int_0^\infty dn_z \frac{1}{\exp[\beta\hbar\omega(n_x + n_y + n_z)] - 1} \qquad (31.47)$$

The oscillator energies are $\varepsilon = \hbar\omega(n_x + n_y + n_z + 3/2)$. In the expression for the mean particle numbers, $\mu = 3\hbar\omega/2$ was used; this applies, if a finite fraction of all particles are in the ground state.

Write the integrand as a geometric series $\sum_{l=1}^\infty \exp(-[...]l)$ and carry out the integration. Determine N_0 as a function of temperature.

31.3   Bose gas in two dimensions

For an ideal Bose gas in two dimensions: Show that there is no Bose-Einstein condensation. To do this, evaluate the relationship between the particle number N and the chemical potential μ and discuss the result for $\mu \to 0$.

32 Ideal Fermi gas

We investigate the ideal Fermi gas for low temperatures. This model is applied to the mobile electrons in a metal. The contribution of the electrons to the specific heat increases linearly with the temperature.

We consider non-relativistic fermions with spin $1/2$ and mass m. The single particle states are characterized by the momentum p and the spin projection s_z. They shall have the energy

$$\varepsilon_p = \frac{p^2}{2m} \tag{32.1}$$

According to (29.31), the mean number of fermions in a single particle state is

$$\overline{n_p} = \frac{1}{\exp\left[\beta(\varepsilon_p - \mu)\right] + 1} \tag{32.2}$$

The thermodynamic energy and the particle number are given by

$$E(T, V, \mu) = \sum_{s_z,\, p} \varepsilon_p\, \overline{n_p}\,, \qquad N(T, V, \mu) = \sum_{s_z,\, p} \overline{n_p} \tag{32.3}$$

As in (30.8), the sum over the discrete momenta is performed as an integral:

$$\sum_{s_z,\, p} \ldots = \frac{2V}{(2\pi\hbar)^3} \int d^3p \,\ldots \tag{32.4}$$

The sum over s_z yields a factor 2. The *ideal Fermi gas* is defined by (32.1)–(32.4). Eliminating the chemical potential μ in (32.3) yields the energy $E(T, V, N)$. From $E(T, V, N)$ one obtains the specific heat c_V and the pressure (29.44), and thus all thermodynamic quantities.

In (32.4) we replace the integration variable p by $\varepsilon = \varepsilon_p = p^2/2m$:

$$\frac{2V}{(2\pi\hbar)^3} \int d^3p \ldots = \frac{2V}{(2\pi\hbar)^3} \int_0^\infty d\varepsilon\, 4\pi \sqrt{2m\varepsilon}\, m \ldots = N \int_0^\infty d\varepsilon\, z(\varepsilon) \ldots \tag{32.5}$$

The density of states

$$z(\varepsilon) = \text{const.} \cdot \frac{V}{N} \sqrt{\varepsilon} \tag{32.6}$$

specifies the number of single particle states per energy interval. The constant contains numerical factors and the mass m. The following calculations are first carried out for an initially unspecified state density $z(\varepsilon)$. The interactions in the real system (such as the electron-lattice interaction) can be partly taken into account by using an effective energy dependent mass in (32.1). This leads to deviations from (32.6).

After introducing the density of states $z(\varepsilon)$, Equations (32.2)–(32.4) are evaluated in the following form:

$$\overline{n(\varepsilon)} = \frac{1}{\exp\left[\beta(\varepsilon - \mu)\right] + 1} \tag{32.7}$$

$$1 = \int_0^\infty d\varepsilon \, z(\varepsilon) \, \overline{n(\varepsilon)} \tag{32.8}$$

$$\frac{E}{N} = \int_0^\infty d\varepsilon \, z(\varepsilon) \, \varepsilon \, \overline{n(\varepsilon)} \tag{32.9}$$

Equation (32.8) determines the chemical potential $\mu(T, v)$; it depends via $\overline{n(\varepsilon)}$ on the temperature and via $z(\varepsilon)$ on $v = V/N$. If one inserts $\mu(T, v)$ into (32.9), one gets $E(T, V, N)$.

For $T \to 0$ or $\beta \to \infty$, the exponential function in the denominator of (32.7) is zero or infinite depending on the sign of $\varepsilon - \mu$. From this follows

$$\overline{n(\varepsilon)} \xrightarrow{T \to 0} \Theta(\mu - \varepsilon) = \begin{cases} 1 & \text{for} \quad \varepsilon < \mu(0, v) \\ 0 & \text{for} \quad \varepsilon > \mu(0, v) \end{cases} \tag{32.10}$$

The function $\Theta(x)$ is called the step or theta function. The value of μ at $T = 0$ is called *Fermi energy*:

$$\varepsilon_{\mathrm{F}} = \frac{p_{\mathrm{F}}^2}{2m} = \mu(0, v) \qquad \text{(Fermi energy)} \tag{32.11}$$

Figure 32.1 on the left shows the occupation numbers (32.10) as a function of the energy. All levels below ε_{F} are occupied, while those above of ε_{F} are empty. This is also called *Fermi sea*; an electron at the sea level has the Fermi energy. In momentum space, all single particle states with $|\boldsymbol{p}| \le p_{\mathrm{F}}$ are occupied; this corresponds to a sphere in momentum space.

The result (32.10) means that for $T \to 0$ the system goes into the lowest possible energy state. Due to the Pauli principle, each of the lowest single particle levels is occupied by just one particle. For a given number of particles N, this results in a occupation boundary p_{F} between occupied and unoccupied levels. We calculate this *Fermi momentum* p_{F} from the condition that just N particles can be accommodated in the states $|\boldsymbol{p}| \le p_{\mathrm{F}}$,

$$N = \sum_{s_z = \pm 1/2} \sum_{|\boldsymbol{p}| \le p_{\mathrm{F}}} 1 = \frac{2V}{(2\pi\hbar)^3} \int_{|\boldsymbol{p}| \le p_{\mathrm{F}}} d^3p = \frac{2V}{(2\pi\hbar)^3} \frac{4\pi}{3} p_{\mathrm{F}}^3 \tag{32.12}$$

This results in

$$p_{\mathrm{F}} = \left(3\pi^2\right)^{1/3} \frac{\hbar}{v^{1/3}}, \qquad \varepsilon_{\mathrm{F}} = \left(3\pi^2\right)^{2/3} \frac{\hbar^2}{2m\, v^{2/3}} \tag{32.13}$$

The ideal Fermi gas can serve as a model for the following systems:

- Atom: electrons of the shell.

- Atomic nucleus: nucleons.

- Metal: electrons in the conduction band.

- Helium liquid: ^{3}He atoms.

- White dwarf: electrons.

The first two cases are known as the shell model (of the atom, or of the atomic nucleus). Due to the small number of particles, a statistical treatment is only possible to a limited extent for these cases.

For the systems listed, we specify the size of the Fermi energy:

$$\varepsilon_F = \frac{p_F^2}{2m} \approx \begin{cases} 10\,\text{eV} & \text{atom, metal } (10^{-8}\,\text{cm}) \\ 10^{-4}\,\text{eV} & {}^3\text{He–liquid } (4 \cdot 10^{-8}\,\text{cm}) \\ 35\,\text{MeV} & \text{atomic nucleus } (10^{-13}\,\text{cm}) \\ 10^6\,\text{eV} & \text{white dwarf } (10^{-11}\,\text{cm}) \end{cases} \tag{32.14}$$

For m the respective mass (electron for atom, metal, white dwarf, ^{3}He atom, nucleon for the atomic nucleus) was used. The approximate size of $v^{1/3}$ is given in the brackets.

In a metal like copper the atoms are placed at regular lattice points. The weakest bound electron of each atom is mobile, that means it can move within the lattice; therefore the metal is a conductor. The mobile electrons in a metal can be described approximately as an ideal Fermi gas. For copper with $v = V/N = 12\,\text{Å}^3$ we estimate the Fermi energy:

$$\varepsilon_F = \left(3\pi^2\right)^{2/3} \frac{\hbar^2}{2m_e v^{2/3}} \approx 7\,\text{eV} \approx 8 \cdot 10^4\,k_B\,\text{K} \tag{32.15}$$

Here, $\hbar^2/(m_e \text{Å}^2) \approx 7.6\,\text{eV}$ is used.

In order to release an electron from the metal, work W of about 2 to 3 eV is required. Form this point of view, an electron at the Fermi edge (at the surface of the Fermi sea) has therefore a negative energy (relative to an electron outside the metal). This can be described by a potential U for the electrons, which is constant and negative in the metal, $U = -\varepsilon_F - W \approx -10\,\text{eV}$, and which disappears outside the metal. Adding such a potential to ε in (32.1) would result in a constant shift of μ.

In the following, we consider the ideal Fermi gas primarily as a model for the electrons in the conduction band of a metal. For the considered temperatures we have

$$k_B T \ll \varepsilon_F \sim 10^5\,k_B\,\text{K} \qquad \text{(metal)} \tag{32.16}$$

The other case $k_B T \sim \varepsilon_F$ can come into question for a ^{3}He liquid.

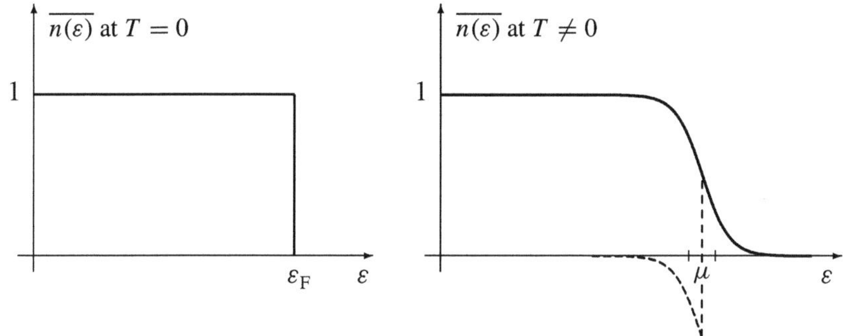

Figure 32.1   Mean occupation number $\overline{n(\varepsilon)}$ of the ideal Fermi gas for $T = 0$ and $T \neq 0$. For $T = 0$, the system is in its quantum mechanical ground state. For $T \neq 0$, the transition from $\overline{n} \approx 1$ to $\overline{n} \approx 0$ takes place in a range of a few $k_{\mathrm{B}}T$ around $\mu \approx \varepsilon_{\mathrm{F}}$. For the electrons in the metal under normal condition $k_{\mathrm{B}}T/\varepsilon_{\mathrm{F}} < 10^{-2}$ holds. Then the "softening of the Fermi edge" would not be visible in the given scale; the right part therefore uses a larger temperature. Here also the difference $\eta(x)$ between the cases $T \neq 0$ and $T = 0$ is shown as a dashed line (which coincides with the solid line for $\varepsilon \geq \mu$). The function $\eta(x)$ is non zero in the vicinity of $\varepsilon = \mu$, and it is an odd function with respect $\varepsilon = \mu$. The energies $\varepsilon = \mu \pm k_{\mathrm{B}}T$ are shown by small marks an the abscissa.

For $k_{\mathrm{B}}T \ll \varepsilon_{\mathrm{F}}$, the mean occupation numbers are sketched on the right in Figure 32.1. For finite temperature, the Fermi edge (border between occupied and unoccupied levels) is smeared out. The transition between $\overline{n} = 1$ and $\overline{n} = 0$ takes place in a range of $k_{\mathrm{B}}T$. This can easily be seen from some special values:

$$
\overline{n(\varepsilon)} = \begin{cases} 0.50 & \varepsilon = \mu \\ 0.50 \pm 0.23 & \varepsilon = \mu \mp k_{\mathrm{B}}T \\ 0.50 \pm 0.45 & \varepsilon = \mu \mp 3k_{\mathrm{B}}T \end{cases} \tag{32.17}
$$

The chemical potential μ depends on the temperature; this dependence will be given below. For $k_{\mathrm{B}}T \ll \varepsilon_{\mathrm{F}}$, however, $\mu \approx \varepsilon_{\mathrm{F}}$ holds.

We consider the difference $\eta(x)$ between the mean occupation number at finite temperature and the theta function $\Theta(\mu - \varepsilon)$,

$$
\eta(x) = \overline{n(\varepsilon)} - \Theta(\mu - \varepsilon) = \frac{1}{\exp(x) + 1} - \Theta(-x), \qquad x = \beta(\varepsilon - \mu) \tag{32.18}
$$

It is easy to show that this function is odd:

$$
\eta(-x) = -\eta(x) \tag{32.19}
$$

In the following, we calculate the specific heat using the *Sommerfeld expansion*, a technique worked out by the German physicist Arnold Sommerfeld for solving certain integrals. We first split integrals of the form (32.8) or (32.9) into two parts,

$$
\int_0^\infty d\varepsilon \; f(\varepsilon) \, \overline{n(\varepsilon)} = \int_0^\mu d\varepsilon \; f(\varepsilon) + \int_0^\infty d\varepsilon \; f(\varepsilon) \left(\overline{n(\varepsilon)} - \Theta(\mu - \varepsilon) \right) \tag{32.20}
$$

Because of (32.16), $\overline{n(\varepsilon)} - \Theta(\mu - \varepsilon)$ is non zero only in a small range around μ. The functions $f(\varepsilon)$ in question (such as $\varepsilon^{1/2}$ or $\varepsilon^{3/2}$) vary only weakly in this range and can therefore be expanded:

$$f(\varepsilon) = f(\mu) + f'(\mu)\,(\varepsilon - \mu) + \frac{f''(\mu)}{2}\,(\varepsilon - \mu)^2 + \ldots \tag{32.21}$$

We insert this expansion into (32.20), where we use the dimensionless variable $x = \beta(\varepsilon - \mu)$ and the function $\eta(x)$ from (32.18):

$$\int_0^\infty d\varepsilon\; f(\varepsilon)\,\overline{n(\varepsilon)} = \int_0^\mu d\varepsilon\; f(\varepsilon) + \frac{1}{\beta}\int_{-\beta\mu}^\infty dx\,\left(f(\mu) + f'(\mu)\,\frac{x}{\beta} + \ldots\right)\eta(x) \tag{32.22}$$

Because of $\beta\mu \gg 1$, the lower limit of the second integral can be replaced by $-\infty$. Because of $\eta(-x) = -\eta(x)$, the integral over the even powers then vanishes; the next contribution in (32.22) would come from the term with $f'''(\mu)\,x^3/\beta^3$. Thus we get

$$\int_0^\infty d\varepsilon\; f(\varepsilon)\,\overline{n(\varepsilon)} = \int_0^\mu d\varepsilon\; f(\varepsilon) + \frac{f'(\mu)}{\beta^2}\int_{-\infty}^\infty dx\, x\, \eta(x) + \mathcal{O}(T^4) \tag{32.23}$$

The term $\mathcal{O}(T^4)$ has the relative size $(k_\mathrm{B}T/\varepsilon_\mathrm{F})^2$ compared to the previous term. This relative magnitude is about 10^{-5} for (32.15) and room temperature. In the following we neglect these higher terms. With

$$\int_{-\infty}^{+\infty} dx\, x\, \eta(x) \overset{(32.19)}{=} 2\int_0^{+\infty} dx\, x\, \eta(x) = 2\int_0^{+\infty} \frac{x}{\exp(x) + 1} = \frac{\pi^2}{6} \tag{32.24}$$

we get

$$\int_0^\infty d\varepsilon\; f(\varepsilon)\,\overline{n(\varepsilon)} = \int_0^\mu d\varepsilon\; f(\varepsilon) + \frac{\pi^2}{6\beta^2}\,f'(\mu) \qquad (k_\mathrm{B}T \ll \varepsilon_\mathrm{F}) \tag{32.25}$$

Using this we evaluate (32.8) and (32.9):

$$1 = \int_0^\mu d\varepsilon\; z(\varepsilon) + \frac{\pi^2}{6\beta^2}\,z'(\mu) \tag{32.26}$$

$$\frac{E}{N} = \int_0^\mu d\varepsilon\; z(\varepsilon)\,\varepsilon + \frac{\pi^2}{6\beta^2}\Big(\mu\,z'(\mu) + z(\mu)\Big) \tag{32.27}$$

For the integral in (32.26) we write

$$\int_0^\mu d\varepsilon\; z(\varepsilon) = \int_0^{\varepsilon_\mathrm{F}} d\varepsilon\; z(\varepsilon) + \int_{\varepsilon_\mathrm{F}}^\mu d\varepsilon\; z(\varepsilon) = 1 + (\mu - \varepsilon_\mathrm{F})\,z(\widetilde{\varepsilon}) \tag{32.28}$$

The value 1 for the first integral follows from (32.26) with $T = 0$. The mean value theorem was used for the second integral; therefore $\widetilde{\varepsilon}$ has some value between μ and ε_F. Inserting (32.28) into (32.26) we see

$$\mu = \varepsilon_\mathrm{F} + \mathcal{O}(T^2) \qquad (k_\mathrm{B}T \ll \varepsilon_\mathrm{F}) \tag{32.29}$$

Therefore, in the correction terms of (32.26) – (32.28) we can use $\mu \approx \varepsilon_F$ and $\widetilde{\varepsilon} \approx \varepsilon_F$ (corrections to this would again result in terms $\mathcal{O}(T^4)$ which we do not take into account). From (32.26) and (32.28) we obtain then

$$\mu - \varepsilon_F = -\frac{\pi^2}{6\beta^2}\frac{z'(\varepsilon_F)}{z(\varepsilon_F)} \tag{32.30}$$

Specifically for (32.6), $z'(\varepsilon_F)/z(\varepsilon_F) = 1/(2\varepsilon_F)$. This results in

$$\mu = \varepsilon_F\left[1 - \frac{\pi^2}{12}\left(\frac{k_B T}{\varepsilon_F}\right)^2\right] \qquad (k_B T \ll \varepsilon_F) \tag{32.31}$$

From (30.12) follows $\mu \to -\infty$ for $T \to \infty$. The temperature dependence of $\mu(T, v)$ is sketched in Figure 32.2.

We evaluate the energy (32.27), calculating the integral as in (32.28), and using $\mu \approx \varepsilon_F$ and $\widetilde{\varepsilon} \approx \varepsilon_F$ in the correction terms:

$$\frac{E}{N} = \underbrace{\int_0^{\varepsilon_F} d\varepsilon\, \varepsilon\, z(\varepsilon)}_{= E_0/N} + (\mu - \varepsilon_F)\, \varepsilon_F\, z(\varepsilon_F) + \frac{\pi^2}{6\beta^2}\left(\varepsilon_F\, z'(\varepsilon_F) + z(\varepsilon_F)\right) \tag{32.32}$$

Here we insert (32.30):

$$E(T, V, N) = E_0 + \frac{\pi^2}{6}\, N\, \frac{z(\varepsilon_F)}{\beta^2} = E_0(V, N) + \frac{\pi^2}{6}\, N\, (k_B T)^2\, z(\varepsilon_F) \tag{32.33}$$

This means that the specific heat increases linearly with the temperature:

$$c_V(T) = \frac{1}{N}\frac{\partial E(T, V, N)}{\partial T} = \frac{\pi^2}{3}\, k_B^2\, T\, z(\varepsilon_F) \tag{32.34}$$

We now evaluate the results specifically for the state density (32.6). First, we consider $T = 0$ and insert $z(\varepsilon) = A\, \varepsilon^{1/2}$ into the particle number condition,

$$1 = \int_0^{\varepsilon_F} d\varepsilon\, z(\varepsilon) = A\int_0^{\varepsilon_F} d\varepsilon\, \sqrt{\varepsilon} = \frac{2}{3}\, A\, \varepsilon_F^{3/2} = \frac{2}{3}\, z(\varepsilon_F)\, \varepsilon_F \tag{32.35}$$

From this follows

$$z(\varepsilon) = \frac{3}{2\,\varepsilon_F}\sqrt{\frac{\varepsilon}{\varepsilon_F}} \quad \text{and} \quad z(\varepsilon_F) = \frac{3}{2\,\varepsilon_F} \tag{32.36}$$

We use this in (32.33) and (32.34):

$$E = E_0 + \frac{\pi^2}{4}\frac{k_B T}{\varepsilon_F}\, N k_B T \tag{32.37}$$

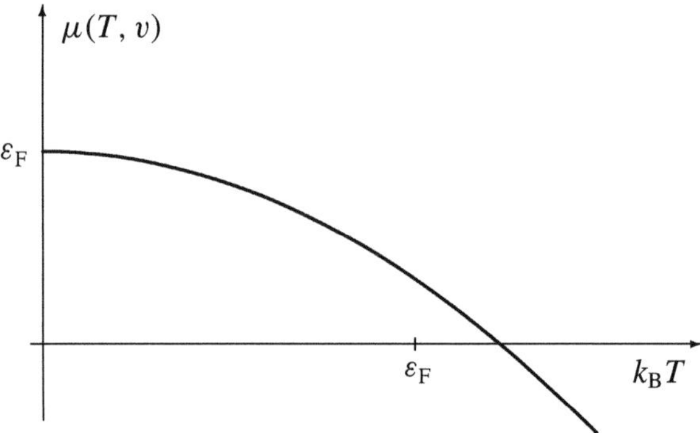

Figure 32.2 Chemical potential (32.31) of the ideal Fermi gas as a function of the temperature ($v = $ const.).

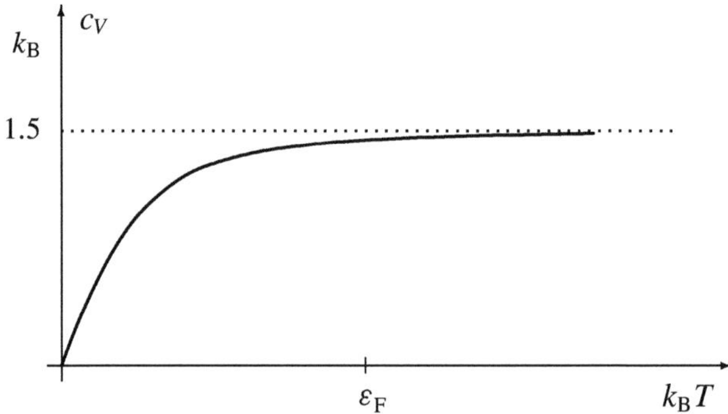

Figure 32.3 Temperature dependence of the specific heat of the ideal Fermi gas. For the mobile electrons in a metal, only the range $k_B T \ll \varepsilon_F$ is relevant.

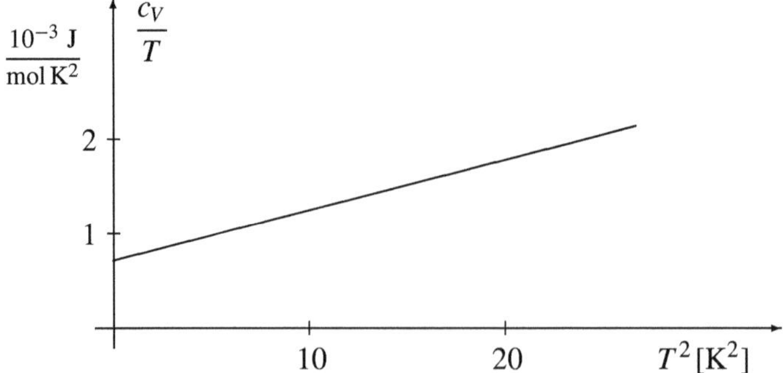

Figure 32.4 The specific heat (32.39) for a metal. A plot of c_V/T over T^2 results in a straight line. By fitting experimental results to this line one obtains experimental values for α and γ in the approach (32.39).

$$\boxed{c_V = \frac{\pi^2}{2} \frac{k_B T}{\varepsilon_F} k_B}$$

$$(32.38)$$

This result can be easily understood looking at Figure 32.1: Because of $k_B T \ll \varepsilon_F$, only a fraction $\mathcal{O}(k_B T/\varepsilon_F)$ of all particles can be excited thermally. On average, the excited particles receive an energy $\mathcal{O}(k_B T)$. This explains (32.37) and (32.38) except for the numerical factors. Compared to the classical expectation $c_V = \mathcal{O}(k_B)$, the specific heat contains an additional factor $k_B T/\varepsilon_F$.

For high temperatures ($k_B T \gg \varepsilon_F$), the classical result $c_V = 3k_B/2$ will be regained. The overall results for the specific heat shown in Figure 32.3. However, when applied to the metal, only the range $k_B T \ll \varepsilon_F$ comes into question.

In order to check the result (32.38) experimentally, one measures the specific heat of a metal. Hereby, the lattice vibrations of the crystal contribute to c_V. For not too high temperatures this contribution is proportional to T^3 (Chapter 33). Now on fits a curve of the form

$$c_V = \alpha T^3 + \gamma T$$

$$(32.39)$$

to the measured specific heat (Figure 32.4); this results in experimental values for α and γ. From the value for γ, the Fermi energy ε_F can be determined according to (32.38), or more generally the state density at the Fermi edge according to (32.34).

Fermi pressure

According to (29.44), the pressure of the non-relativistic Fermi gas is

$$P(T, V, N) = \frac{2\,E(T, V, N)}{3\,V} = \frac{2\,E_0(V, N)}{3\,V} + \frac{\pi^2}{6} \frac{N k_B T}{V} \frac{k_B T}{\varepsilon_F}$$

$$(32.40)$$

Using (32.36), we obtain for the ground state energy (32.32):

$$\frac{E_0(V, N)}{N} = \int_0^{\varepsilon_F} d\varepsilon\, z(\varepsilon)\, \varepsilon = \frac{3}{5}\,\varepsilon_F$$

$$(32.41)$$

In contrast to the ideal classical gas or the ideal Bose gas, the pressure of the Fermi gas does not approach zero for $T \to 0$. It rather approaches the finite value

$$P_{\text{Fermi}} = P(0, V, N) = \frac{2}{5} \frac{N}{V} \varepsilon_F = \frac{(3\pi^2)^{2/3}}{5} \frac{\hbar^2}{m\, v^{5/3}}$$

$$(32.42)$$

This *Fermi pressure* is the reason for the relative *incompressibility* of ordinary solid or liquid matter. Of course, the internal structure of this matter is very different from an ideal Fermi gas of electrons. The decisive prerequisites for (32.42) are the *uncertainty principle* and the *Pauli principle*. These are independent of the specific structure of the condensed matter (liquid or solid) and lead to a Fermi pressure of this size. In simplified terms, due to the Pauli principle, one electron only has effectively only $v = V/N$ as an available volume. Because of the uncertainty principle, it then has a minimum energy of the size $\hbar^2/(m\,v^{2/3})$. To halve the volume of condensed matter ($v \to v/2$), this kinetic energy of the electrons must be increased by about 60%. According to (32.15), this requires this requires an energy of about $4\,\text{eV}\,(Z_1 + Z_2 + \dots)$ per molecule, where Z_1, Z_2,... are the atomic numbers of the atoms of the molecule. This energy exceeds vastly the energy set free in an explosive chemical reaction (approximately 1 eV per molecule in oxyhydrogen gas). It is therefore relatively difficult (energy consuming) to compress condensed matter.

Finally, it should be mentioned that the Fermi pressure plays a central role in some stellar models. In a stellar equilibrium the matter pressure must balance the gravitational pressure. We briefly present some stellar equilibria:

- For our Earth, it is the one just described incompressibility of ordinary matter that prevents a further compression by gravity.

- In our Sun, the kinetic pressure of the plasma balances the gravitational pressure. This kinetic pressure can be expressed by $P = N k_B T / V$; where $T \approx 5 \cdot 10^7$ K in the interior of the Sun. This temperature is maintained by the fusion of hydrogen to helium. The Fermi pressure plays no role for this stellar equilibrium.

- The final stage of the Sun (mass $M_\odot$, radius $R_\odot$) could be a star made of helium, in which the atoms are crashed by the gravitational pressure (forming an electron soup with α-particles). It is then the Fermi pressure of the electrons that balances the gravitational pressure, and preserves the star from further compression. This is the star model of the *white dwarf* (see Chapter 47 in [3]). For a mass $M \approx M_\odot$, the equilibrium condition yields a (dwarf) radius $R \approx 10^{-2}\,R_\odot$. Such stars often have temperatures in the range $10^4 \dots 10^5$ K and therefore have a "white" spectrum. Because of $k_B T \ll \varepsilon_F$, the temperature plays practically no role for the stellar equilibrium.

- A star can be so massive and therefore so compact that the kinetic energy of the electrons exceeds $1.5\,m_e c^2$. Then the conversion into neutrons according to the reaction $p + e^- \to n + \nu_e$ is energetically favored. The resulting neutron star can be described as an equilibrium between the Fermi pressure of the neutrons and the gravitational pressure. Neutron stars are observed as *pulsars*; for a mass $M \approx M_\odot$ they have a radius R of about 5 to 10 km.

Exercises

32.1 Mean velocities in the Fermi gas

In the ideal Fermi gas at $T = 0$ all states up to the Fermi level at $p_F = m\,v_F$ are occupied; we consider spin 1/2 fermions. Express the mean velocity values $\bar{v}$ and $\overline{v^2}$ by the Fermi velocity v_F.

32.2 Relativistic ideal Fermi gas

Let the density of an ideal Fermi gas be so high that $\hbar/(V/N)^{1/3} \gg mc$ holds. Then most momenta are highly relativistic, $p \gg mc$, and one can use the approximation $\varepsilon \approx cp$. For this case, derive the Fermi momentum and the Fermi energy. For $T \approx 0$, determine the energy $E_0(V, N)$ and the pressure $P(V, N)$ does the system.

32.3 Current from hot cathode

In order to escape from a metal, electrons must overcome a potential barrier of height V_0 (relative to the Fermi energy ε_F). The electrons shall be treated as an ideal Fermi gas. Calculate the current density of the emitted electrons at the temperature T. This emission of electrons is called Richardson effect.

32.4 Pauli's paramagnetism

The electrons of a metal are considered as an ideal Fermi gas with the state density $z(\varepsilon)$. In an external magnetic field B, the single particle energies are

$$\varepsilon_\pm = \varepsilon \mp \mu_B B$$

The upper signs applies if the magnetic moment is parallel to the field. It is assumed that $\mu_B B \ll \varepsilon_F$. Calculate the number of parallel $(+)$ and antiparallel $(-)$ magnetic moments,

$$N_\pm = \sum \overline{n_+(\varepsilon)}$$

for $T \approx 0$. For $T \approx 0$ the mean occupation numbers $\overline{n_+(\varepsilon)}$ become Θ functions. Determine the magnetization $M(B) = \mu_B(N_+ - N_-)/V$.

32.5 Temperature dependent correction for paramagnetism

In the magnetic field B, the electrons of an ideal Fermi gas have the energies $\varepsilon_\pm = p^2/(2m) \mp \mu_B B$. Calculate the number $N_\pm$ of parallel $(+)$ and antiparallel $(-)$ magnetic momenta using the *Sommerfeld technique* up to the order T^2; one may use $k_B T \ll \varepsilon_F$ and $\mu_B B \ll \varepsilon_F$. Determine the magnetization $M(T, B) = \mu_B (N_+ - N_-)/V$.

33 Phonon gas

In the crystal lattice of a solid, the translational degrees of freedom lead to displacements of the atoms from their equilibrium positions. The atoms (or molecules) will then oscillate around their equilibrium position; and the coupling of these oscillations leads to lattice waves. The quantized excitations of the lattice waves are called phonons. The statistical treatment of the phonon gas (as an ideal Bose gas) leads to a characteristic temperature dependence of the specific heat. For low temperatures, the specific heat increases with the third power of the temperature.

Linear chain

The vibration of a lattice atom is coupled to the vibrations of the neighboring atoms. The first step is therefore to determination of the natural oscillations of the coupled system. For this we consider a one dimensional crystal model, the linear chain sketched in Figure 33.1. The equilibrium positions of the atoms are $x_n = a \cdot n$; where a denotes the lattice constant. Initially, we restrict the deflections to the x direction. The displacement of the nth atom from its equilibrium position is denoted by $q_n(t)$. For small deflections, the restoring force is proportional to the relative displacement $q_{n+1} - q_n$ between two neighboring atoms. This means that the force acting on the nth atom is

$$F_n = f\left(q_{n+1} - q_n\right) - f\left(q_n - q_{n-1}\right) \tag{33.1}$$

with a (spring) constant f. To solve the equations of motion

$$m\,\ddot{q}_n = F_n = f\left(q_{n+1} + q_{n-1} - 2q_n\right) \tag{33.2}$$

we use the approach

$$q_n(t) = A\,\exp\left[i(\omega t + kna)\right] \tag{33.3}$$

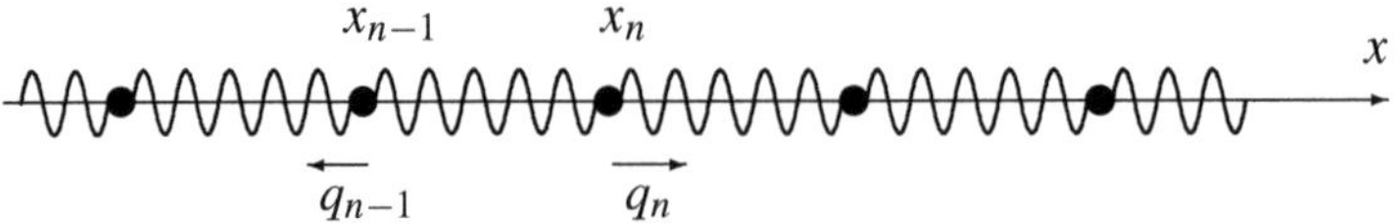

Figure 33.1 The one dimensional chain presents a simple model for lattice vibrations. The atoms with mass m have the equilibrium positions $x_n = a \cdot n$; the length a corresponds to the lattice constant of a crystal. The forces between the atoms are assumed to be harmonic.

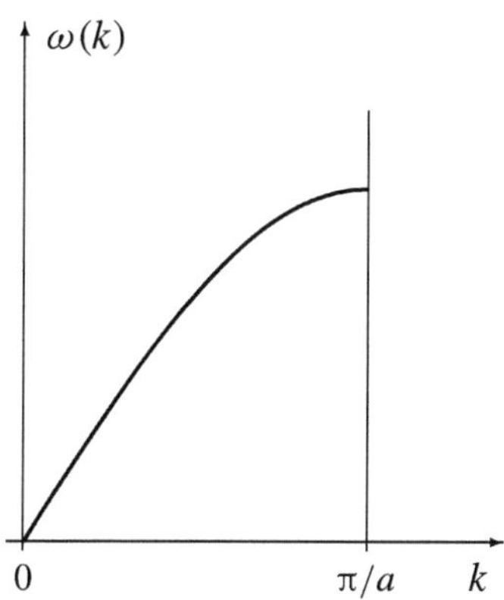

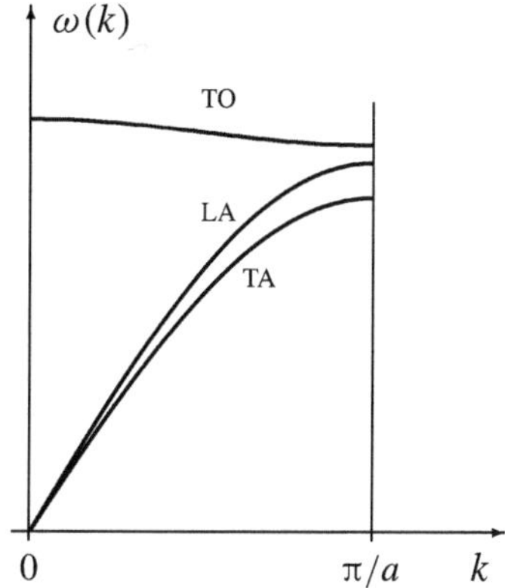

Figure 33.2 Dispersion relation $\omega = \omega(k)$ of the linear chain (left). In the real crystal there are deviations from this, similar to those sketched in the right part. Transverse and longitudinal waves (TA and LA, A stands for acoustic) have slightly different dispersion relations. In an ionic crystal there are also transverse optical waves (TO) with a finite $\omega(0)$.

We allow for a complex amplitude A with the proviso that the physical solution is equal to the real part of (33.3). The parameters k and ω are real. We substitute (33.3) into (33.2):

$$-m\,\omega^2 = f\left[\exp(ika) + \exp(-ika) - 2\right] = -4f\,\sin^2\left(\frac{ka}{2}\right) \qquad (33.4)$$

This yields the *dispersion relation*

$$\omega = \omega(k) = \sqrt{\frac{4f}{m}}\,\sin\left(\frac{|k|\,a}{2}\right) \qquad (33.5)$$

We restrict the frequency to positive values, $\omega > 0$. In (33.3), k and $k + 2\pi/a$ give the same solution. Therefore we can restrict the wave number k by

$$-\frac{\pi}{a} < k \le \frac{\pi}{a} \qquad (33.6)$$

Figure 33.2 sketches the dispersion relations for the linear chain and for a real crystal. Because $\omega(-k) = \omega(k)$, the graph can be restricted to the range $0 \le k \le \pi/a$.

With $A = C\exp(i\varphi)$ (with real C and φ) the real part of (33.3) becomes $q_n(t) = C\cos(\omega t + kna + \varphi)$. With $x = x_n = na$ we write this as a function of x and t:

$$q(x,t) = q_n(t) = C\cos(\omega t + kx + \varphi) \qquad (33.7)$$

This shows that the natural oscillations $q_n(t)$ can be considered as *waves* with the amplitude $q(x,t)$. The wave number $k = 2\pi/\lambda$ determines the wavelength λ. For large wavelengths ($\lambda \gg a$), many neighboring atoms oscillate in the same direction, i.e. in phase. At the smallest wavelength $\lambda_{\min} = 2\pi/k_{\max}$ neighboring atoms oscillate in antiphase; because for $k_{max} a = \pi$, Equation (33.7) yields $q_{n+1}/q_n = -1$.

We now consider a *finite* chain of length $L = Ma$ consisting of $M \gg 1$ atoms. For the finite chain we must specify boundary conditions. The simplest is the periodic boundary condition

$$q_1 = q_{1+M} \qquad\qquad (33.8)$$

One can imagine that the chain is closed to a circle so that the $(M+1)$-th atom is the first atom again. From this condition and (33.3) follows $\exp(ikaM) = 1$ or

$$k_i = \frac{2\pi}{aM} i \qquad i = 0, \pm 1, \pm 2, \ldots, \pm \frac{M-2}{2}, +\frac{M}{2} \qquad (33.9)$$

Taking into account the restriction (33.6), we obtain exactly M discrete k-values. Since the system has M degrees of freedom, there are also be M natural oscillations. The value $k_i = 0$ means a translation of the entire chain and can therefore be omitted. From (33.9) we see that the possible natural oscillations have equidistant k-values. The gap between two neighboring values is

$$\Delta k = \frac{2\pi}{L} \qquad\qquad (33.10)$$

Also if (33.8) is replaced by a physical boundary condition one obtains equidistant k-values (Exercise 33.1).

Crystal lattice

We generalize the previous results for a three-dimensional crystal lattice. We imagine that each atom of the linear chain is replaced by a two-dimensional lattice of atoms (y-z-plane). Then the solution (33.7) describes a displacement of this y-z-plane in the x-direction. In the three-dimensional crystal, this is a density wave or *sound wave*. The step from the chain to the crystal does not change the dispersion relation. One could now identify m and f in (33.5) as the mass and spring constant of the individual planes; thereby the ratio f/m remains unchanged.

The group velocity of the wave for low frequencies defines the *speed of sound*

$$c_S = \left.\frac{d\omega}{dk}\right|_{k=0} = \sqrt{\frac{f}{m}}\, a \qquad\qquad (33.11)$$

For ordinary sound $k \ll k_{max} = \pi/a \sim 1/\text{Å}$ holds. For example, iron with $c_S = 5000\,\text{m/s}$ and the audible frequency $\nu = \omega/2\pi = 5\,\text{kHz}$ leads to the wavelength $\lambda = 1\,\text{m}$, i.e. $k \sim 10^{-10}\, k_{max}$.

A general wave is of the form $A \exp[i(\boldsymbol{k}\cdot\boldsymbol{r} - \omega t)]$. With $\boldsymbol{k} = k\,\boldsymbol{e}_x$ we obtain from this the wave (33.7); the wave vector $\boldsymbol{k}$ points in the x direction. The deflections described by $q(x, t)$ also point in the x direction. A wave in which the wave vector and the deflection are parallel is called *longitudinal*.

For the chain in Figure 33.1, in addition to the displacements in the x-direction, we can also consider deflections in two perpendicular directions. Also this leads to wave solutions of the type (33.7), where the deflections q_n are perpendicular to the wave vector $\boldsymbol{k} = k\boldsymbol{e}_x$; such wave is called *transversal*. For these oscillations, too, the linear chain can be supplemented to form a three-dimensional crystal. In the three-dimensional crystal, three directions of the wave vector $\boldsymbol{k}$ are possible. For each $\boldsymbol{k}$ value there is one longitudinal and two transverse waves. This specification is referred to as the *polarization* of the wave.

We assume a cubic lattice with the lattice constant a and the volume is $V = L^3$. We take into account all three polarization directions, the distances $\Delta k = 2\pi/L$ and the maximum π/a for the $|k|$ values. With this we obtain the number of oscillation modes:

$$\sum_{\text{pol}} \sum_{k} 1 = 3 \int_{-\pi/a}^{\pi/a} \frac{dk_x}{\Delta k} \int_{-\pi/a}^{\pi/a} \frac{dk_y}{\Delta k} \int_{-\pi/a}^{\pi/a} \frac{dk_z}{\Delta k} \, 1 = \frac{3V}{a^3} = 3N \tag{33.12}$$

Since the N atoms can be displaced in three directions, the system has $3N$ degrees of freedom and thus $3N$ natural oscillations (or eigenmodes).

We facilitate the counting of the eigenmodes by using the simplified linear dispersion relation

$$\omega = \omega(k) \approx c_{\text{S}} |\boldsymbol{k}|, \qquad c_{\text{S}} = c_{\text{l}} \approx c_{\text{t}} \tag{33.13}$$

The actual dispersion relations will deviate from this, especially in the upper range (see Figure 33.2). In addition, the slopes c_{l} and c_{t} for the longitudinal and transverse branches are slightly different in a real crystal. We further assume that all $\boldsymbol{k}$ directions are equivalent. Then the limitation (33.6) for Cartesian components becomes $|\boldsymbol{k}| \le k_{\max}$ with a yet to be determined upper limit $k_{\max} = \mathcal{O}(\pi/a)$. The number of natural oscillations now results from

$$3N = \sum_{\text{pol}} \int \frac{d^3k}{\Delta k^3} = \frac{3V}{(2\pi)^3} \int_0^{k_{\max}} 4\pi k^2 \, dk = \frac{3V}{2\pi^2} \frac{1}{c_{\text{S}}^3} \int_0^{\omega_{\text{D}}} d\omega \, \omega^2 = \frac{V}{2\pi^2 c_{\text{S}}^3} \omega_{\text{D}}^3 \tag{33.14}$$

The integral limit is the *Debye frequency*

$$\omega_{\text{D}} = c_{\text{S}} \, k_{\max} = c_{\text{S}} \left(\frac{6\pi^2}{v} \right)^{1/3} \approx 3.9 \, \frac{c_{\text{S}}}{a} \tag{33.15}$$

As an example, we insert $c_{\text{S}} = 5 \cdot 10^3 \, \text{m/s}$ and $a = 2 \cdot 10^{-8} \, \text{cm}$ and get $\omega_{\text{D}} \approx 10^{14} \, \text{s}^{-1}$.

Using (33.14) simplifies the sums occurring in the statistical treatment. The sum over the discrete $\boldsymbol{k}$ values and polarization directions are performed as

$$\sum_{\text{pol}} \sum_{k} \ldots = \frac{9N}{\omega_{\text{D}}^3} \int_0^{\omega_{\text{D}}} d\omega \, \omega^2 \ldots = 3N \int_0^{\infty} d\omega \, z_{\text{D}}(\omega) \ldots \tag{33.16}$$

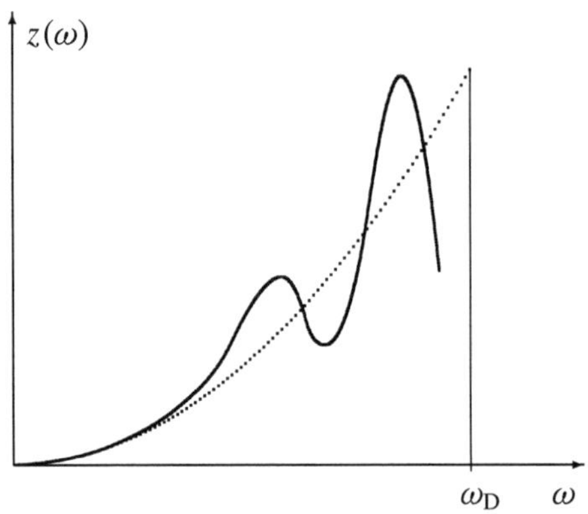

Figure 33.3 Possible form of the frequency spectrum $z(\omega)$ of the lattice vibrations in a real crystal. The Debye model with $z_\mathrm{D} \propto \omega^2$ (dotted) is based on the simplified dispersion relation $\omega = c_\mathrm{S} k$.

This *Debye model* is characterized by the *frequency spectrum*

$$z_\mathrm{D}(\omega) = \begin{cases} \dfrac{3\,\omega^2}{\omega_\mathrm{D}^3} & \omega \leq \omega_\mathrm{D} \\[2mm] 0 & \omega > \omega_\mathrm{D} \end{cases} \qquad \text{(Debye model)} \qquad (33.17)$$

The frequency spectrum $z(\omega)$ is the density of states of the natural oscillations, i.e. the number of modes per ω interval.

In a real crystal, there are deviations from the Debye model, comparable to those sketched in Figure 33.3. In any case, the function $z(\omega)$ must be normalized, $\int d\omega\, z(\omega) = 1$, because the number $3N$ of degrees of freedom or natural oscillations is fixed.

Phonons

Each of the $3N$ natural oscillations of a crystal lattice is characterized by its wave vector and polarization:

$$j = (\boldsymbol{k}, \text{pol}) = (\boldsymbol{k}, m)\,, \qquad j = 1, 2, ..., 3N \qquad (33.18)$$

The three possible polarization directions (one longitudinal and two transverse waves) are counted by $m = 1, 2, 3$. We assume $\omega = c_\mathrm{S}|\boldsymbol{k}|$ and the spectrum (33.17) for the natural frequencies.

Classically, the energy of a natural oscillation can have any value. Quantum mechanically, however, only discrete energy values are allowed:

$$e_j = \hbar\,\omega_j \left(n_j + \frac{1}{2} \right) = \hbar\,\omega(k) \left(n_{\boldsymbol{k}}^m + \frac{1}{2} \right) \qquad (33.19)$$

The quantum numbers $n_j = n_{\boldsymbol{k}}^m$ can adopt the values $0, 1, 2, \ldots$. The assumption (33.1) of harmonic restoring forces is, however, only valid for small deflections from the rest position, and therefore only for not too high quantum numbers n_j. A real crystal disintegrates for too high vibrational energies. This means that for a sufficiently high temperature the crystal melts.

A microstate r of the lattice is now characterized by the oscillation quantum numbers $n_j = n_k^m$:

$$r = (n_1, n_2, \ldots, n_{3N}) = \{n_k^m\} \tag{33.20}$$

Since the natural oscillations are independent of each other, the energy of this microstate is equal to

$$E_r = \sum_{j=1}^{3N} e_j = \sum_{m,k} \hbar\,\omega(k)\left(n_k^m + \frac{1}{2}\right) = E_0(V) + \sum_{m,k} \varepsilon_k\, n_k^m \tag{33.21}$$

Here

$$\varepsilon_k = \hbar\,\omega(k) \tag{33.22}$$

denotes the energy of an oscillator quantum. In the volume $V = L^3$, the k-values are a multiple of $\Delta k = 2\pi/L$. From this follows $\varepsilon_k \approx \hbar c_s k \propto V^{-1/3}$.

In (33.21), the energy E_0 of the zero-point oscillations was written in front of the sum. The energy E_0 may also include other energy contributions not considered here; then $E_0(V) = E(T = 0, V)$ is the ground state energy of the lattice. In a realistic description, the volume dependence of $E_0(V)$ accounts for the (relative) incompressibility of the lattice. The second part in (33.21) also depends on V; it leads to a temperature dependent contribution of the pressure.

Since the total number of natural oscillations is $3N$, the result will depend on N. However, the number N of atoms in the crystal is a constant and not an external parameter such as the volume.

Except for the contribution E_0, Equation (33.21) is of the form (29.17) of an ideal quantum gas. Because of

$$n_k^m = 0, 1, 2, \ldots \tag{33.23}$$

we are dealing with an *ideal Bose gas*. In this perspective, we no longer speak of n_k^m oscillation quanta of the oscillator, but of n_k^m bosons in the single particle state $j = (k, m)$ with the energy ε_k. These bosons are called *phonons*; they represent the quantized waves of the lattice. The comparison with (29.17) shows that the three polarizations $m = 1, 2, 3$ correspond to the three spin quantum numbers $s_z = 0, \pm 1$. In addition to the energy ε_k and the momentum $p = \hbar k$, the spin 1 can be assigned to a phonon.

The system of lattice oscillations can be treated as an ideal Bose gas. This perspective is advantageous because we can then treat the system analogously to the other quantum gases. It is also the starting point for the treatment of other processes such as electron-phonon scattering.

In contrast to the Bose gas of ^{4}He atoms, the total number of phonons is not fixed:

$$N_{\text{ph}} = \sum_{m,k} n_k^m \neq \text{const.} \tag{33.24}$$

The mean number $\overline{N_{ph}}$ turns out to be a function of the temperature, Exercise 33.3 Since the energy eigenvalues $E_r(V)$ do not depend on N_{ph}, the chemical potential of the phonons is zero:

$$\mu = \overline{\frac{\partial E_r(V)}{\partial N_{ph}}} = 0 \qquad \text{(phonons)} \tag{33.25}$$

The mean occupation numbers (29.39) become

$$\overline{n_k^m} = \overline{n_k} = \frac{1}{\exp(\beta\,\varepsilon_k) - 1} \tag{33.26}$$

From this we get the energy

$$E(T, V) = \overline{E_r} = E_0(V) + \sum_{m,k} \varepsilon_k\,\overline{n_k} \tag{33.27}$$

Here we insert (33.16), (33.17) and $\varepsilon_k = \hbar\omega(k)$,

$$
\begin{aligned}
E(T, V) &= E_0(V) + 3N \int_0^{\omega_D} d\omega\, \frac{3\,\omega^2}{\omega_D^3}\, \frac{\hbar\omega}{\exp(\beta\hbar\omega) - 1} \\
&= E_0(V) + \frac{9N}{(\hbar\omega_D)^3}\,(k_B T)^4 \int_0^{x_D} dx\, \frac{x^3}{\exp(x) - 1}
\end{aligned}
\tag{33.28}
$$

We use the dimensionless quantities

$$x = \frac{\hbar\omega}{k_B T} \quad \text{and} \quad x_D = \frac{\hbar\omega_D}{k_B T} = \frac{T_D}{T} \tag{33.29}$$

and the Debye temperature $T_D = \hbar\omega_D/k_B$. For $x_D \ll 1$ we expand the integrand:

$$
\begin{aligned}
E(T, V) &= E_0(V) + \frac{9Nk_B T}{x_D^3} \int_0^{x_D} dx\, x^3 \left(\frac{1}{x} - \frac{1}{2} + \frac{x}{12} + \cdots \right) \\
&= E_0(V) + 3Nk_B T \left(1 - \frac{3}{8}\frac{T_D}{T} + \frac{1}{20}\frac{T_D^2}{T^2} + \cdots \right)
\end{aligned}
\tag{33.30}
$$

From this follows the heat capacity of the Debye model for high temperatures:

$$C_V = \frac{\partial E(T, V)}{\partial T} = 3Nk_B \left(1 - \frac{1}{20}\frac{T_D^2}{T^2} \right) \qquad (T \gg T_D) \tag{33.31}$$

The classical limit $C_V = 3Nk_B$ is called *Dulong-Petit law*. According to the equipartition theorem, each canonical variable that occurs quadratically in the Hamilton function carries the mean energy $k_B T/2$. An oscillator has two such canonical variables; it gives the contribution k_B to the heat capacity. In the crystal there are $3N$ independent natural oscillations (or oscillators), i.e. $C_V = 3Nk_B$.

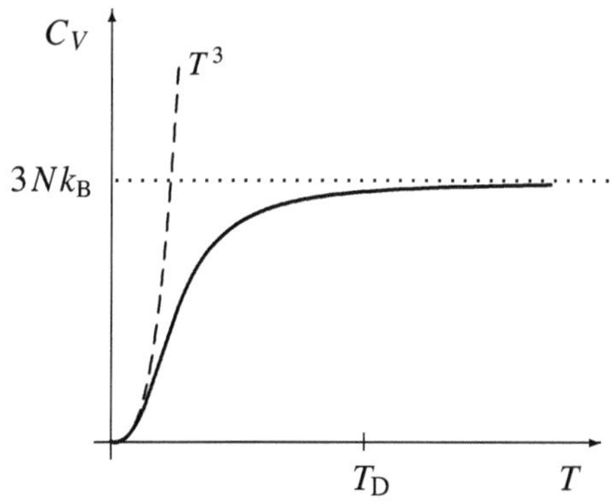

Figure 33.4 The lattice contribution to the specific heat as a function of the temperature. The limiting cases for low (dashed) and high (dotted) temperatures are shown.

For $x_D \gg 1$ or $T \ll T_D$ the upper limit of the integral can be set to infinity. With

$$\int_0^\infty dx \, \frac{x^3}{\exp(x) - 1} = \sum_{n=1}^\infty \int_0^\infty dx \, x^3 \exp(-nx) = 6 \sum_{n=1}^\infty \frac{1}{n^4} = \frac{\pi^4}{15} \qquad (33.32)$$

we get

$$E(T, V) = E_0(V) + \frac{3\pi^4}{5} \frac{N(k_B T)^4}{(\hbar \omega_D)^3} \qquad (T \ll T_D) \qquad (33.33)$$

From this follows the heat capacity of the Debye model for low temperatures:

$$C_V = \frac{\partial E(T, V)}{\partial T} = \frac{12\pi^4}{5} N k_B \frac{T^3}{T_D^{\,3}} \qquad (T \ll T_D) \qquad (33.34)$$

The T^3–behavior can be understood as follows: For low temperatures all modes $\hbar\omega \lesssim k_B T$ are thermally excited. The number of these modes is proportional to $\int d^3k \propto k^3 \propto \omega^3 \propto T^3$. Each mode yields a contribution of the size $\mathcal{O}(k_B)$ to the heat capacity.

The entire course of the specific heat, as it results from (33.28), is sketched in Figure 33.4. The transition between the T^3–behavior and the classical limit occurs in the temperature range between $T = 0$ and T_D.

In principle, we have linked ω_D to microscopic quantities (33.15, 33.11). The force constant f is determined by the potential between the atoms of the crystal. In practice, T_D in (33.34) is regarded as a parameter that is adapted to the experimental data. This yields for example

$$T_D = \begin{cases} 105 \text{ K} \\ 343 \text{ K} \end{cases} \qquad T_m = \begin{cases} 601 \text{ K} & \text{for lead} \\ 1357 \text{ K} & \text{for copper} \end{cases} \qquad (33.35)$$

The melting temperature T_m is also shown. Lead is a soft metal compared to copper; this is reflected by the lower temperature values. For $T \to T_m$ the lattice vibrations become so strong that they dissolve the crystal; the considered model becomes invalid. As can be seen in Figure 33.4, the classical limit $c_V = 3k_B$ is almost reached for T_D. The approximation (33.31) can therefore also be used for $T \sim T_D$, i.e. beyond the specified range of validity.

The Debye model accounts for two aspects correctly: The total number of possible modes is $3N$, and for small ω, $z(\omega) \propto \omega^2$ holds. In this respect the limiting cases (33.31) and (33.34) are realistic. In the range between these cases, it will generally come to deviations between the calculated and measured $c_V(T)$. These deviations allow conclusions about the deviation of the frequency spectrum $z(\omega)$ from the Debye model (Figure 33.3).

The longitudinal and transverse waves considered here are also called *acoustic*; the ratio $d\omega/dk$ for $k \to 0$ is the speed of sound. In addition, there may be also so-called *optical* oscillations, which are can be excited by electromagnetic fields (light). An optical oscillation may occur in a lattice of ions (for example table salt, NaCl). In this case the positive and negative ions vibrate against each other. This can be taken into account by generalizing the Debye model to the Einstein model, Exercise 33.4. In contrast to the acoustic oscillations, the optical oscillations have a finite energy for $k \to 0$. Figure 33.2 on (right part) schematically shows the acoustic branches and the optical branch of the dispersion relation.

Exercises

33.1 Oscillation modes of the linear chain

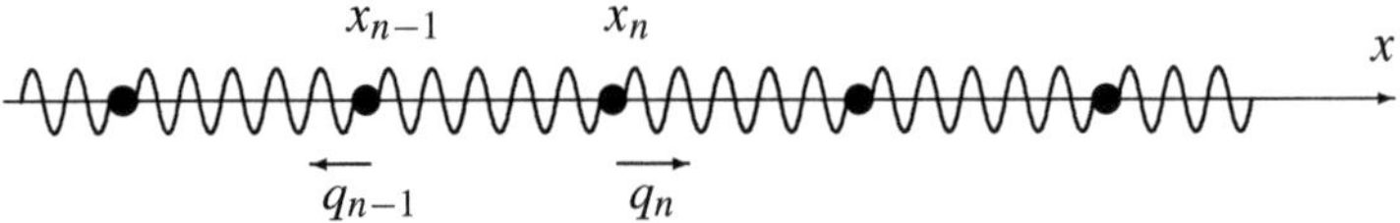

The rest positions of the masses of a linear chain are $x_n = na$. The deflections $q_n(t) = q(x_n, t)$ obey the equation of motion $m\,\ddot{q}_n = f\,(q_{n+1} + q_{n-1} - 2q_n)$. The chain shall consist of $N + 1$ masses (with $N \gg 1$). Specifically, these are either the masses at $x = 0$, $x = a$, $x = 2a, \ldots, x = Na = L$ for a finite chain, or $N + 1$ neighboring masses of an infinite chain; in both cases we consider an interval of length $L = Na$. Specify the possible discrete k-values for

- Periodic boundary conditions $q(x + L, t) = q(x, t)$
- Physical boundary conditions $q(0, t) = q(L, t) = 0$

Justify that it does not matter for the statistical counting which of these boundary conditions are used.

33.2   Specific heat of the phonon gas for low temperatures

The lattice vibrations of a crystal yields the energy

$$E(T, V) = E_0(V) + \frac{9N}{(\hbar\omega_{\mathrm{D}})^3} (k_{\mathrm{B}}T)^4 \int_0^{x_{\mathrm{D}}} dx \, \frac{x^3}{\exp(x) - 1} \tag{33.36}$$

where $x_{\mathrm{D}} = \hbar\omega_{\mathrm{D}}/(k_{\mathrm{B}}T) = T_{\mathrm{D}}/T$. For low temperatures, $T \ll T_{\mathrm{D}}$, one gets the known behavior $C_V \propto T^3$. Calculate the leading correction to this limiting case.

33.3   Average phonon number in the Debye model

Express the mean phonon number for the Debye model by an integral. Evaluate the integral for high and low temperatures from.

33.4   Einstein model

1. Calculate the specific heat of lattice vibrations using the approach $z_{\mathrm{E}}(\omega) = \delta(\omega - \omega_{\mathrm{E}})$ for the frequency spectrum. With a suitably chosen frequency ω_{E}, this *Einstein model* may describe the contribution of the optical phonons.

2. For a NaCl crystal (with N sodium and chlorine atoms each), the phonon spectrum can be approximated by

$$z(\omega) = z_{\mathrm{D}}(\omega) + \delta(\omega - 2\,\omega_{\mathrm{D}}) \tag{33.37}$$

i.e. by a Debye spectrum for the acoustic phonons and an Einstein term for the optical phonons. Calculate the specific heat for low and high temperatures.

34 Photon gas

We study systems in which matter at temperature T is in equilibrium with electromagnetic radiation. Such a system is, for example, the plasma on the surface of the Sun. This electromagnetic radiation can be treated as an ideal photon gas. This model leads to Planck's radiation formula and to the Stefan-Boltzmann law. The specific heat of the photon gas increases with the third power of the temperature.

For the theoretical treatment, we consider a cavity resonator in which standing electromagnetic waves can be excited. Each such wave represents an independent oscillation mode of the system, and can be quantized accordingly. The quanta of these waves are called photons. We determine how many quanta of the individual oscillation modes are excited in statistical equilibrium. This yields the thermodynamic energy of the photon gas.

The electromagnetic waves are solutions of Maxwell's equations in vacuum (source free space). These equations yield the wave equation

$$\left(\Delta - \frac{1}{c^2} \frac{\partial^2}{\partial t^2} \right) E(r, t) = 0 \tag{34.1}$$

for the electric field $E(r, t)$. The approach

$$E(r, t) = E_0 \sin(k \cdot r + \varphi) \sin(\omega t + \psi) \tag{34.2}$$

results in the dispersion relation

$$\omega^2 = c^2 k^2 \quad \text{or} \quad \omega = \omega(k) = c\,|k| = ck \tag{34.3}$$

where c is the speed of light. The electric field (34.2) represents a standing wave; the magnetic field follows from the Maxwell equation $\partial B/\partial t = -c \operatorname{rot} E$. The general solution of the free Maxwell equations can be written as a superposition of these solutions.

From the Maxwell equation $\operatorname{div} E = 0$ follows the restriction

$$E_0 \cdot k = 0 \quad \text{or} \quad E_0 \perp k \tag{34.4}$$

Specifically for $k = k e_z$, the solution (34.2) becomes

$$E(z, t) = (E_{0x} e_x + E_{0y} e_y) \sin(kz + \varphi) \sin(\omega t + \psi) \tag{34.5}$$

with $\omega = ck$. This means that there are two polarizations of the wave. For $E_{0y} = 0$ the wave is linearly polarized in the x-direction; then E points in the x-direction and B points in the y direction. Since the vectors E and B are perpendicular to the wave vector k, the electromagnetic waves are *transversal*. In contrast to the sound waves in the previous chapter, there are no longitudinal waves.

We consider electromagnetic waves inside a cavity. The cavity shall have the cubic volume $V = L^3$; its walls are made of metal. At the walls, the tangential component of the electric field must disappear. For (34.5) this means $E(0, t) = E(L, t) = 0$. From this follows $\varphi = 0$ and $kL = i\,\pi$, i.e. the discrete k values

$$k_i = \frac{\pi}{L}\, i \quad \text{with} \quad i = 1, 2, 3,\ldots \tag{34.6}$$

For counting the modes, we can restrict i to positive values because $\sin(kz)$ and $\sin(-kz)$ result in the same solution. The sum over the k_i values can be replaced by an integral:

$$\sum_{k_i} \ldots = \frac{L}{\pi} \int_0^{\infty} dk \,\ldots = \frac{L}{2\pi} \int_{-\infty}^{\infty} dk \,\ldots = \frac{1}{\Delta k} \int_{-\infty}^{\infty} dk \,\ldots \tag{34.7}$$

where

$$\Delta k = \frac{2\pi}{L} \tag{34.8}$$

We now consider the vector k and the sum over all possible momentum directions:

$$\sum_{m=1}^{2} \sum_{k} \ldots = 2 \int_{-\infty}^{\infty} \frac{dk_x}{\Delta k} \int_{-\infty}^{\infty} \frac{dk_y}{\Delta k} \int_{-\infty}^{\infty} \frac{dk_z}{\Delta k} \ldots = \frac{2V}{(2\pi)^3} \int d^3k \,\ldots \tag{34.9}$$

For this result, the shape of the volume V is irrelevant. In contrast to lattice waves, the number of electromagnetic modes is not finite; there is no maximum wave number. The sum over m stands for the two possible polarization directions of the wave.

The cavity merely serves as an conceptual aid for counting the number of oscillation modes. Here we have not presented the complete expressions for cavity oscillations (this may be found in Chapter 21 in [2]). For the statistically relevant excitations $\hbar\omega \sim k_B T$, the walls of the cavity are insignificant because of $\overline{k} \gg k_{\min} = \Delta k$. The result depends on the value of V only but not on the shape of the volume.

As solutions of the free Maxwell equations, the electromagnetic oscillations are harmonic and decoupled from each other. Each mode represents a classical harmonic oscillator. We take into account the quantization by using the discrete energy values

$$e_j = \hbar\omega(k) \left(n_k^m + \frac{1}{2} \right), \qquad j = (k, m) \tag{34.10}$$

Here $\omega = ck$, and n_k^m may adopt the values $0,1,2,\ldots$.

We briefly sketch the formal path that leads to this quantization. For the classical field, the time dependent amplitudes $A_j(t)$ of the individual oscillation modes can be chosen as generalized coordinates of the Lagrange function; in (34.5) $A(t) = E_{0x}\sin(\omega t + \psi)$ represents such an amplitude. From the Lagrange function

$$L(A_1, A_2, ..., , \dot{A}_1, \dot{A}_2, ...) = \text{const.} \cdot \sum_j \left(\dot{A}_j^2 - \omega_j^2 A_j^2 \right) \qquad (34.11)$$

follow the equations of motion $\ddot{A}_j = -\omega_j^2 A_j$ with the solutions $A_j(t) = a_j \sin(\omega_j t + \psi_j)$. The Lagrange function L is used to determine the Hamilton function $H(A_1, A_2, ..., P_{A_1}, P_{A_2}, ...)$; one can choose the constant in (34.11) such that H is equal to the energy of the cavity oscillation. Then we turn from the Hamilton function to the Hamilton operator. The Hamilton operator describes independent oscillators and therefore has the energy eigenvalues $E_r = \sum e_j$ with the e_j from (34.10).

A microstate r of the cavity is defined by the oscillation quantum numbers $n_j = n_k^m$ of all standing waves:

$$r = (n_1, n_2, n_3 \ldots) = \left\{ n_k^m \right\} \qquad (34.12)$$

The quantum numbers $n_j = n_k^m$ can adopt the values 0, 1, 2, ... Since the natural oscillations (the standing waves) are independent of each other, the energy of the microstate is

$$E_r(V) = \sum_{j=1}^{\infty} e_j = \sum_{m,k} \hbar \omega(k) \left(n_k^m + \frac{1}{2} \right) = E_0 + \sum_{m,k} \varepsilon_k \, n_k^m \qquad (34.13)$$

Here we have introduced the energy of an oscillator quantum,

$$\varepsilon_k = \hbar \omega(k) \qquad (34.14)$$

In the volume $V = L^3$ the k-values are a multiples of $\Delta k = 2\pi/L$; from this follows $\varepsilon_k = \hbar c k \propto V^{-1/3}$.

Except for the energy E_0 of the zero-point oscillations, (34.13) is of the form (29.17) of the ideal quantum gas. Because of

$$n_k^m = 0, 1, 2, \ldots \qquad (34.15)$$

it is an ideal *Bose gas*. In this perspective, we do no longer speak of n_k^m quanta of the oscillator, but of n_k^m bosons in the single particle state $j = (k, m)$ with energy ε_k. These bosons are called *photons*. A photon is characterized by its momentum $p = \hbar k$ and by its spin 1. The two polarization options correspond to the spin projections $\pm \hbar$ in the direction of k; these are all possible spin states (Chapter 20 of [2]).

The photons could merely be a conceptual construction or way of speaking. However, many experiments (such as the Compton effect) show that photons are real particles with definite values for energy, momentum and spin.

The number of photons in the radiation cavity is not fixed,

$$N_{\mathrm{ph}} = \sum_{m,k} n_k^m \neq \text{const.} \tag{34.16}$$

The mean number $\overline{N_{\mathrm{ph}}}$ turns out to be a function of the temperature, Exercise 34.1. Since the Hamilton operator and the energy eigenvalues $E_r(V)$ do not depend on N_{ph}, the chemical potential disappears,

$$\mu = \frac{\overline{\dfrac{\partial E_r(V)}{\partial N_{\mathrm{ph}}}}}{} = 0 \qquad \text{(photons)} \tag{34.17}$$

With this the mean occupation number (29.39) becomes

$$\overline{n_k^m} = \overline{n_k} = \frac{1}{\exp(\beta \varepsilon_k) - 1} \tag{34.18}$$

The zero point energy in (34.13), $E_0 = \sum \hbar\omega(k)/2$, is infinite because the number of oscillation modes is not is not limited. Therefore, we consider only the measurable energy difference

$$E'(T, V) = \overline{E_r} - E_0 = E(T, V) - E(0, V) = \sum_{m,k} \varepsilon_k \, \overline{n_k} \tag{34.19}$$

The difference formation is a rule for handling the infinite result for $E(T, V)$; in this way we obtain a physically meaningful result. The procedure is questionable in itself, because it consists of subtracting two infinite quantities, $E(T, V)$ and $E_0 = E(0, V)$, from each other. In quantum electrodynamics, however, this is a common and successful method, which we use here in a particularly simple form.

Although the cavity oscillations represent a completely different physical system, the occurring structures are largely analogous to lattice oscillations. Comparing the phonon and photon gas we note some points:

- In the first case, the wave velocity is the speed of sound (phonons), in the second case the speed of light (photons).

- For photons, the dispersion relation is exactly linear ($\omega \propto k$). For phonons, this only applies approximately. In addition, the dispersion relation for phonons is characterized by a maximum value ω_{D}. As a consequence, the relation $E \propto T^4$ applies to phonons only for low temperatures, for photons it holds exactly.

- For photons, arbitrarily large n_k are possible, for phonons, the crystal dissolves if the excitations become too large. In other words, the harmonicity of the oscillation applies exactly for the electromagnetic field, but only approximately for lattice waves.

- There is one longitudinal and two transverse lattice waves, but only two transverse electromagnetic waves. In the particle picture this means that the spin 1 of the phonon has three possible settings, but the spin 1 of the photon has only two.

- The energy of the zero-point oscillations is finite for phonons, for photons it is infinite.

We calculate the energy E':

$$E'(T, V) = \sum_{m,k} \varepsilon_k \overline{n_k} \overset{(34.9)}{=} \frac{2V}{(2\pi)^3} \int d^3k \, \frac{\hbar ck}{\exp(\beta\hbar ck) - 1}$$

$$= \frac{V}{\pi^2 \beta^4 \hbar^3 c^3} \int_0^\infty dx \, \frac{x^3}{\exp(x) - 1} \overset{(33.32)}{=} \frac{\pi^2 V (k_B T)^4}{15 \, \hbar^3 c^3} \tag{34.20}$$

With the Stefan-Boltzmann constant

$$\sigma = \frac{\pi^2 k_B^4}{60 \, \hbar^3 c^2} = 5.67 \cdot 10^{-8} \, \frac{\mathrm{W}}{\mathrm{m^2 \, K^4}} \tag{34.21}$$

the result becomes

$$E'(T, V) = \frac{4\sigma}{c} \, V T^4 \tag{34.22}$$

The heat capacity of a cavity with the volume V is then

$$\boxed{C_V = \frac{16\,\sigma}{c} \, V T^3} \tag{34.23}$$

Planck's radiation formula

The energy (34.20) can be written as an integral over the frequencies ω,

$$\frac{E'(T, V)}{V} = \int_0^\infty d\omega \, u(\omega) \tag{34.24}$$

Using $\omega = ck$, the comparison with (34.20) defines the spectral distribution $u(\omega)$ of the energy density,

$$\boxed{u(\omega) = \frac{\hbar}{\pi^2 c^3} \frac{\omega^3}{\exp(\hbar\omega/k_B T) - 1} \qquad \text{Planck's radiation formula}} \tag{34.25}$$

This is also the radiation distribution of a *black body*. If the cavity resonator is provided with a small hole, radiation with the distribution $u(\omega)$ can escape from it. The hole shall be small so that the resulting disturbance of the equilibrium in the

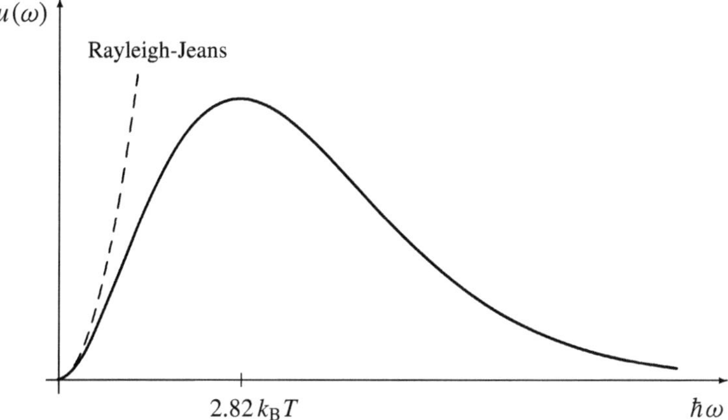

Figure 34.1 Planck's radiation formula: If matter of temperature T is in equilibrium with electromagnetic radiation, it emits radiation with this frequency distribution.

cavity is negligible. The term "black" comes from the fact that such a hole appears black from the outside.

Planck's formula is shown in Figure 34.1. For small ω, the function $u(\omega)$ increases proportionally to ω^2, for large ω, it decreases exponentially. The increase $u(\omega) \propto \omega^2$ is determined by the number of possible modes, i.e. by the factor k^2 in $d^3k = 4\pi k^2 \, dk$. For high ω, the exponential function dominates in the denominator of (34.25); thus $u(\omega) \propto \exp(-\hbar\omega/k_B T)$. The Planck's formula has a maximum that results from

$$\frac{du(\omega)}{d\omega} = 0 \quad \Longrightarrow \quad \beta\hbar\omega = 3\left[1 - \exp(-\beta\hbar\omega)\right] \tag{34.26}$$

The numerical solution of the equation $x/3 = 1 - \exp(-x)$ is $x \approx 2.82$, therefore

$$\hbar\,\omega_{\max} = 2.82\,k_B T \qquad \text{(Wien's displacement law)} \tag{34.27}$$

This relation describes how the maximum of the radiation distribution shifts with temperature.

Planck's radiation formula was of particular importance in the development of quantum physics. The result of a classical statistical treatment is obtained from the equipartition theorem or formally from (34.25) by the limit "$\hbar \to 0$",

$$u(\omega) = \frac{\hbar}{\pi^2 c^3} \frac{\omega^3}{\exp(\beta\hbar\omega) - 1} \quad \xrightarrow{\hbar \to 0} \quad u_{\text{RJ}}(\omega) = \frac{k_B T \,\omega^2}{\pi^2 c^3} \tag{34.28}$$

This result is known as the Rayleigh-Jeans law. It implies an "ultraviolet catastrophe": Each classically possible oscillation, even that with a very high frequency or small wavelength, obtains the mean energy $k_B T$. Since the density of possible oscillations increases with ω^2, the high frequencies (ultraviolet) lead to an infinite total energy, $\int d\omega\, u_{\text{RJ}} = \infty$. Note that this infinite result concerns the *measurable* energy difference between T and $T = 0$. Such a result is physically nonsensical; of course it is also in contradiction to the observed distribution. This situation led Planck to the ad hoc assumption that the energy of each oscillation can only occur in discrete units (i.e. quanta) that are proportional to the frequency. The proportionality constant $\hbar$ of these energy quanta $\hbar\omega$ could then be determined from comparing (34.25) with measured radiation distributions.

Stefan-Boltzmann law

We calculate the radiation power of a black body. We consider the energy current density

$$\frac{E'c}{V} = \frac{\text{energy}}{\text{area} \cdot \text{time}} \tag{34.29}$$

If this current density passes through an area f, the transferred power is equal to fcE'/V. In the cavity wall there shall be a small opening with the area f. Only part of the power fcE'/V passes through this hole because the energy density E'/V consists of photons with statistically distributed velocity directions. We denote the speed of a photon by v (where $|v| = c$). Of the photons immediately before the opening, only those contribute to an outgoing current for which $\theta = \arccos(v \cdot n/v)$ is between 0 and $\pi/2$; here n is the normal to the opening. For the radiant power (energy/time), only the velocity component perpendicular to the area f of the opening counts, i.e. $v \cdot n = c \cos\theta$. This means that through the opening with the area f the electromagnetic power

$$P_{\text{em}} = \frac{E'c}{V} \, f \, \frac{1}{4\pi} \int_0^{2\pi} d\phi \int_0^1 d(\cos\theta) \, \cos\theta = \frac{E'c f}{4V} \tag{34.30}$$

is emitted. With (34.22), Equation (34.30) becomes the *Stefan-Boltzmann law*

$$\boxed{P_{\text{em}} = \sigma \, f \, T^4 \qquad \text{Stefan-Boltzmann law}} \tag{34.31}$$

This is the power (energy/time) emitted by a black body (or by a cavity hole).

Applications

In the derivation, we have referred to a cavity resonator. Inside the cavity the free Maxwell equations hold. The resonator yields particularly simple boundary conditions. The actual applications of the results go, however, far beyond this artificial scenario. Whenever matter is in equilibrium with electromagnetic radiation, the radiation can be described by a photon gas. The matter can lead to deviations from Planck's formula (for example, absorption lines); physically, the presence of matter is usually necessary to specify a temperature.

In the treatment of the photon gas, only the properties of of electromagnetic waves and basic statistical assumptions were used. Because of the generality of these assumptions, Planck's formula and the Stefan-Boltzmann law can be applied to many systems. We present some examples:

- If iron is heated from $1\,000$ K to $2\,000$ K, it appears first glowing red and then glowing white. The emitted light is described by Planck's radiation formula. In both cases only the upper tail of the Planck distribution extends to the visible region, In the first case, the lowest visible frequencies (red) are seen. In the second case also other colors contribute leading to a white appearance.

- In the Sun's surface, the plasma is in equilibrium with electromagnetic radiation. The measurement of the radiation spectrum determines the surface temperature of the Sun (about $5\,800$ K). This method is also used to determine the surface temperature of other stars.

- The energy flux density of solar radiation hitting the Earth is $1.36\,\text{kW}/\text{m}^2$; this quantity is called solar constant.

  Averaged over time (day-night) and position (latitude) and weather (atmosphere, clouds) the energy flux density which reaches the Earth's surface is about $I_0 = 340\,\text{W}/\text{m}^2$. This input must be radiated back into space one the average (otherwise the Earth would continuously heat up; the Earth's internal heat can be neglected in this energy balance). The Stefan-Boltzmann law $I_0 = \sigma\,T_0{}^4$ results in a first, rough estimate for the mean temperature $T_{\mathrm{E}} \approx T_0$ of the Earth's surface. The natural greenhouse effect caused by a non polluted atmosphere, increases the temperature T_0 by about 30 degrees.

- Greenhouse effect: The burning of fossil fuels increases the CO_2 content of the atmosphere. Carbon dioxide absorbs light stronger in the infrared range than in the visible range. This means qualitatively

$$\text{absorption by } CO_2 = \begin{cases} \text{large} & \hbar\omega \sim 1000\,k_{\mathrm{B}}\text{K} \\ \text{small} & \hbar\omega \sim 16\,000\,k_{\mathrm{B}}\text{K} \end{cases} \tag{34.32}$$

The Earth has a temperature T_{E} of about 300 K, so that the maximum of its radiation is at $\hbar\omega_{\max} \sim 1000\,k_{\mathrm{B}}\text{K}$. The solar surface has a temperature of about $5\,800$ K, so that the maximum of its radiation at lies $\hbar\omega_{\max} \sim 16\,000\,k_{\mathrm{B}}\text{K}$. For both numbers we have used Wien's displacement law (34.27).

The light absorbed in the CO_2 layer is re-emitted in other directions. This effectively results in partial back radiation or reflection. If the CO_2 layer reflects 10% of the Earth's thermal radiation, then the Earth's surface must, in equilibrium, radiate 10% more heat (Figure 34.2). According to (34.31), the fourth power of the Earth's surface temperature T_{E} must increase by 10%, i.e. T_{E} increases by about 2.5% or $\Delta T_{\mathrm{E}} \approx 7.5$ K.

Other trace gases (such as methane) strengthen this greenhouse effect. Glass panes of a greenhouse causes an increase in temperature in a analogous way.

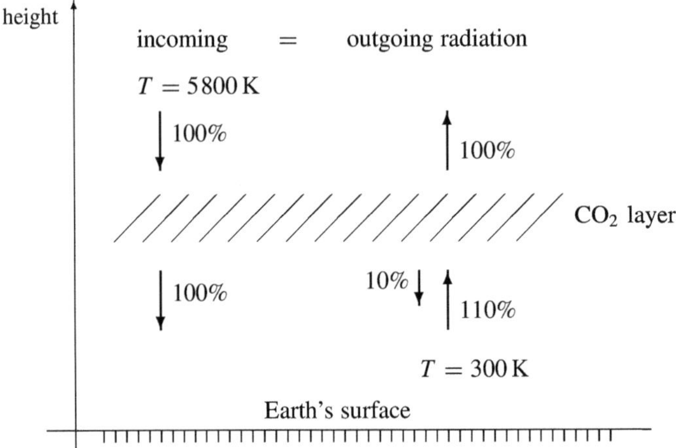

Figure 34.2 A greenhouse effect is caused by a layer of gas that absorbs solar radiation (5800 K) less than the thermal radiation (300 K) from the Earth's surface.

- Cosmic background radiation: The entire cosmos is filled with a cavity radiation that belongs to the temperature 2.73 K. In 1965, Penzias and Wilson measured the intensity of this radiation (for some wavelengths) and derived such a temperature from this. In the following years larger wavelength range ($\lambda = 70\,\text{cm}\ldots 0.1\,\text{cm}$) were explored. This confirmed that the radiation obeys Planck's radiation formula. Also the isotropy of the radiation was experimentally verified. This isotropy distinguishes the photon gas from the radiation of specific sources.

Such radiation was predicted at the end of the 1940s on the basis of a cosmological model that extrapolated the current expansion of the cosmos to earlier times. According to this model, matter in the cosmos used to be denser and correspondingly hotter. The radiation is a remnant from the time when matter was ionized and in equilibrium with radiation. This was around the time $t \approx 4 \cdot 10^5$ years, which is to be compared to $t_{\text{today}} \approx 2 \cdot 10^{10}$ years. The extrapolation to earlier times ultimately leads to a singularity (big bang), for which $t = 0$ is set. As the matter formed neutral atoms with decreasing temperature, there was a decoupling of the photon gas and the matter. In the further course the temperature of the Planck's formula of this photon gas decreases in parallel with the expansion of the cosmos to its current value.

Radiation pressure

Because of (34.6), $\varepsilon_k = ck \propto V^{-1/3}$ holds. From $E'_r(V) = \sum \varepsilon_k(V)\, n_k$ we then obtain for the pressure

$$P = -\overline{\frac{\partial E'_r(V)}{\partial V}} = \frac{E'}{3V} \tag{34.33}$$

In a kinetic approach, the pressure is determined by the momentum transfers of the reflected gas particles; this picture can also be used for a photon gas.

As a numerical example, we determine the pressure of a radiation with the energy flux density $j = c E'/V = 1\,\text{kW/m}^2$ (comparable to the solar constant):

$$P \sim \frac{j}{c} = \frac{1\,000\,\text{W}}{\text{m}^2}\,\frac{1}{3 \cdot 10^8\,\text{m/s}} \approx 3 \cdot 10^{-6}\,\frac{\text{N}}{\text{m}^2} = 3 \cdot 10^{-11}\,\text{bar} \qquad (34.34)$$

The pressure of solar radiation is of this size. The formula (34.33), however, applies to the isotropic pressure of a radiation equilibrium. The pressure of directed radiation (such as the solar radiation) differs from this by numerical factors.

Dispersion relation and specific heat

The reason for the T^3-dependence of the specific heat heat in the phonon and photon gas lies in the common dispersion relation $\varepsilon_k \propto k$. We discuss this relationship in general for bosons with $\mu = 0$.

To do this, we assume a dispersion relation with a certain power behavior for $k \to 0$,

$$\varepsilon_k = \hbar\omega = a k^\nu \qquad (k \to 0,\ \nu > 0) \qquad (34.35)$$

For bosons with $\mu = 0$ and for small temperatures, the following applies

$$E = \sum_{m,k} \varepsilon_k\,\overline{n_k} \propto \int_0^\infty dk\,k^2\,\frac{a k^\nu}{\exp(\beta a k^\nu) - 1} \qquad (34.36)$$

For $T \to 0$, only the lowest energy states contribute; therefore, irrespective of the range of validity of the dispersion relation, the upper integral limit can be chosen to be infinite. We introduce the dimensionless variable x:

$$x = \beta a k^\nu, \qquad k \propto (x T)^{1/\nu}, \qquad dk \propto T^{1/\nu} x^{1/\nu - 1}\,dx \qquad (34.37)$$

This turns the energy into

$$E \propto T^{3/\nu+1} \int_0^\infty dx\,\frac{x^{3/\nu}}{\exp(x) - 1} \qquad (34.38)$$

The integral yields a dimensionless number, so that

$$E \propto T^{3/\nu+1} \quad \text{and} \quad C_V \propto T^{3/\nu} \qquad (34.39)$$

We have seen some examples for this, namely (i) the ideal Bose gas with $\nu = 2$ and $C_V \propto T^{3/2}$ and (ii) the phonon gas (for low temperatures) and the photon gas, both with $\nu = 1$ and $C_V \propto T^3$.

If the relevant excitations for $k \to 0$ have a finite energy Δ (such as the optical phonons in Figure 33.2), then the exponential factor in $\overline{n}$ dominates the behavior for $T \to 0$. From this follows

$$C_V \propto \exp(-\Delta/k_\text{B}T) \qquad \left(T \to 0,\ \hbar\omega \overset{k\to 0}{\longrightarrow} \Delta\right) \qquad (34.40)$$

A contribution of this form results, for example, from the vibrations in a diatomic gas or the optical phonons in an ion crystal.

Exercises

34.1   Average number of photons in the radiation cavity

Give the mean number of photons in a given radiation cavity (volume V, temperature T).

34.2   Temperature difference Europe–equator

Estimate the temperature difference between mid-latitudes (45°) and the tropics (0°) on the Earth's surface, which results from the geometrically caused difference or the solar radiation.

34.3   Range of visible light in the Planck formula

Visible light has wavelengths in the range of $\lambda = 3800\,\text{Å}$ to $7800\,\text{Å}$. By what factor are the associated frequencies above the maximum $\omega_{\max}$ of the distribution of red glowing ($T \approx 1000$ K) or white glowing ($T \approx 2000$ K) iron?

34.4   Surface temperature of the Sun

The solar constant $I_S = 1.36\,\text{kW/m}^2$ determines the intensity of solar radiation on the Earth. The distance Earth-Sun is about eight light minutes, and the radius of the Sun is $R_\odot \approx 7 \cdot 10^5$ km. Which temperature $T_\odot$ has the surface of the Sun?

VI Phase transition

35 Classification

In Part VI we deal with phase transitions. After a general introduction we investigate simple models for ferromagnetism, for the gaseous-liquid transition and for the λ transition of liquid helium. Finally, we consider critical phenomena in the framework of Landau theory and derive relations between critical exponents.

Overview

Figure 35.1 lists some phases that occur in nature. We are looking at homogeneous substances consisting of a specific kind of atoms or molecules; composite probes such as wood are excluded. The obvious phases at room temperature are gas, liquid and solid. For high temperatures, a gas can be ionized leading to a plasma of ions and free electrons. In addition, there are numerous other phase transitions, for example graphite-diamond, white tin$-$grey tin, paramagnet$-$ferromagnet, liquid$-$superfluid and normal conductor$-$superconductor.

The three phases gaseous, liquid and solid can be characterized by the relative spatial positions of the molecules. In the gaseous state, the average distance between a molecule and the next one is large compared to the molecule size. The distance is then also large compared to the range of the interaction; the interaction plays a minor role. In the liquid state, the distance is comparable to the range of the attractive interaction between the molecules. The interaction can no longer be regarded as a small disturbance. But the molecules are still movable and displaceable against each other (even over larger distances). The equilibrium state of a solid is generally a crystal, i.e. a spatially periodic arrangement of the molecules. The individual molecules can only oscillate around their equilibrium positions in the crystal lattice.

Some substances have phases that do not fit easily into this scheme. For example, ordinary glass is solid and homogeneous, but actually without a crystal structure; it is a a "frozen liquid" (also to be taken literally in terms of the manufacturing process). This means that the molecules have a time independent spatial distribution similar to that found in a liquid at a certain point in time. Another structure that does not fit into the simple liquid-solid scheme is that of liquid crystals. In this case the deformability of the material is that of a liquid; however, over large ranges (containing many molecules) there is a spatially periodic structure.

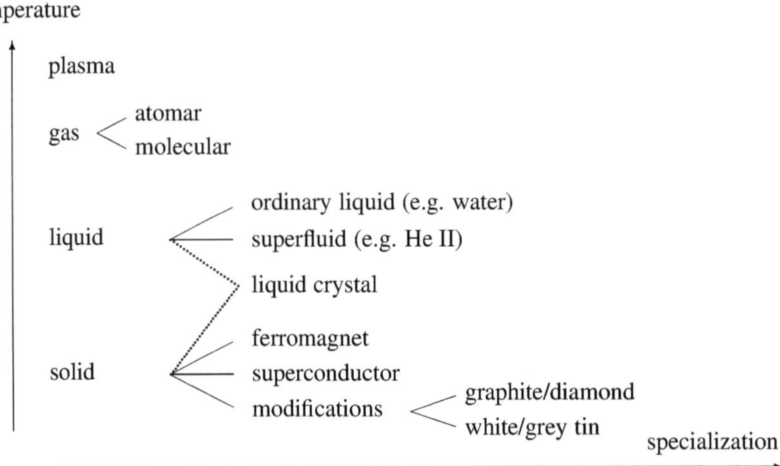

Figure 35.1   Overview of some selected phases that occur in nature. The qualitative temperature scale refers to the phases listed on the left.

For a given P and T, the phase with the lowest chemical potential $\mu(T, P)$ is present at equilibrium, (20.17). The *approach* to equilibrium, i.e. to the state with minimum μ, can take a very long time, possibly even an infinite time. Therefore, systems can exist stably in phases that are not the equilibrium phases. We refer to such a state as *quasi-equilibrium*; in these cases one may have a partial equilibrium in which the degrees of freedom with finite relaxation time (for example the lattice vibrations) are in equilibrium. A quasi-equilibrium state is only *metastable*, as it does not correspond to the absolute minimum of $G = N\mu(T, P)$. The transition to the absolute minimum is hindered by energy barriers. These barriers cannot be overcome by the available thermal energy (or only with a low probability). We give some examples of quasi-equilibrium states:

- Under normal conditions, graphite is the actual equilibrium state of carbon. Nevertheless, nobody needs to fear that his diamonds will turn into graphite on their own; the metastable diamond phase is in fact stable. Another largely stable modification of carbon are the fullerenes (molecules consisting of a large number of carbon atoms such as C_{60}).

- Above $13.2\,^\circ$C, the equilibrium of tin is the so-called β phase, a shiny silvery metal. Below this temperature, the equilibrium state is gray tin (α-tin, powdery). In practice, however, the white form is also stable at low temperatures; tin foil remains shiny white in the refrigerator, at least for a while. The transition between these modifications can easily be achieved experimentally—in contrast to the production of diamonds from graphite.

- The "frozen liquid" glass is not in its deepest (crystalline) state, but it is quite stable.

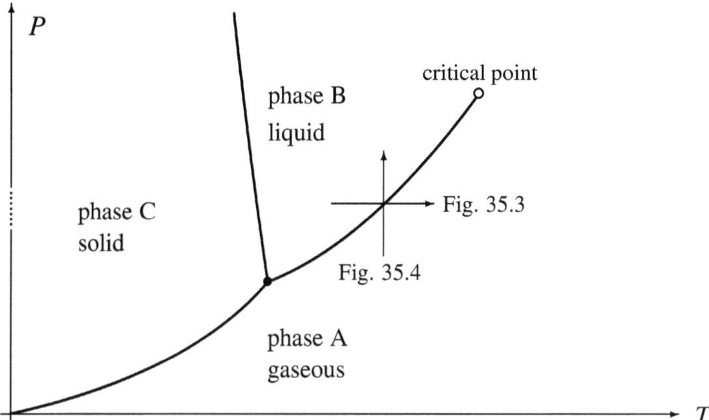

Figure 35.2   Phase diagram of a simple substance. Figures 35.3 and 35.4 show the behavior of μ when crossing the vapor pressure curve (arrows) at constant pressure or at constant temperature.

- Solids are usually in (stable) polycrystalline form, and not in the actual equilibrium state of a defect-free single crystal (mono-crystal).

The transition between two phases can be discrete or continuous. For this discussion, we refer to the simple phase diagram in Figure 35.2. This diagram is typical in that the structure shown occurs for almost all pure substances; however, there are often additional subdivisions. In Chapter 14 and 21, Figure 35.2 was shown as a phase diagram for water. In fact, it is a simplification, because there are several modifications (crystal forms) of ice which are separated by lines in the phase diagram.

If the phase transition is discrete, then the phases are separated by a line in the phase diagram; this case is discussed below. The gaseous and liquid phases are separated by the vapor pressure curve , which ends in a *critical point*. At the critical point this transition becomes continuous. And a liquid and a gaseous state can be connected to each other along a path that leads around the critical point. In the range of high temperatures and high pressures ($T > T_{cr}$ or $P > P_{cr}$), there is no longer a well-defined distinction between the gaseous and the liquid phase.

For technically achievable pressures, there is no such critical point for the solid-liquid curve there is no critical point. However, at sufficiently high pressures the atomic structures will dissolve (as in a white dwarf), so that this curve does not continue indefinitely.

Chemical potential

We investigate the behavior of the chemical potential at the discrete boundary line between two phases. Some properties have already discussed in Chapter 21, in particular the equilibrium condition and the Clausius-Clapeyron equation.

We consider a system with two phases, A and B. There shall be an equilibrium with respect to heat and volume exchange, which implies $T = \text{const.}$ and $P = \text{const.}$ in the entire sample. We denote the chemical potentials of the two phases by μ_A and μ_B. Since the structures of the two phases are different, $\mu_A(T, P)$ and $\mu_B(T, P)$ are different functions.

At equilibrium, the chemical potential $\mu(T, P)$ is minimal, (20.17). Therefore, of the two possible phases, the one with the smaller chemical potential will be present:

$$\mu(T, P) = \begin{cases} \mu_A(T, P) & \text{if } \mu_A \leq \mu_B \\ \mu_B(T, P) & \text{if } \mu_B \leq \mu_A \end{cases} \tag{35.1}$$

This alternative separates areas in the T-P-plane. The boundary line is given by

$$\boxed{\mu_A(T, P) = \mu_B(T, P) \quad \Longrightarrow \quad \text{transition curve } P = P(T)} \tag{35.2}$$

This relation makes the existence of discrete transition lines plausible. On the transition curve, the condition for an equilibrium under exchange of heat, volume and particles is fulfilled. This implies that the two phases can coexist. Specifically for the transition liquid to gaseous, the transition curve is given by the vapor pressure $P_v(T)$ or the boiling temperature $T_b(P)$.

We investigate in more detail how μ changes at the transition A $\leftrightarrow$ B. Figure 35.2 shows arrows indicating two simple ways of crossing the boundary line between the phases: One changes the temperature at constant pressure, or the pressure at constant temperature. At the transition line, the chemical potentials are equal; as a rule, this will not apply to their derivatives (along the selected path):

$$\left(\frac{\partial \mu_A}{\partial T}\right)_P \neq \left(\frac{\partial \mu_B}{\partial T}\right)_P \quad \text{at } T_{tr}, \quad \text{or} \quad \left.\frac{\partial \mu}{\partial T}\right|_{T_{tr}^+} \neq \left.\frac{\partial \mu}{\partial T}\right|_{T_{tr}^-} \tag{35.3}$$

Here $T_{tr}^{\pm}$ denotes a temperature immediately above or below the transition (index tr). The jump in $\partial \mu / \partial T$ implies a jump $\Delta s = s(T_{tr}^+, P) - s(T_{tr}^-, P)$ in the entropy, Figure 35.3, i.e. a transformation enthalpy. For a phase transition induced by an increase in pressure (vertical arrow in Figure 35.2) on gets an analogous jump in volume (Figure 35.4). A transition with a jump in the first partial derivative of μ is referred to as a *1st order phase transition*. The qualitative behavior discussed here applies also to solid-liquid and solid-gas transitions.

If we follow the transition line in Figure 35.2 towards the critical point, the jumps in entropy and volume approach zero. Immediately behind the critical point, the transition becomes is continuous. The following then applies at the critical point

$$\mu_A = \mu_B, \quad \left(\frac{\partial}{\partial T}\right)_P (\mu_A - \mu_B) = 0, \quad \left(\frac{\partial^2}{\partial T^2}\right)_P (\mu_A - \mu_B) \neq 0 \tag{35.4}$$

The partial derivatives with respect to the pressure behave accordingly. Such a transition is referred to as a 2nd order phase transition. This designation leads to the

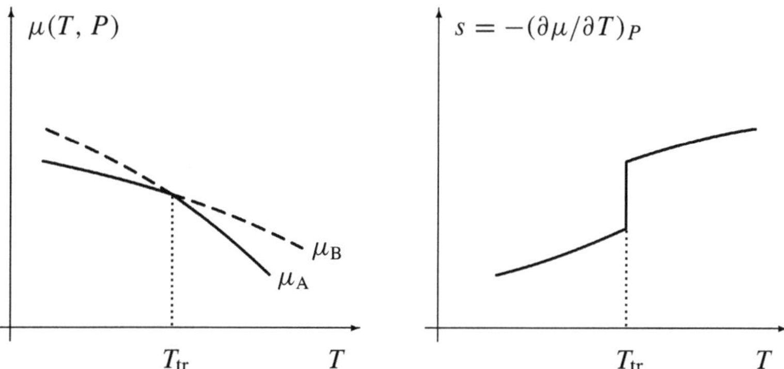

Figure 35.3 The chemical potentials of phases A and B as a function of temperature (left). At equilibrium, the system is in the phase with the lower potential, so the solid line shows the actual chemical potential μ. An increase in temperature at constant pressure results in the transition B $\rightarrow$ A (horizontal arrow in Figure 35.2). If the slopes of μ_A and μ_B are different at the transition point, then the entropy has a jump (right). The jump in the entropy means a finite transformation enthalpy and a δ-functional behavior of c_P (Figure 35.5 left).

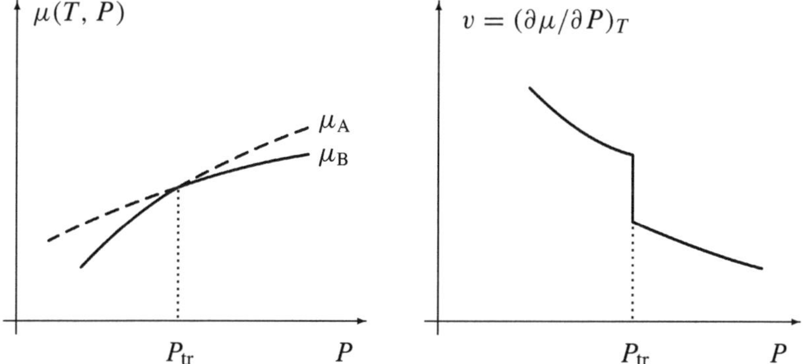

Figure 35.4 The chemical potentials of the phases A and B as a function of pressure (left). At equilibrium, the system is in the phase with the lower potential, so the solid line shows the actual chemical potential μ. Increasing the pressure at a constant temperature results in the transition A $\rightarrow$ B (vertical arrow in Figure 35.2). If the slopes of μ_B and μ_A are different at the transition point, then the volume displays a jump (right).

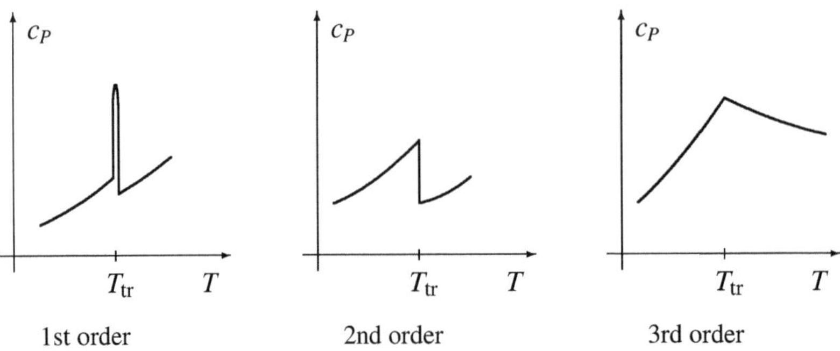

<table>
<tr><td>1st order</td><td>2nd order</td><td>3rd order</td></tr>
</table>

Figure 35.5 Classification of phase transitions according to Ehrenfest: If the first, second or third derivative of the chemical potential at the transition temperature has a jump, it is a phase transition 1st, 2nd or 3rd order phase transition.

Ehrenfest classification: If the nth derivative of $\mu(T, P)$ has a jump, it is a phase transition of the nth order, i.e.

$$\left(\frac{\partial^m}{\partial T^m}\right)_P (\mu_A - \mu_B) \begin{cases} = 0 & (m < n) \\ \neq 0 & (m = n) \end{cases} \qquad \begin{array}{l} \text{(transition of } n\text{th order} \\ \text{according to Ehrenfest)} \end{array} \qquad (35.5)$$

For $n = 1$, 2 and 3, the corresponding behavior of the specific heat is sketched in Figure 35.5.

The theoretical treatment of the transition usually begins with the choice of a macroscopic quantity that changes in a characteristic way at the transition point. This can be, for example, the volume for the liquid-gas transition, or the magnetization for the ferromagnetic transition. This appropriately selected parameter is called *order parameter* (Chapter 39). Today, the usual designation of transitions follows from the behavior of this order parameter ψ at the transition:

$$\overline{\psi} = \begin{cases} \text{discontinuous} & \text{transition of 1st order} \\ \text{continuous} & \text{transition of 2nd order} \end{cases} \qquad (35.6)$$

For the transitions discussed above (liquid-gas and transition at the critical point) this classification agrees with that by Ehrenfest (1st and 2nd order). For the Bose-Einstein condensation, the order parameter (condensate fraction N_0/N in Figure 31.3) changes continuously. This holds also for the order parameter $v - v_{cr}$ of the liquid-gaseous transition at the critical point. According to (35.6), these are therefore a 2nd order phase transition.

Microscopic calculation

We discuss the possibility of calculating the phases of a substance microscopically. The basic procedure is clear: Starting from the Hamilton operator $H(V, N)$ of the system, one determines the energy eigenvalues E_r of the microstates r, and from this the partition function

$$Z(T, V, N) = \sum_r \exp\left(-\frac{E_r(V, N)}{k_B T}\right) \tag{35.7}$$

In this way, the free energy $F = -k_B T \ln Z$ and all other thermodynamic quantities can be determined. From $F(T, V, N)$ follows in particular the free enthalpy $G = N\mu(T, P)$ and thus the chemical potential $\mu(T, P)$. The transition lines of the phase diagram are then determined as those lines where this $\mu(T, P)$ has discontinuous first (or higher) derivatives. This shows in principle how the existence of the phases and their position in the P-T diagram can be derived from the Hamiltonian operator.

If the first (or a higher) order derivative of μ is discontinuous this also applies to the respective derivatives of F; specifically for a 1st order phase transition, $S = -\partial F/\partial T$ has a jump. So at the transition point, the first (or a higher) derivative of $Z(T, V, N)$ must be discontinuous; the next higher derivative is then singular.

The principle determinability of the phase transitions from (35.7) gives rise to the following problem: Each term on the right-hand side of (35.7) is an analytic function that can be differentiated any number of times with respect to T (for $T \neq 0$). Must this then not also hold for a sum of such terms? In this case there could be no singularities in the higher derivatives of Z; i.e. there would be no phase transitions. Then the possibility of the statistical treatment of equilibrium states via (35.7) would generally be called into question.

The solution to this apparent contradiction is that a *infinite* sum of analytical functions can result in a singular function. We demonstrate this with a simple example. The function

$$f_v(x) = \sum_{n=1}^{\infty} \frac{\exp(-xn)}{n^v} \tag{35.8}$$

depends on the argument x and a parameter v. In Exercise 35.1 it is shown that $f_{1/2}$ for $x \to 0$ behaves like

$$f_{1/2} = \sqrt{\frac{\pi}{x}} + \text{(terms that are finite for } x \to 0) \tag{35.9}$$

Note for this example:

- Every single term on the right-hand side of (35.8) is arbitrarily often differentiable with respect to x. This then also applies to a finite sum of such terms. For $v = 1/2$ the infinite sum results, however, results in a function that is singular at $x = 0$.

- The functions f_v with $v = m + 1/2$ have a singularity in the m-th derivative. This follows from $f_v' = -f_{v-1}$.

- The treatment of the ideal Bose gas led to the functions $f_\nu(x)$ with $x = -\beta\mu$. With $z = \exp(-x)$ as argument, the function $f_\nu(x)$ becomes the generalized Riemann function $g_\nu(z)$, (31.10). The derivative of $g_{3/2}(z) = f_{3/2}(x)$ results in a function that is singular at $\mu = 0$; at this point the phase transition occurs in the ideal Bose gas. The example under consideration is therefore relevant for this particular phase transition.

For the singularity properties it is essential that (35.7) and (35.8) are *infinite* sums. This implies:

1. The partition function cannot be approximated by a finite sum.

2. A phase transition can only occur in an infinite system.

An infinite system corresponds to the *thermodynamic limit*

$$N \to \infty, \qquad V \to \infty, \qquad v = \frac{V}{N} = \text{const.} \tag{35.10}$$

We have already referred to this limit for the Bose-Einstein condensation in the ideal Bose gas (Chapter 31).

For a finite system, the singularities (such as the δ function, the jump or the kink in Figure 35.5) are smeared out. Every concrete system is finite, but for $N = \mathcal{O}(10^{24})$ the width of the blurring of the singularity is smaller than the measurement accuracy.

If we could calculate the partition function (35.7) with sufficient accuracy for a realistic system, this would determine a potential phase transition and its properties. Due to the structure of the phases, it can be expected that the phases correspond to different solution types (classes of microstates). In (35.7), all possible microstates have to be counted. But depending on the value of T and V, different solution can dominate the partition function to a greater or lesser extent.

The calculation of the existence and properties of the phases from (35.7) is a extremely difficult problem that cannot be solved in general; the discussion of the emergence of singularities already indicates this. However, there are special models that can be solved exactly and whose solution contains a phase transition. The ideal Bose gas from Chapter 31 is such a model. In this case, the partition function could be evaluated exactly, and that leads to a phase transition (Bose-Einstein condensation). In the following chapters we will deal with models that can describe phase transitions but which cannot be derived (at least not completely) microscopically. Examples are the Weiss model of ferromagnetism and the van der Waals gas.

Exercises

35.1 Singularity due to infinite sum

Derive the behavior of the function $g(x)$ for $x \to 0^+$:

$$g(x) \equiv \sum_{n=1}^{\infty} \frac{\exp(-xn)}{\sqrt{n}} \quad \overset{x\to 0^+}{\longrightarrow} \quad \sqrt{\frac{\pi}{x}} \qquad (x > 0)$$

This is an example of how an infinite sum of analytic terms can be singular. Use Euler's summation formula

$$\sum_{n=n_0}^{n_1} f(n) = \int_{n_0}^{n_1} dn\, f(n) + \frac{f(n_0) + f(n_1)}{2} - \frac{f'(n_0) - f'(n_1)}{12} \pm \ldots$$

and the substitution $y^2 = xn$.

36 Ferromagnetism

For an ideal spin system, the magnetization is proportional to the external magnetic field; the system is paramagnetic (Chapter 26). In realistic systems, the interaction between the spins can result in a magnetization even without an external field; the material is then called ferromagnetic. At high temperatures, a ferromagnetic material becomes paramagnetic again. The phase transition paramagnetic-ferromagnetic takes place at a critical temperature T_{cr}. Below T_{cr} there is a spontaneous magnetization. The free energy is considered as a function of the temperature, the external field and the magnetization. The discussion of spin systems has model character for the treatment of phase transitions.

Weiss model

We construct a simple model for the ferromagnetism. In a crystal, there shall be at each lattice site an unpaired electron with the magnetic moment $\boldsymbol{\mu} = -g\mu_B \boldsymbol{s}$. Here, $\mu_B = e\hbar/2m_e c$ is the Bohr magneton[1]. For the g-factor we use the value for free electrons, $g \approx 2$. The z component of a spin can adopt the values $s_z = \pm 1/2$ (the factor $\hbar$ is contained in μ_B). As a model system, we consider $i = 1, 2, ..., N$ independent spins. In an external magnetic field $\boldsymbol{B}$, the Hamilton operator of this ideal spin system is then

$$H_0 = -\sum_i \boldsymbol{\mu}_i \cdot \boldsymbol{B} = 2\sum_i \mu_B\, \hat{\boldsymbol{s}}_i \cdot \boldsymbol{B} \tag{36.1}$$

For $\boldsymbol{B} = B\boldsymbol{e}_z$ the microstates are defined by the quantum numbers $s_{z,i}$, i.e. $r = (s_{z,1}, ..., s_{z,N})$ or $r = (\pm 1/2, ..., \pm 1/2)$.

In a real crystal, the spins of these electrons interact with each other. The main contribution to this is not the magnetic dipole-dipole interaction, but the Coulomb interaction in interplay with the *exchange symmetry*. This will be explained in the following. Thereby a Hamilton operator for the interacting system is set up.

The real spatial wave functions of two neighboring electrons are denoted by $\phi_a(\boldsymbol{r}_1)$ and $\phi_b(\boldsymbol{r}_2)$; ϕ_a and ϕ_b are functions localized at adjacent lattice sites. The spin state $|SS_z\rangle$ for the total spin $\boldsymbol{S}$ of the two electrons can be symmetric ($S = 1$)

---

[1] The chemical potential μ_B of a phase B is not used in this chapter.

or antisymmetric ($S = 0$). The wave function $\Psi(1, 2)$ of the two electrons must be antisymmetric in total, i.e.

$$\Psi = \psi(r_1, r_2)\,|SS_z\rangle = \frac{1}{\sqrt{2}} \cdot \begin{cases} \left[\phi_a(r_1)\,\phi_b(r_2) + \phi_a(r_2)\,\phi_b(r_1)\right]|00\rangle \\ \left[\phi_a(r_1)\,\phi_b(r_2) - \phi_a(r_2)\,\phi_b(r_1)\right]|1S_z\rangle \end{cases} \tag{36.2}$$

For the normalization, we have assumed $\langle\phi_a|\phi_b\rangle \approx 0$. We calculate the Coulomb energy E_C of the two electrons:

$$E_C = \left\langle \Psi \left| \frac{e^2}{r_{12}} \right| \Psi \right\rangle = \underbrace{\int d^3r_1 \int d^3r_2\, |\phi_a(r_1)|^2\, |\phi_b(r_2)|^2\, \frac{e^2}{r_{12}}}_{=\,I_0}$$

$$\pm \underbrace{\int d^3r_1 \int d^3r_2\, \phi_a(r_1)\,\phi_b(r_1)\,\phi_a(r_2)\,\phi_b(r_2)\, \frac{e^2}{r_{12}}}_{=\,I/2} \tag{36.3}$$

The first integral I_0 is the direct Coulomb energy, the second one is referred to as the *exchange integral*. The exchange integral is positive and of the size $I/2 = \mathcal{O}(\text{eV}/10) = \mathcal{O}(10^3\,k_\text{B}\text{K})$. Depending on the total spin, the result is

$$E_C = \begin{cases} I_0 + I/2 & (S = 0) \\ I_0 - I/2 & (S = 1) \end{cases} \tag{36.4}$$

From $\hat{\boldsymbol{S}}^2 = (\hat{\boldsymbol{s}}_1 + \hat{\boldsymbol{s}}_2)^2 = \hat{\boldsymbol{s}}_1^2 + \hat{\boldsymbol{s}}_2^2 + 2\,\hat{\boldsymbol{s}}_1 \cdot \hat{\boldsymbol{s}}_2$ follows

$$\langle SS_z|\,\hat{\boldsymbol{S}}^2\,|SS_z\rangle = S(S + 1) = 2s(s + 1) + 2\langle SS_z|\,\hat{\boldsymbol{s}}_1 \cdot \hat{\boldsymbol{s}}_2\,|SS_z\rangle \tag{36.5}$$

For $s = 1/2$ we get

$$\langle 00|\,\hat{\boldsymbol{s}}_1 \cdot \hat{\boldsymbol{s}}_2\,|00\rangle = -\frac{3}{4} \quad \text{and} \quad \langle 1S_z|\,\hat{\boldsymbol{s}}_1 \cdot \hat{\boldsymbol{s}}_2\,|1S_z\rangle = \frac{1}{4} \tag{36.6}$$

This means that we can express the Coulomb energy (36.4) in the form

$$E_C = I_0 - I\left(\langle \hat{\boldsymbol{s}}_1 \cdot \hat{\boldsymbol{s}}_2\rangle + \frac{1}{4}\right) \tag{36.7}$$

The brackets $\langle\ldots\rangle$ denote the quantum mechanical expectation values (36.6). The energy contribution (36.7) is obtained for all neighboring spins. We take this spin-spin interaction into account by an additional term to the Hamilton operator (36.1):

$$\boxed{H = 2\sum_i \mu_\text{B}\,\hat{\boldsymbol{s}}_i \cdot \boldsymbol{B} - I\sum_{\{i,\,j\}} \hat{\boldsymbol{s}}_i \cdot \hat{\boldsymbol{s}}_j \qquad \text{Heisenberg model}} \tag{36.8}$$

The constant contributions in (36.7) play no role. The index $\{i, j\}$ means that we sum over all neighboring pairs, whereby each pair is counted once. The localized spatial wave functions ϕ_a and ϕ_b have a noticeable overlap only for neighboring lattice locations; therefore only these contributions are taken into account.

The Hamilton operator (36.8) defines the *Heisenberg model*. If $\hat{s}_i \cdot \hat{s}_j$ replaced by $\hat{s}_{z,i}\,\hat{s}_{z,j}$, the result is called *Ising model*. The Ising model is starting point for many studies of phase transitions. In the one- and two-dimensional lattice it can be solved analytically, in three dimensions numerically.

We solve the Heisenberg model using the Weiss approximation. Here, the sum over the spin operators $\hat{s}_j$ of the neighbors of any lattice site is approximated by the mean value of all spins:

$$\sum_{\{i,\,j\}} \hat{s}_i \cdot \hat{s}_j = \sum_i \hat{s}_i \cdot \sum_{j:\ \text{neighbors of } i} \hat{s}_j \approx \left(\sum_i \hat{s}_i\right) \cdot \nu\,\overline{s} \qquad (36.9)$$

Here ν stands for the number of nearest neighbors, for example $\nu = 6$ in the cubic lattice. Formally, $\overline{s}$ is the statistical mean value of the quantum mechanical expectation value $\langle\,\hat{s}_j\,\rangle$. In the *mean field approximation* (36.9) (also called molecular field approximation), the influence of the other particles on a selected particle is described by an effective mean field. In this approximation, we obtain $H \approx H_{\text{eff}}$ with the effective Hamilton operator

$$H_{\text{eff}} = 2 \sum_i \mu_{\text{B}}\,\hat{s}_i \cdot \left(\boldsymbol{B} - \frac{\nu I\,\overline{s}}{2\,\mu_{\text{B}}}\right) = 2 \sum_i \mu_{\text{B}}\,\hat{s}_i \cdot \boldsymbol{B}_{\text{eff}} \qquad (36.10)$$

In the effective field

$$\boldsymbol{B}_{\text{eff}} = \boldsymbol{B} - \frac{\nu I\,\overline{s}}{2\,\mu_{\text{B}}} = \boldsymbol{B} + W\boldsymbol{M} \qquad (36.11)$$

the magnetization $\boldsymbol{M}$ and a numerical factor W occur:

$$\boldsymbol{M} = \frac{N}{V}\,\overline{\mu} = -2n\mu_{\text{B}}\,\overline{s}, \qquad W = \frac{\nu I}{4n\mu_{\text{B}}^2} \qquad (36.12)$$

With $I/2 = \mathcal{O}(\text{eV}/10)$, $\nu = 6$, $\mu_{\text{B}} = e\hbar/(2m_e c)$ and $n = (3\,\text{Å})^{-3}$ we estimate the size of W:

$$W = \frac{\nu I}{4n\mu_{\text{B}}^2} \approx \frac{6 \cdot \mathcal{O}(\text{eV}/10)}{\underset{\text{magnetic interaction}}{\underbrace{}}} \sim 10^3 \qquad (36.13)$$

The result $W \gg 1$ means that the exchange interaction $\pm I/2$ in (36.3) is much stronger than the magnetic interaction ($\sim 2n\mu_{\text{B}}^2$) between neighboring magnetic moments.

Magnetization

The effective Hamilton operator H_{eff} has the same form as the Hamilton operator (36.1) for N independent spins. Therefore, the partition function can be calculated in the same way as in Chapter 26; whereby the external field $\boldsymbol{B}$ is replaced by the effective field $\boldsymbol{B}_{\text{eff}}$. For the sake of simplicity, we set $\boldsymbol{B} = B\,\boldsymbol{e}_z$ and consequently $\boldsymbol{M} = M\,\boldsymbol{e}_z$. The result for the magnetization M is known from (26.11):

$$M = n\,\mu_{\text{B}}\,\tanh\left(\beta\mu_{\text{B}}B_{\text{eff}}\right) = M_0\,\tanh\left[\beta\mu_{\text{B}}(B + WM)\right] \tag{36.14}$$

This is an *implicit* solution; because the magnetization to be calculated is part of the effective field B_{eff}. In this way, the Weiss model takes into account the mutual interaction of the spins.

For discussing (36.14) we consider two limiting cases. For a strong external field B, the magnetization approaches the maximum value M_0 exponentially:

$$\frac{M}{M_0} \approx 1 - 2\,\exp(-2\beta\mu_{\text{B}}B) \qquad (B \gg WM) \tag{36.15}$$

For $M = M_0 = n\,\mu_{\text{B}}$ all spins are aligned. For $M \ll M_0$ we expand (36.14):

$$\frac{\mu_{\text{B}}}{k_{\text{B}}T}\,(B + WM) = \operatorname{artanh}\left(\frac{M}{M_0}\right) = \frac{M}{M_0} + \frac{1}{3}\left(\frac{M}{M_0}\right)^3 + \dots \tag{36.16}$$

We neglect the higher order terms and solve for B:

$$B = W\left(\frac{T}{T_{\text{cr}}} - 1\right)M + \frac{k_{\text{B}}T}{3\,\mu_{\text{B}}}\left(\frac{M}{M_0}\right)^3 \qquad (M \ll M_0) \tag{36.17}$$

Here we have introduced the critical temperature T_{cr},

$$k_{\text{B}}T_{\text{cr}} = \mu_{\text{B}}^2\,n\,W \overset{(36.13)}{=} \frac{vI}{4} \sim 10^3\,k_{\text{B}}\,\text{K} \tag{36.18}$$

This is a transition temperature between the phases with and without spontaneous magnetization. Previously, we have denoted the transition temperature by T_{tr}. For 2nd phase transitions (as here, or as for the liquid- gas transition at the critical point), the transition temperature is usually denoted as the critical (index cr) temperature T_{cr}.

Figure 36.1 shows the relationship between B and M graphically. For $T > T_{\text{cr}}$, M is initially proportional to B; for $B \to \infty$, the magnetization M approaches the saturation value M_0. The initially paramagnetic behavior is similar to that of independent spins. The situation changes qualitatively for $T < T_{\text{cr}}$. There are B-values with three solutions for this; they result in Figure 36.1 as intersections of a horizontal line $B = \text{const.}$ with the cubic curve (36.17). Because of $\boldsymbol{M} \parallel \boldsymbol{B}$, only the solutions in the first and third quadrants in Figure 36.1 are indeed solutions. The solution in the first quadrant ends at $B = 0$ with a finite M value, which we call *spontaneous magnetization* M_{s}. This magnetization occurs below T_{cr} without an external field, i.e. by itself or spontaneously. Such behavior is referred to as *ferromagnetism*. The solution in the third quadrant ends accordingly at $-M_{\text{s}}$.

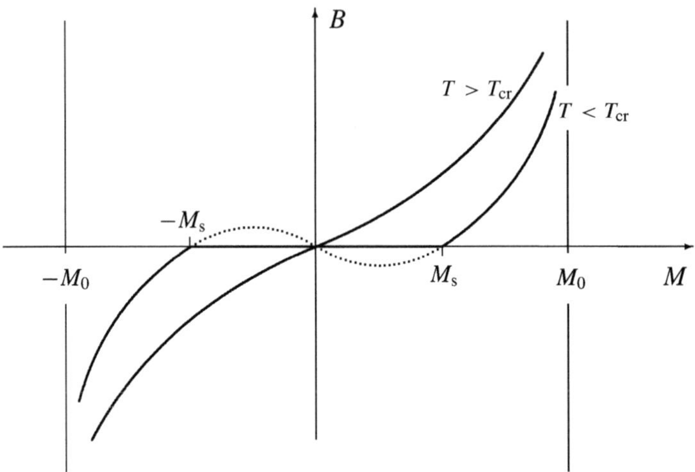

Figure 36.1 Relationship between the magnetization M and the applied magnetic field B in the Weiss model. Because of $\boldsymbol{M} \parallel \boldsymbol{B}$ the dotted parts of the curve do not constitute a solution.

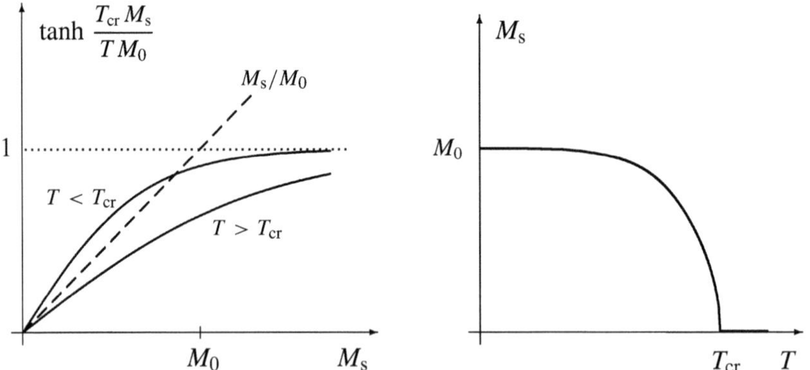

Figure 36.2 In the left part the two sides of Equation (36.19) are displayed. The intersection point is the solution $M_s(T)$, which is shown on the right.

No direction is distinguished for $B = 0$, so that all directions of M_s (with fixed $|M_s|$) are possible. The projection $M_s \cdot e_z$ can adopt all possible values between $-M_s$ and M_s; the two solutions $-M_s$ and $+M_s$ can therefore be connected in Figure 36.1 by a horizontal straight line.

From (36.14) with $B = 0$ follows the strength of the spontaneous magnetization:

$$\boxed{\;\frac{M_s}{M_0} = \tanh\left(\frac{T_{cr}\,M_s}{T\,M_0}\right) \qquad \text{spontaneous} \atop \text{magnetization}\;} \tag{36.19}$$

This equation is solved graphically in Figure 36.2. The following applies

$$M_s = M_s(T) \approx M_0 \cdot \begin{cases} 0 & (T \geq T_{cr}) \\[2mm] \sqrt{\dfrac{3\,(T_{cr} - T)}{T_{cr}}} & (T \to T_{cr}^{-}) \\[2mm] 1 - 2\,\exp\left(-\dfrac{2\,T_{cr}}{T}\right) & (T \ll T_{cr}) \end{cases} \tag{36.20}$$

The spontaneous magnetization is sketched on the right in Figure 36.2 as a function of temperature. Above T_{cr} is M_s is equal to zero, below it is non zero. The behavior of the system changes qualitatively at the temperature T_{cr}; the system undergoes a discrete phase transition at T_{cr}. The quantity M_s is the obvious order parameter for characterizing this transition. Since the order parameter changes continuously at T_{cr}, we are dealing with a phase transition of 2nd order.

The magnetic behavior can be described by the material parameter *magnetic susceptibility* $\chi_m = (\partial M / \partial B)_T$. For $B \to 0$ and $T > T_c$, we obtain $M \to 0$. Then χ_m is determined by the linear term in (36.17):

$$\chi_m = \left(\frac{\partial M}{\partial B}\right)_T \overset{B \to 0}{=} \frac{T_{cr}/W}{T - T_{cr}} \qquad \text{(Curie-Weiss law, } T \geq T_{cr}) \tag{36.21}$$

This *Curie-Weiss law* applies to a ferromagnet; it may be compared to the Curie law (26.15) for a paramagnet. For paramagnetic materials at room temperature the susceptibility χ_m is of the size 10^{-4}. In contrast to this, the susceptibility (36.21) is of size 1 for $T - T_{cr} = 1$ K.

The magnetic susceptibility (36.21) becomes singular at the transition point; an arbitrarily weak field may then cause a finite change in the magnetization. This applies also applies to $T \to T_{cr}^{-}$ and $B \parallel M$ (Exercise 36.1).

In the three-dimensional case, each direction of M_s represents a possible equilibrium state (for $T < T_{cr}$ and $B = 0$). If one now applies a weak magnetic field B, the equilibrium state is the one with $M_s \parallel B$. This means that for $T < T_{cr}$ a weak field can cause a finite change in magnetization. In this respect, the susceptibility in three dimensions is infinite for $T < T_{cr}$. However, this is only the case in our model. A real ferromagnet displays a resistance to a change in direction of M_s; therefore the susceptibility remains finite.

Weiss domains

The best-known example of a ferromagnet is iron with the transition temperature $T_{cr} = 1041$ K. If, in absence of an external magnetic field, iron is cooled below T_{cr} it comes to a spontaneous magnetization in finite ranges, the *Weiss domains*. The temperature dependence is similar to that shown in Figure 36.2 on the right. The global magnetization $\langle M \rangle$ —averaged over several Weiss' domains— can disappear, however; despite $T < T_{cr}$ the iron does not appear as a magnet. By an external magnetic field the magnetization directions of different Weiss' domains can be aligned. For weak fields, this alignment and the resulting global magnetization is approximately proportional to the field:

$$\langle M \rangle \approx \chi_{\text{ferro}} \, B \tag{36.22}$$

For this, the susceptibility can assume large values, such as $\chi_{\text{ferro}} \sim 10^6$. For strong fields one approaches the saturation magnetization M_0. The alignment of the Weiss domains requires a finite amount of energy. As a result of such an alignment, a piece of iron becomes permanent magnet. In addition, the Weiss domains lead to hysteresis effects.

Free energy

We want to investigate the behavior of the specific heat at the phase transition. To do this, we first determine a special form of the free energy.

The magnetic field is an external parameter $x = B = \boldsymbol{B}\, \boldsymbol{e}_z$ of the Hamilton operator (36.8). In the following we assume that other external parameters do play no role; in particular the volume V shall be constant. The force generalized force X associated with $x = B$ is the magnetic moment $V M$:

$$X = -\overline{\frac{\partial E_r(B)}{\partial B}} = -2\mu_B \sum_i \overline{s_{z,i}} = N\overline{\mu} = VM \tag{36.23}$$

The energy eigenvalues of the Hamilton operator (36.8) are to be used for this. According to the 1st and 2nd law of TD, $dE = T\,dS - \sum_i X_i\,dx_i = dE_0 - VM\,dB$. Compared to the case without a magnetic field (dE_0), the energy dE and thus all other thermodynamic potentials receive the additional term $-VM\,dB$. The free energy can then be expressed in the form

$$dF = dF_0 - VM\,dB = dF_0 - V d(BM) + VB\,dM \tag{36.24}$$

This yields $F = F_0 - VBM + V \int B\,dM$. In this integral we insert $B = B(M)$ from (36.17) and carry out the integration:

$$\begin{aligned}
\mathcal{F}(T, B, M) &= F_0(T) - VMB + \frac{VW}{2}\frac{T - T_{cr}}{T_{cr}} M^2 + \frac{V k_B T}{12\,\mu_B}\frac{M^4}{M_0^3} \\
&= F_0(T) + V\left[a\,(T - T_{cr})\, M^2 + u\, M^4 - MB \right]
\end{aligned} \tag{36.25}$$

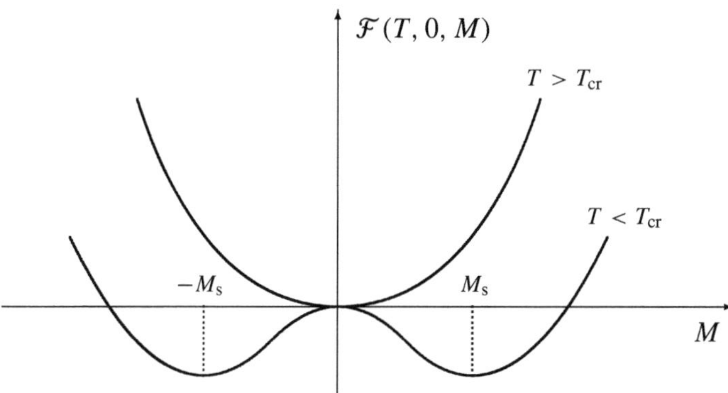

Figure 36.3 Free energy $\mathcal{F}(T, B, M)$ of the Weiss model as a function of M for $B = 0$. At T_{cr} the sign of the quadratic term changes. For $T > T_{\mathrm{cr}}$ the minimum is $M = 0$, for $T < T_{\mathrm{cr}}$, on the other hand, the minimum occurs at a finite value $M = M_{\mathrm{s}}$.

Here we used the font $\mathcal{F}$ to point out that this is not yet the equilibrium free energy F. The equilibrium energy depends alone on T and B (and not, as $\mathcal{F}$ on T, B and M). The equilibrium free energy $F(T, B)$ can be obtained by inserting the equilibrium magnetization $M = M(T, B)$ into $\mathcal{F}(T, B, M)$. The more general expression $\mathcal{F}(T, B, M)$ includes non-equilibrium states.

The magnetic field shall have no influence on the degrees of freedom not explicitly treated here; then F_0 does not depend on B. When $\int B\,dM$ is performed, the temperature is kept constant (the change of F with temperature change is taken into account in dF_0). In the neighborhood of T_{cr}, the coefficients a and u in (36.25) can be assumed to be constants.

The following discussion is limited to temperatures close to T_{cr}, and to a weak magnetic field (because we have used (36.17). The free energy (36.25) has model character (Chapter 39) for phase transitions.

In the free energy expression $\mathcal{F}(T, B, M)$ any values of the magnetization M are permitted. For a given T and x (here $x = B$), the free energy at equilibrium is minimal (Chapter 17). From this follows the condition for the equilibrium value $M(T, B)$ of the magnetization:

$$\mathcal{F} = \text{minimum} \quad \longrightarrow \quad \left(\frac{\partial \mathcal{F}}{\partial M}\right)_{T,B} = 0 \quad \longrightarrow \quad M = M(T, B) \qquad (36.26)$$

Figure 36.3 shows $\mathcal{F}$ for $B = 0$ as a function of M. For $T > T_{\mathrm{cr}}$, $\mathcal{F}$ has only one minimum at $M = 0$. For $T < T_{\mathrm{cr}}$, the sign of the quadratic term in (36.25) reverses, and for small M this term initially dominates over the one with M^4. This results in two minima, namely at $M = M(T, 0) = \pm M_{\mathrm{s}}(T)$. The minima correspond to the possible equilibrium states of the system. Without the restriction to the z-direction we obtain a continuum of minima (all directions of M_{s}). The minimum determines the equilibrium magnetization. For $B \neq 0$, the one minimum in Figure 36.3 would shift downwards, the other upwards. The equilibrium state is then that of the absolute minimum.

By inserting $M = M(T, B)$ into the free energy (36.25), we obtain the equilibrium free energy:

$$F(T, B) = \mathcal{F}\big(T, B, M(T, B)\big) \tag{36.27}$$

The function $\mathcal{F}(T, B, M)$ is analytical in all variables (all derivatives are continuous). The function $F(T, B)$, on the other hand, is not analytic for $T = T_{\mathrm{cr}}$. For example, $M(T, 0) = M_{\mathrm{s}}(T)$ has a kink at T_{cr}. Inserting $M(T, B)$ into $\mathcal{F}$ results in a function F that has a jump in the second derivative at T_{cr}.

Spontaneous symmetry breaking

The solution of $\mathcal{F} = $ minimal displays for $T < T_{\mathrm{cr}}$ a *spontaneous symmetry breaking*. This means that a solution lacks a certain symmetry which the Hamilton operator of the system possesses. The Hamilton operator (36.8) is invariant under rotations; it does not distinguish any direction. The equilibrium state for $T > T_{\mathrm{cr}}$ has this symmetry; it does not prefer an direction. For $T < T_{\mathrm{cr}}$, on the other hand, the solution is given by M_{s}; and this magnetization distinguishes a specific direction. This distinguished direction *breaks* (violates) the rotational symmetry. This breaking occurs *spontaneously*, i.e. without external influence. The symmetry operations (here rotations) transforms a solution into another solution (i.e. it changes the directions of M_{s}).

Landau developed a general theory (Chapter 39) for phase transitions of second order by assuming an expression for free energy analogous to (36.25). Such a free energy then depends on the order parameter (like $\mathcal{F}$ depends on M).

Specific heat

For $B = 0$ and $T \approx T_{\mathrm{cr}}$ we insert the equilibrium solution (36.20),

$$M(T, 0) = M_{\mathrm{s}}(T) = M_0 \cdot \begin{cases} 0 & (T \geq T_{\mathrm{cr}}) \\ \sqrt{3\,(T_{\mathrm{cr}} - T)/T_{\mathrm{cr}}} & (T \to T_{\mathrm{cr}}^-) \end{cases} \tag{36.28}$$

into the general free energy form $\mathcal{F}$ of (36.25):

$$F(T, B = 0) = \mathcal{F}(T, 0, M_{\mathrm{s}}(T)) = F_0(T) + \begin{cases} 0 \\ -K\,(T - T_{\mathrm{cr}})^2 \end{cases} \tag{36.29}$$

The upper line applies to $T \geq T_{\mathrm{cr}}$, the lower line to $T \to T_{\mathrm{cr}}^-$. The constant K shall be determined in Exercise 36.2. We calculate the heat capacity for a constant, vanishing field B:

$$C_B(T) = -T \left(\frac{\partial^2 F}{\partial T^2}\right)_B \overset{(B=0)}{=} C_0(T) + \begin{cases} 0 & (T \geq T_{\mathrm{cr}}) \\ 2K\,T_{\mathrm{cr}} & (T \to T_{\mathrm{cr}}^-) \end{cases} \tag{36.30}$$

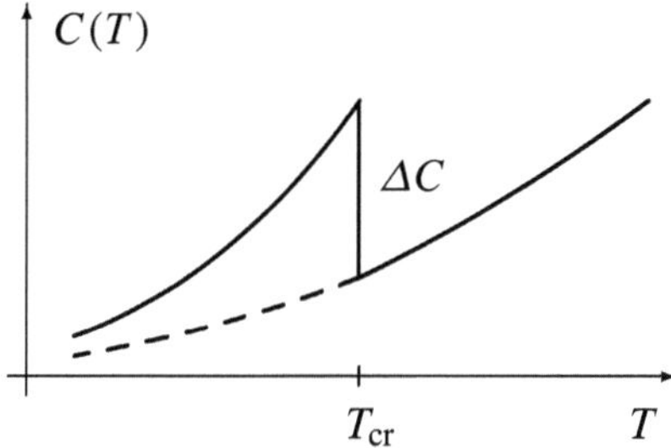

Figure 36.4 In the Weiss model, the heat capacity displays a jump at the transition temperature.

This heat capacity is sketched in Figure 36.4; it has a jump at T_{cr}. The exact solution of the Heisenberg model (36.8), has a divergent heat capacity; the Weiss approximation falls short of this result. For iron, the experimental specific heat at the transition point diverges logarithmically.

Exercises

36.1 Free energy in the Weiss model

In the Weiss model, the general form of free energy can be written as

$$\mathcal{F}(T, B, M) = F_0(T) + V\left[a\,(T - T_{\mathrm{cr}})\,M^2 + u\,M^4 - MB \right]$$

The condition $\partial\mathcal{F}/\partial M = 0$ yields the equilibrium magnetization $\overline{M}$. Determine $\overline{M_0}$ for $B = 0$, and $\overline{M_0} + \delta M$ for a weak field $B = \delta B$. Calculate the magnetic susceptibility $\chi_{\mathrm{m}} = \delta M/\delta B$ for $T > T_{\mathrm{cr}}$ and $T < T_{\mathrm{cr}}$ in first order in the small quantities.

36.2 Specific heat in the Weiss model

In the Weiss model, the general expression for free energy $\mathcal{F}(T, B, M)$ is

$$\mathcal{F}(T, B, M) = F_0(T) - VMB + \frac{VW}{2}\frac{T - T_{\mathrm{cr}}}{T_{\mathrm{cr}}}M^2 + \frac{Vk_{\mathrm{B}}T}{12\,\mu_{\mathrm{B}}}\frac{M^4}{M_0^3}$$

Here W is the Weiss factor, $k_{\mathrm{B}}T_{\mathrm{cr}} = \mu_{\mathrm{B}}^2\,n\,W$ defines the critical temperature, $M_0 = N\mu_{\mathrm{B}}/V$ is the saturation magnetization, and $F_0(T)$ contains the non-magnetic contributions. Express all parameters in $\mathcal{F} - F_0(T)$ by k_{B}, T_{cr} and M_0. For $B = 0$ and $T \approx T_{\mathrm{cr}}$, determine the equilibrium value $M_{\mathrm{S}}(T)$ and the behavior (jump!) of the heat capacity.

37 Van der Waals gas

The van der Waals gas is investigated as a model for the phase transition gaseous–liquid. Van der Waals introduced this model in 1873 in his doctoral thesis. The thermal equation of state is supplemented by the so-called Maxwell construction. Then the van der Waals model explains the existence of a vapor pressure curve $P_v(T)$, which ends at a critical point, and a number of properties of the gaseous–liquid phase transition.

The *van der Waals gas* is defined by the equations of state

$$P \;=\; P(T, v) \;=\; \frac{k_B T}{v - b} - \frac{a}{v^2} \tag{37.1}$$

$$E \;=\; E(T, V, N) \;=\; N e(T) - N \frac{a}{v} \tag{37.2}$$

Here $v = V/N$, and a and b are positive parameters. We adopt the thermal equation of state, also known as the van der Waals equation, from (28.29). In the caloric equation of state (28.37), we have replaced the volume-independent term $3 N k_B T / 2$ by $N e(t)$. This allows for rotational or vibrational contributions of polyatomic molecules.

We have derived the equations (37.1) and (37.2) in Chapter 28 for a dilute classical gas. Thereby, the parameters a and b were linked to the microscopic interaction $w(r)$. The lowering of the pressure by a/v^2 is caused by the attractive part of the interaction. The finite intrinsic volume of the atoms (i.e. the repulsive part of the interaction) means that effectively only the volume $v - b$ is available per particle.

The derivation in Chapter 28 was done under the condition of a sufficient dilution (in particular $v \gg b$). We will now ignore this prerequisite. Rather, we consider (37.1) and (37.2) as phenomenological equations and discuss them for any values of T and v. For moderate dilution, the microscopic derivation from Chapter 28 can be used as a plausibility argument for the equations of state. In the limit $v \to b$, the prerequisites of this derivation are blatantly violated. However, equation (37.1) is constructed in such a way that it produces a plausible behavior for $v \to b$ (namely $P \to \infty$). Taking into account the finite intrinsic volume of the atoms, this corresponds to the behavior of the ideal gas ($P = k_B T / v \to \infty$ for $v \to 0$).

Figure 37.1 shows various isotherms in the P-v diagram. Compared to the ideal gas with $P = k_B T / v$, the isotherms are shifted to the right by b, i.e. $k_B T / v \to k_B T / (v - b)$. In addition, the term $-a/v^2$ leads to a lowering of the

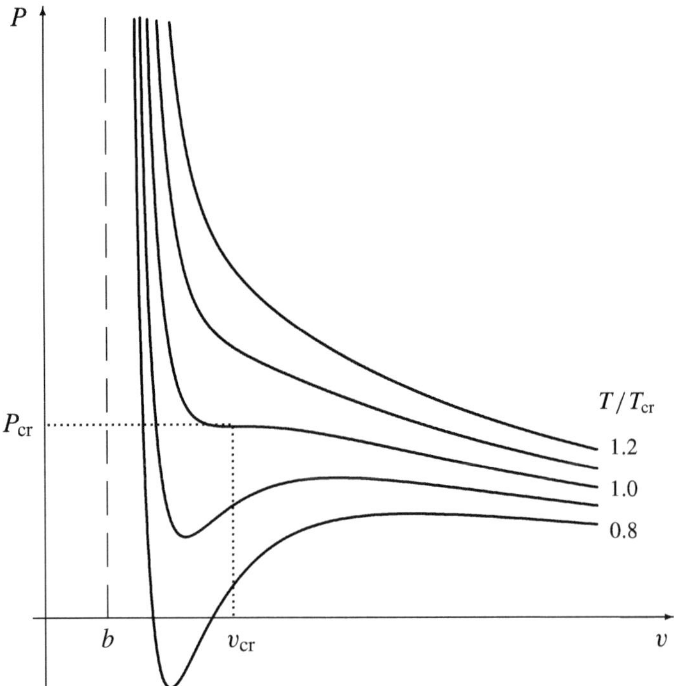

Figure 37.1 Isotherms of the van der Waals gas for some values of T/T_{cr}. For $T > T_{\mathrm{cr}}$ the isotherms decrease monotonically, for $T < T_{\mathrm{cr}}$ they have a minimum and a maximum.

pressure. For high temperatures, this term is small compared to $k_{\mathrm{B}}T/(v-b)$ and causes only a slight small deformation of the monotonically decreasing isotherm. For lower temperatures, the term $-a/v^2$ leads to a down bending of the isotherm with decreasing v; the isotherm exhibits a maximum. Because of $P \to \infty$ for $v \to b$, the isotherm must eventually go up again. This explains that there is a minimum between the maximum and $v = b$.

As the isotherms transform continuously into one another, there is exactly one isotherm for which the minimum and maximum coincide. This isotherm then has a horizontal inflection point, the function is stationary at this point. The temperature of this specific isotherm is denoted by T_{cr}, the coordinates of the horizontal inflection point are denoted by P_{cr} and v_{cr}. The index cr stand for critical. As will be seen this corresponds to the critical point in the phase diagram shown in Figure 35.2.

The pressure P and temperature T are usually specified experimentally. The volume $v(T, P)$ of the van der Waals gas results in Figure 37.1 as the intersection of the horizontals $P = $ const. with the isotherm T. For an isotherm with minimum and maximum, there can be three intersection points, i.e. three solutions for v. This raises the question of what these solutions mean and which solution describes the equilibrium state.

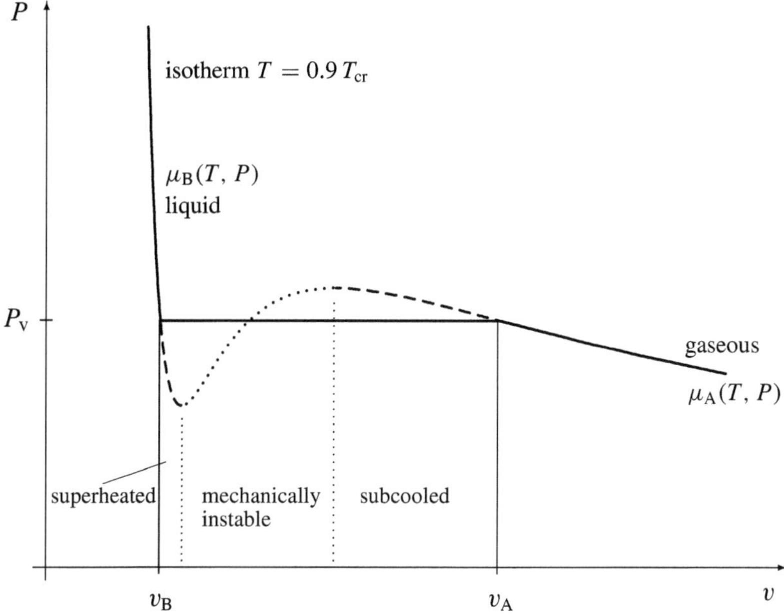

Figure 37.2 Phase transition in the van der Waals model: The left part of the isotherm corresponds to the liquid phase, the right part corresponds to the gaseous phase. The horizontal part describes the transition from the liquid to the gaseous phase; along the horizontal line, the gas content increases from 0% to 100%. The position of the horizontal is determined by the rule that the isotherm (dashed and dotted lines) encloses equal areas with this line.

Maxwell construction

The compressibility must always be positive:

$$\kappa_T = -\frac{1}{V}\left(\frac{\partial V}{\partial P}\right)_T > 0\,, \quad \text{also} \quad \left(\frac{\partial P}{\partial V}\right)_T < 0 \tag{37.3}$$

A system with $\kappa_T < 0$ would be mechanically unstable because there is no restoring force against a small volume fluctuation. Therefore, in Figure 37.1 all isothermal sections with a positive slope are excluded as non-physical.

Figure 37.2 shows a single isotherm in the P-v diagram. If we exclude the area with a positive slope (dotted line) a horizontal line between minimum and maximum still has two intersection points. We denote the left-hand solution by $v_{\mathrm{B}}(T, P)$ and the right one by $v_{\mathrm{A}}(T, P)$. On the two parts of the isotherm we denote the chemical potential by

$$\mu_{\mathrm{A}}(T, P) = \mu(v_{\mathrm{A}}(T, P), T)\,, \qquad \mu_{\mathrm{B}}(T, P) = \mu(v_{\mathrm{B}}(T, P), T) \tag{37.4}$$

For sufficiently low pressure, there is only one solution in Figure 37.2 solution, namely the one with μ_{A}. Here the isotherm differs only slightly from that of the ideal gas; this part of the solution therefore describes the gaseous phase. For sufficiently high pressure, there is also only one solution, namely the one with μ_B. For this part of the isotherm $v = \mathcal{O}(b)$ holds; this characterizes the liquid phase, in which the volume per particle is of the same size as the intrinsic volume. We can therefore assign the phases *gaseous* and *liquid* to these two branches of the isotherm.

For an isotherm with maximum and minimum, we show that the equation

$$\mu_A(T, P_v) = \mu_B(T, P_v) \tag{37.5}$$

has exactly one solution. This solution determines the pressure $P = P_v(T)$ at which the two phases are in equilibrium.

We use the relation $\mu = G/N = F/N + PV/N = f + Pv$; where $f = F/N$ is the free energy per particle. This turns (37.5) into

$$f_B - f_A = P_v(v_A - v_B) \tag{37.6}$$

The difference $f_B - f_A$ can alternatively be determined by an integration along the isotherm:

$$f_B - f_A = f(T, v_B) - f(T, v_A) = \int_{v_A}^{v_B} dv \, \frac{\partial f(T, v)}{\partial v} = \int_{v_B}^{v_A} dv \, P(T, v) \tag{37.7}$$

Here, $P(T, v)$ from (37.1) is to be used, i.e. the dashed and dotted curve in Figure 37.2. This curve does not describe any equilibrium states, but it can be used to calculate the difference in free equilibrium energies. From the last two equations follows

$$P_v(v_A - v_B) = \int_{v_B}^{v_A} dv \, P(T, v) \tag{37.8}$$

In the P-v diagram in Figure 37.2, the left side is the rectangular area below the horizontal $P = P_v(T)$. The right side, on the other hand, is the area under the isotherm $P(T, v)$ (dashed and dotted curve). This implies that the two areas enclosed by the isotherm and the horizontal line must be equal. For an isotherm with a maximum and a minimum, this *Maxwell construction* determines the position of the horizontal $P = P_v(T)$.

For $P = P_v$, the equality $\mu_A = \mu_B$ holds. The behavior of the chemical potentials in the vicinity of this point follows from the Duhem-Gibbs relation (20.15),

$$\frac{\partial \mu_A(T, P)}{\partial P} = v_A = v_{\text{gas}} \quad \text{and} \quad \frac{\partial \mu_B(T, P)}{\partial P} = v_B = v_{\text{liquid}} \tag{37.9}$$

Because of $v_{\text{gas}} > v_{\text{liquid}}$, the slope of μ_A is larger than that of μ_B. This explains the behavior shown in Figure 37.3 on the left. The behavior of $v(T, P)$ results from the jump $v_{\text{gas}} - v_{\text{liquid}}$ and the condition $\partial v/\partial P < 0$, Figure 37.3 right.

The equilibrium condition $\mu(T, P) = \text{minimal}$ means

$$\mu(T, P) = \begin{cases} \mu_A(T, P) & (P \leq P_v(T)) \\ \mu_B(T, P) & (P \geq P_v(T)) \end{cases} \tag{37.10}$$

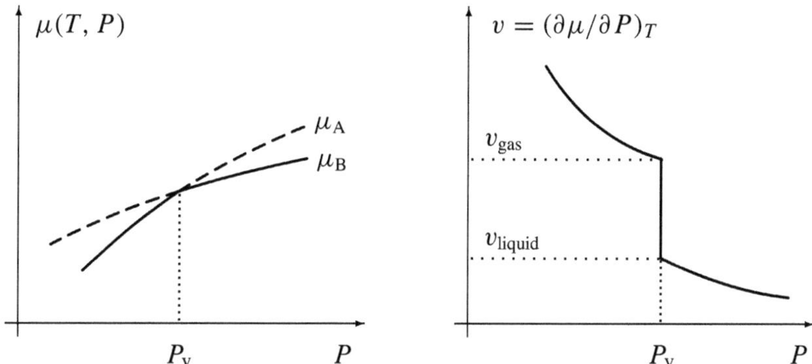

Figure 37.3   The chemical potential as a function of the pressure (T = const.). At the transition or vapor pressure P_v the chemical potentials of the two phases have the same value but different slopes (left). At equilibrium, the phase with the lower μ will be present. The derivative of the chemical potential (solid line in the left part) yields the volume v shown on the right.

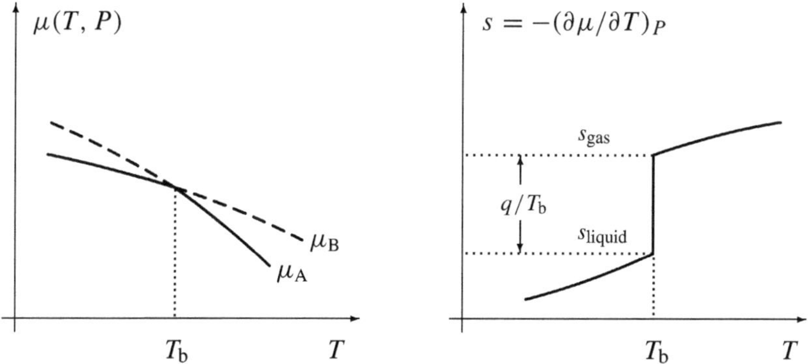

Figure 37.4   The chemical potential as a function of the temperature (P = const.). At the transition or boiling temperature $T_{tr} = T_b$, the chemical potentials of the two phases have the same value but different slopes (left). At equilibrium, the phase with the lower μ will be present. The derivative of the chemical potential (solid line left) results in the entropy s shown on the right. The jump in entropy at the transition point determines the latent heat q (or enthalpy of transformation).

The physical isotherm is the solid line shown in Figure 37.2. The following quasi-static process runs along this isotherm: We start in gas phase A and slowly increase the pressure at a constant temperature. The system moves along the right-hand part of the isotherm. At $P = P_v$ the phases A and B are in equilibrium. An infinitesimal pressure increase from $P_v - \epsilon$ to $P_v + \epsilon$ (with arbitrarily small ϵ) leads to the continuous conversion of initially 100% gas into 100% liquid. In the process, the volume changes from v_A to v_B, and the enthalpy of vaporization (also called latent heat) is transferred to the heat bath (which is guarantees $T = $ const.). The equilibrium states, that the system passes through are given by the horizontal line in Figure 37.2. If the pressure is further increased the system moves along the left, steep part of the isotherm. The change in the chemical potential in the considered process is shown in Figure 37.3 on the left.

In the multiple solution region, the theoretical isotherm must be replaced by the horizontal $P = P_v(T)$, Figure 37.2. With this rule called "Maxwell construction", the isotherms (37.1) are modified in a well-defined and justified manner. In the range of the horizontal line, we have already excluded the isotherm section with positive slope (dotted line) as non-physical due to mechanical instability. In addition, we now also exclude the dashed sections (negative slope) because they have a higher chemical potential (higher than the minimum chemical potential). Therefore, they do not correspond to equilibrium states, i.e. they are thermodynamically unstable. Such non-equilibrium states can, however, be reached temporarily under certain conditions. These metastable states are referred to as superheated liquid (boiling delay) or supercooled gas.

Enthalpy of vaporization

The enthalpy of vaporization is the (latent) heat that the system absorbs on the way from the temperature $T_b - \epsilon$ to $T_b + \epsilon$ at constant pressure. (For the reverse process one talks of the enthalpy of fusion). The transition takes place at the boiling temperature, $T_{tr} = T_b$. With $dH = T\,dS + V\,dP$ and $dP = 0$ we obtain

$$Q = \int_{T_b-\epsilon}^{T_b+\epsilon} T\,dS = \int_{T_b-\epsilon}^{T_b+\epsilon} dH = H_A - H_B \qquad (37.11)$$

The enthalpies H_A and H_B are taken at the considered point T, P of the vapor pressure curve. According to (37.2), we get the enthalpy per particle

$$h = \frac{H}{N} = \frac{E + PV}{N} = e(T) - \frac{a}{v} + Pv \qquad (37.12)$$

This yields the enthalpy of vaporization $q = Q/N$ per particle:

$$q = h_A - h_B = \frac{a}{v_B} - \frac{a}{v_A} + P(v_A - v_B) \approx \frac{a}{v_{\text{liquid}}} + P\,v_{\text{gas}} \qquad (37.13)$$

For $v_{\text{gas}} \gg v_{\text{liquid}}$ we have simplified the result. The term a/v_{liquid} is the energy that must be supplied to remove the atoms from the attractive range of mutual interaction. The second term $P\,v_{\text{gas}}$ is the expansion work to be performed at constant pressure.

The van der Waals model yields a finite enthalpy of vaporization, i.e. a jump in entropy. This and $\partial s/\partial T > 0$ explains the behavior of s sketched in Figure 37.4 on the right. The left part shows the dependence of the chemical potential on the temperature. According to the classification of Chapter 35, this is a 1st order phase transition.

Critical point

For an isotherm with a minimum and a maximum, the Maxwell construction determines the pressure at which the phase transition occurs; this is the vapor pressure curve $P_{\text{v}}(T)$. As discussed for Figure 37.1, the minimum and maximum of the isotherm move closer together for increasing temperature. There is then exactly one isotherm $T = T_{\text{cr}}$ at which they coincide. For $T \to T_{\text{cr}}$ (and $T \le T_{\text{cr}}$), the length of the horizontal in Figure 37.2 approaches zero. This means

$$v_{\text{gas}} - v_{\text{liquid}} = v_{\text{A}} - v_{\text{B}} \overset{T \to T_{\text{cr}}}{\longrightarrow} 0\,, \qquad q = T\,(s_{\text{A}} - s_{\text{B}}) = h_{\text{A}} - h_{\text{B}} \overset{T \to T_{\text{cr}}}{\longrightarrow} 0 \quad (37.14)$$

For $T = T_{\text{cr}}$, the derivative $\partial \mu(T, P)/\partial T$ becomes continuous, and the 1st order phase transition (for $T < T_{\text{cr}}$) becomes a 2nd order transition at the critical point. Above the critical point ($T > T_{\text{cr}}$ or $P > P_{\text{cr}}$) there is no longer a discrete phase transition; the structure then changes continuously from gaseous to liquid. This means that the vapor pressure curve ends at $T = T_{\text{cr}}$. This end point is called *critical point*; the variables at this point are labeled with the index cr for *critical*. The van der Waals model explains the existence of a critical point.

We determine the isotherm for which the maximum and minimum coincide. At this point, this isotherm has a horizontal inflection point which is defined by the following three equations:

$$P(T, v) = \frac{k_{\text{B}} T}{v - b} - \frac{a}{v^2}\,, \qquad \left(\frac{\partial P}{\partial v}\right)_T = 0\,, \qquad \left(\frac{\partial^2 P}{\partial v^2}\right)_T = 0 \qquad (37.15)$$

These three equations define the three variables P, T and v (Exercise 37.1):

$$v_{\text{cr}} = 3b\,, \qquad k_{\text{B}} T_{\text{cr}} = \frac{8}{27}\frac{a}{b}\,, \qquad P_{\text{cr}} = \frac{a}{27\,b^2} \qquad (37.16)$$

The quantities a and b depend on the interaction between the atoms or molecules. This connection was given by Equation (28.26) for a dilute gas.

From (37.16) follows

$$\frac{P_{\text{cr}}\,v_{\text{cr}}}{k_{\text{B}} T_{\text{cr}}} = \frac{3}{8} = 0.375 \qquad (37.17)$$

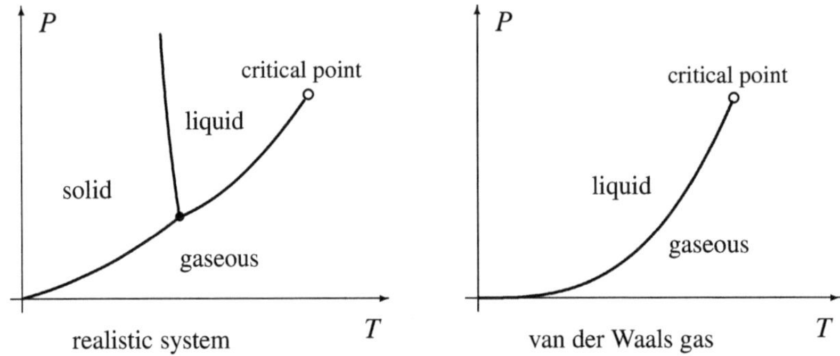

Figure 37.5 Comparison of the phase diagrams of a realistic system and of the van der Waals gas. The van der Waals model is able to explain the existence of the vapor pressure curve, of the critical point, and of the phase transition gaseous $\leftrightarrow$ liquid.

The experimental values of real systems are usually somewhat smaller:

$$\left(\frac{P_{\text{cr}}\, v_{\text{cr}}}{k_{\text{B}}\, T_{\text{cr}}} \right)_{\text{exp}} = \begin{cases} 0.230 & \text{water (H}_2\text{O)} \\ 0.291 & \text{argon (Ar)} \\ 0.292 & \text{oxygen (O}_2\text{)} \\ 0.304 & \text{hydrogen (H}_2\text{)} \\ 0.308 & \text{helium (}^4\text{He)} \end{cases} \tag{37.18}$$

If T, P and v are set in relation to the critical quantities,

$$P^* = \frac{P}{P_{\text{cr}}}, \qquad T^* = \frac{T}{T_{\text{cr}}}, \qquad v^* = \frac{v}{v_{\text{cr}}} \tag{37.19}$$

then (37.1) yields the dimensionless equation

$$P^* + \frac{3}{v^{*2}} = \frac{8\, T^*}{3\, v^* - 1} \tag{37.20}$$

Many gases deviate from this form of the van der Waals equation by less than 10%; in this respect, the van der Waals gas is a remarkably good model. The generality in the form (37.20) is also known as the *law of corresponding states*: Relative to the critical variables, different systems behave similarly.

Summary

Figure 37.5 compares the phase diagram of the van der Waals gas with that of real substances. The van der Waals model explains the existence of a vapor pressure curve and of the critical point. The transition gaseous-liquid is reproduced qualitatively correct (2nd order at the critical point, 1st order below and no phase transition above). From Figure 37.1 and the Maxwell construction, it can be seen that at higher temperatures the transition pressure P_{v} is also higher. In other words, the vapor pressure curve has a positive slope, $dP_v/dT > 0$. However, there is no solid phase in this model; the vapor pressure curve looses its meaning for low temperatures.

Overall, the van der Waals gas is a remarkable model that reproduces essential features of the phase transition liquid–gaseous.

Exercises

37.1   Dimensionless van der Waals equation

Determine the critical values v_{cr}, T_{cr} and P_{cr} for which the following three equations are fulfilled:

$$P(T, v) = -\frac{a}{v^2} + \frac{k_B T}{v - b}, \qquad \left(\frac{\partial P}{\partial v}\right)_T = 0, \qquad \left(\frac{\partial^2 P}{\partial v^2}\right)_T = 0 \qquad (37.21)$$

Write on the van der Waals equation for the quantities $P^* = P/P_{cr}$, $T^* = T/T_{cr}$ and $v^* = v/v_{cr}$.

37.2   Van der Waals equation for nitrogen

For nitrogen, the experimental critical values are $T_{cr} = 126.2\,\text{K}$ and $P_{cr} = 33.9\,\text{bar}$. At normal pressure $P_0 \approx 1\,\text{bar}$, nitrogen boils at $T_0 \approx 77.4\,\text{K}$. The vaporization enthalpy is $q_{exp} = 5.6\,\text{kJ/mol}$.

    The van der Waals model shall be applied to this phase transition: First determine the numerical values of the volumes v_A^* (gaseous) and v_B^* (liquid) at which the transition occurs. Check whether the Maxwell construction is fulfilled for P_0 and T_0, and calculate the vaporization enthalpy q.

37.3   Energy and entropy of the van der Waals gas

Show that the energy per particle for the van der Waals gas of the form

$$e(T, v) = \frac{E(T, V, N)}{N} = u(T) - \frac{a}{v} \qquad (37.22)$$

where $u(T)$ is an unknown function. From this, determine the entropy $s(T, v)$ per particle. Show that this entropy can be used to calculate the transformation enthalpy

$$q = T\left(s_A - s_B\right) = h_A - h_B = \frac{a}{v_B} - \frac{a}{v_A} + P(v_A - v_B)$$

37.4   Dieterici gas

A gas satisfies the so-called Dieterici equation of state:

$$P \exp\left(\frac{\alpha}{v R T}\right) (v - \beta) = R T \qquad \text{(Dieterici gas)}$$

Here $v = V/v$ is the volume per mole, and α and β are parameters. Determine the critical values P_{cr}, T_{cr} and v_{cr}. Express the Dieterici equation by the dimensionless variables $P^* = P/P_{cr}$, T^*/T_{cr} and v^*/v_{cr}. Sketch some isotherms in the P-v diagram.

38 Liquid helium

We consider the λ-transition of liquid ^{4}He. Some of the remarkable properties associated with this transition can be understood in the ideal Bose gas model (IBG) of Chapter 31.

Helium has the stable isotopes ^{4}He and ^{3}He. From naturally occurring helium (atmosphere, natural gas deposits) the noble gases ^{4}He and ^{3}He can be produced technically; ^{3}He occurs in very small traces only. At normal pressure, these gases condense at the temperatures 4.2 K and 3.2 K, respectively. Unlike all other known substances, helium remains liquid even at $T \to 0$ (under normal pressure). This is due to the rather weak interaction between the atoms. A realistic description of this interaction is the Lennard-Jones potential

$$w(r) = \epsilon \left(\frac{r_0^{12}}{r^{12}} - 2\, \frac{r_0^6}{r^6} \right) \qquad (\epsilon = 10.2\, k_{\mathrm B}\, \mathrm{K}, \ \ r_0 = 2.87\,\text{Å}) \tag{38.1}$$

This potential is shown in Figure 38.1

We consider the Schrödinger equation for two helium atoms with the potential (38.1). A bound state (localized around the potential minimum) would mean that there are stable He$_2$ molecules. For such a state, the relative motion had to be restricted to a range around the minimum of the potential. Figure 38.1 indicates a value $\Delta x \approx 1$ Å for this range. Then the uncertainty principle implies a kinetic energy of the size

$$\Delta E_{\mathrm{kin}} \approx \frac{1}{2 m_{\mathrm r}} \frac{\hbar^2}{\Delta x^2} = \frac{1}{4 m_{\mathrm n} c^2} \left(\frac{\hbar c}{e^2} \right)^2 \left(\frac{e^2}{\text{Å}} \right)^2 \approx 12\, k_{\mathrm B}\, \mathrm{K} \tag{38.2}$$

The reduced mass of the two helium atoms is approximately equal to two nucleon masses, $m_{\mathrm r} \approx 2\, m_{\mathrm n}$. For the evaluation we used $m_{\mathrm n} \approx 1\,\mathrm{GeV}/c^2$, $\hbar c/e^2 \approx 137$, $e^2/\text{Å} \approx 14.4\,\mathrm{eV}$ and $1\,\mathrm{eV} \approx 1.2 \cdot 10^4\, k_{\mathrm B}\mathrm{K}$.

The kinetic energy (38.2) has the same size as the attractive strength of the potential ($w_{\mathrm{min}} = -\epsilon \approx -10\, k_{\mathrm B}\mathrm{K}$). This is the reason why there are no stable He$_2$ molecules. There is a weakly bound so-called dimer. This dimer is a bound with an energy $1\,\mu\mathrm{eV}$ and has a large extension, $\langle r \rangle \approx 52$ Å. This state is readily broken off thermodynamically

The weakness of the attractive interaction explains that helium remains liquid under normal pressure even for $T \to 0$. Solid helium is only formed under high pressure; it is then the softest solid known.

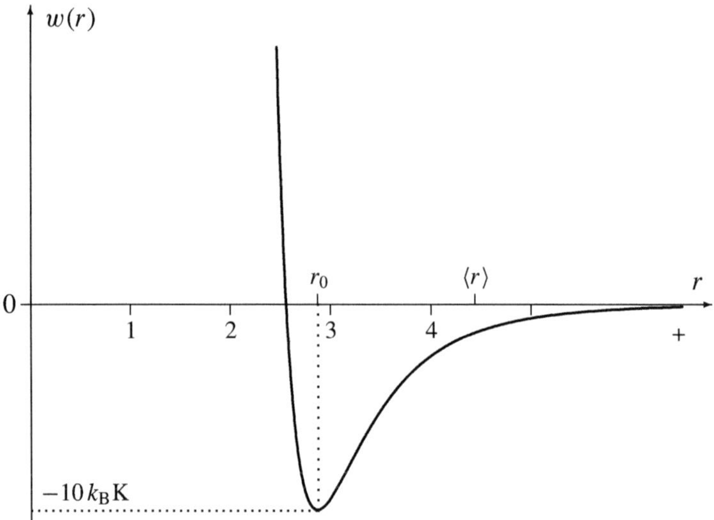

Figure 38.1 Potential between two helium atoms at a distance of r. The minimum lies at $r_0 = 2.87\,\text{Å}$. The average distance between two atoms in the liquid is $\langle r \rangle \approx 4.44\,\text{Å}$.

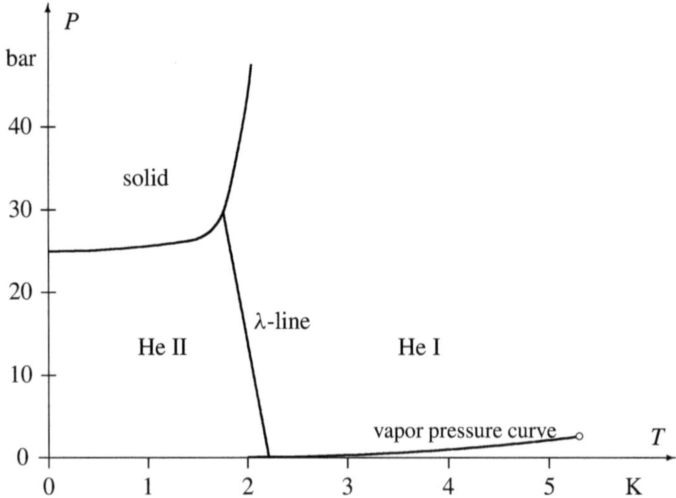

Figure 38.2 Phase diagram of ^{4}He. The vapor pressure curve starts at $T = 0$ and $P = 0$ and ends at the critical point at about 2.2 bar and 5.3 K. In this chapter discusses the λ transition, i.e. the transition from the normal fluid He I to the superfluid He II.

λ transition

Figure 38.2 shows the phase diagram of ^{4}He. There is a remarkable phase transition at $T_\lambda = 2.17\,\text{K}$ (at normal pressure). The transition displays is a singularity of the specific heat at T_λ, and a number of unusual properties below T_λ. Because of very different properties, a distinction is made between He II (below T_λ) and He I (above T_λ). These properties include in particular the superfluidity of He II.

In the ^{3}He liquid, such a phase transition does not occur. It is therefore evident that this transition is associated with the different exchange symmetry, i.e. with the fact that ^{4}He atoms are bosons and ^{3}He atoms are fermions. The interaction potentials between the atoms are practically the same in both cases. The different mass could at most lead to a shift in the transition temperature.

Since the exchange symmetry and not the interaction is decisive for this transition is decisive, it makes sense to compare these systems (^{4}He and ^{3}He liquids) to the ideal Bose gas (IBG) and the ideal Fermi gas. In these models, the degrees of freedom of the translation are treated quantum mechanically and statistically, whereby the interaction is neglected. In a liquid the degrees of freedom of translation are preserved (but not in a solid). Compared to the ideal Bose or Fermi gas the translational motion in the ^{4}He or ^{3}He liquid is influenced by the interactions.

The IBG (Chapter 31) actually shows a phase transition at approximately the right temperature, but the ideal Fermi gas (Chapter 32) does not. In the following, we discuss how the IBG may explain some essential properties of the λ transition and how the He II phase can be understood.

It should be noted that ^{3}He below $2.7\cdot10^{-3}\,\text{K}$ also becomes superfluid, too. This phase transition has other causes, however. Basically, for superproperties, such as those that occur in lasers, superconductivity and superfluidity, there is a *macroscopic wave function*, i.e. a quantum mechanical wave function that is adopted by macroscopically many particles. This is only possible for bosons, i.e. for photons in laser light or for ^{4}He atoms in superfluid helium. In the case of superconductivity the electrons join together to form pairs (Cooper pairs) and thus bosons. A comparable mechanism also leads to the superfluidity of ^{3}He. In these cases, the transition temperature is linked to the strength of the pair interaction. In liquid ^{4}He, however, the transition temperature T_λ is determined by the condition that the thermal wavelength becomes comparable to the mean particle distance.

Ideal Bose gas

We discuss some properties of the ideal Bose gas model (IBG) which can be compared with properties of liquid ^{4}He.

The IBG yields the transition temperature (31.14),

$$k_\text{B}T_\text{cr} = \frac{2\pi}{[\zeta(3/2)]^{2/3}}\frac{\hbar^2}{m\,v^{2/3}} \approx 3.13\,k_\text{B}\,\text{K} \qquad (\text{with } v = 46\,\text{Å}^3 \text{ for }^4\text{He}) \quad (38.3)$$

In Chapter 31 it was shown that below T_{cr} a finite fraction of all particles goes to the lowest level. This part of the liquid is called condensate. We summarize some of its properties.

We denote the condensate density by $\varrho_0 = m\, N_0/V$. The entire density splits up into a condensed and a non-condensed part:

$$\varrho = \varrho_0 + \varrho_{n.c.} \quad \text{with} \quad \frac{\varrho_0}{\varrho} = \frac{N_0}{N} \overset{(31.25)}{=} 1 - \left(\frac{T}{T_{cr}}\right)^{3/2} \tag{38.4}$$

The temperature dependence of the condensate fraction ϱ_0/ϱ is sketched in Figure 38.3 on the left.

All condensed particles have the same wave function; in this sense, the wave function is *macroscopic*. In Chapter 31, the wave function $\psi_0(r)$ of the condensed particles was the lowest solution in the box enclosing the gas. In the following, we want to investigate motions of the condensate and allow for a more general (time independent) wave function $\psi_0(r)$. Any quantum mechanical wave function $\psi_0(r)$ can be expressed by two real fields, $\phi(r)$ and $S(r)$,

$$\psi_0(r) = \phi(r)\, \exp\left[i\, S(r)\right] \tag{38.5}$$

The condensate density

$$\varrho_0(r) = m\, \left|\phi(r)\right|^2 \tag{38.6}$$

is independent of the phase S. We calculate the current density of the condensate:

$$j_0(r) = \frac{\hbar}{2i}\left(\psi_0^*\, \nabla\psi_0 - \psi_0\, \nabla\psi_0^*\right) = \hbar\,\left|\phi(r)\right|^2 \operatorname{grad} S(r) = \varrho_0\, v_0(r) \tag{38.7}$$

A flow of condensed density can therefore be characterized by a position dependent phase $S(r)$. The velocity field of the flow is

$$v_0(r) = \frac{\hbar}{m}\, \operatorname{grad} S(r) \tag{38.8}$$

In the following, we assume that the condensate density ϱ_0 is homogeneous:

$$\frac{\varrho_0}{\varrho} = \frac{N_0/V}{N/V} = 1 - \left(\frac{T}{T_{cr}}\right)^{3/2} \tag{38.9}$$

Since all condensed particles have the same wave function ψ_0, they move *coherently*. For the IBG condensate density and its motion, we note:

1. STABILITY: A state populated with many bosons has a particular stability. This is due to the fact that the scattering of one particle does not change the velocity field $v_0(r)$, simply because the remaining $N_0 - 1$ particles remain in the state ψ_0 with the phase $S(r)$ and with $v_0(r) = (\hbar/m)\operatorname{grad} S(r)$. When a non-condensed particle enters state ψ_0, it takes on the phase of the other particles.

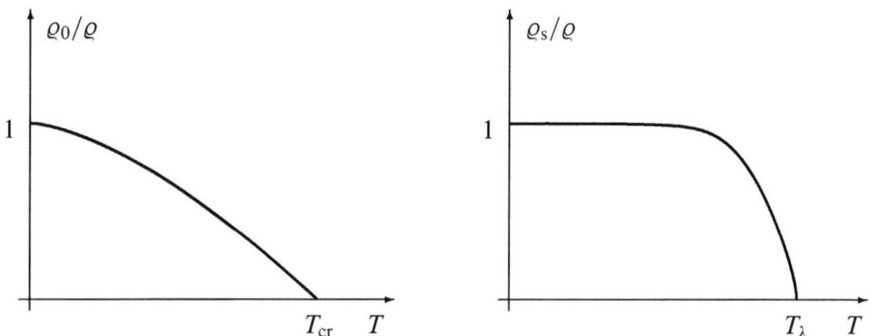

Figure 38.3   Temperature dependence of the condensate density (left) in ideal Bose gas and of the superfluid density (right) in liquid ^{4}He.

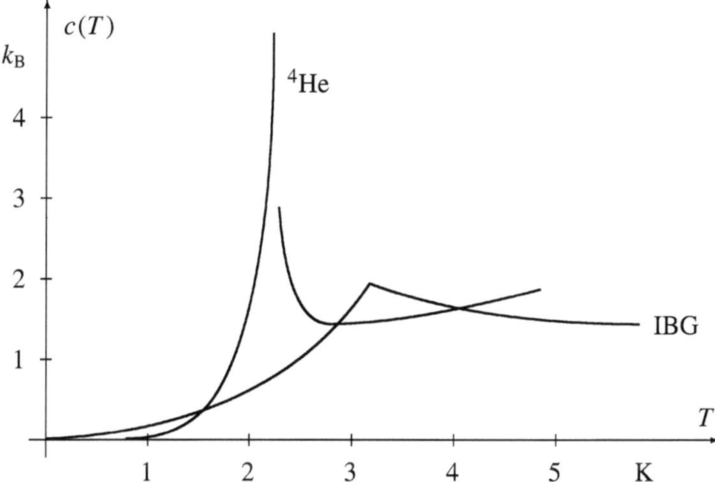

Figure 38.4   The specific heats of the ideal Bose gas and of liquid ^{4}He show a qualitative similarity. Quantitatively, however, the behavior is quite different at the phase transition as well as for low temperatures. The name λ-point was chosen because of the λ-like form of the specific heat of ^{4}He.

The true statistical equilibrium implies $v_0 \equiv 0$, because no direction is distinguished. Because of the mentioned stability, however, there can be a metastable state with a flow field $v_0(r) \neq 0$. The theoretical investigation of the stability is difficult because all possible modes that lead to a decay of the flow field must be taken into account. In this respect, we restrict ourselves to qualitative statements on the stability.

2. VORTEX FREEDOM: It follows from (38.8) that the velocity field v_0 of the condensate flow is vortex-free:

$$\mathrm{rot}\, v_0(r) = 0 \tag{38.10}$$

This is a consequence of the description of the condensate by a macroscopic wave function.

3. ENTROPY: Since all condensed particles are in the same state, the number Ω_0 of possible microstates of these N_0 particles is one. Therefore, the condensate therefore does not contribute to the entropy:

$$S_0 = S(\varrho_0) = k_B \ln \Omega_0 = 0 \tag{38.11}$$

Helium

Specific heat

Figure 38.4 compares the specific heats of ^{4}He with that of the ideal Bose gas. The specific heat of ^{4}He has a jump and a logarithmic term; whereby c_V and c_P are similar to each other. For c_P an approximate logarithmic behavior has been observed over many orders of magnitude of the relative temperature t, i.e. ($c_P \approx -A \ln|t| + \ldots$ in the range $|t| = |T - T_\lambda|/T_\lambda = 10^{-2} \ldots 2 \cdot 10^{-9}$).

The specific heat c_V of the IBG shows a qualitative similarity with that of liquid helium, but it is quantitatively significantly different. This also applies to low temperatures, where the quasiparticles occurring in the helium liquid determine the specific heat (last section of this chapter).

Two-fluid model

Below the λ transition, helium exhibits a number of unusual properties, which can be understood quantitatively if one assumes that the density of the liquid consists of a superfluid and a normal part, i.e.

$$\varrho = \varrho_s(T) + \varrho_n(T) \tag{38.12}$$

This approach is referred to as *two-fluid model*. The two parts are not separated spatially, but they differ significantly in their behavior. In particular, the superfluid density ϱ_s can flow without resistance through narrow capillaries. As a result of this and other experiments, the division (38.12) is defined experimentally; the densities ϱ_s and ϱ_n are therefore measurable variables. The temperature dependence of the superfluid density is sketched in Figure 38.3 on the right.

For the superfluid density and its motion, one finds experimentally:

1. VISCOSITY AND STABILITY: The superfluid density has vanishing viscosity. The concept of viscosity η (internal friction) is introduced in Equation (43.41). The viscosity η of ^{4}He is determined by the normal density alone:

$$\eta_s = 0, \qquad \eta = \eta_n \tag{38.13}$$

This means that the superfluid density flows without resistance through narrow capillaries, and such a superfluid flow field can be persistent. In a normal liquid, an existing flow field decays due to internal friction.

A superfluid flow field represents a metastable state. In the exact equilibrium state, no direction is distinguished, so $v_s \equiv 0$ must hold. The decay of a metastable superfluid flow field depends decisively on the velocity v_s. A superfluid flow is only possible below a critical velocity v_{crit} (where $v_{crit} \sim 10^2$ cm/s).

2. VORTEX FREEDOM: The velocity field of a superfluid flow is vortex-free,

$$\text{rot } v_s(r) = 0 \tag{38.14}$$

3. ENTROPY: Within the scope of the measurement accuracy (1 to 2 % of the total entropy), the entropy of the superfluid component disappears:

$$S_s = S(\varrho_s) = 0, \qquad S_n = S(\varrho_n) = S \tag{38.15}$$

The entire entropy is therefore carried by the normal density.

In the following, we present some key experiments for the mentioned properties (38.12) – (38.15) of He II.

Andronikashvili's experiment

A rotating pendulum made of metal disks hangs from a torsion thread in a helium liquid (Figure 38.5). The disks are arranged so densely that viscous liquid is carried along with the pendulum and thereby contributes to the mass of the pendulum. For He II the normal density then contributes to this effective pendulum mass, but not but the superfluid density. From this, Andronikashvili determined in 1946 the temperature dependence of the apportionment (38.12). The result for the temperature dependence for the superfluid density is shown in Figure 38.3 on the right.

This experiment confirms and quantifies the apportionment $\varrho_s + \varrho_n$, (38.12) with (38.13).

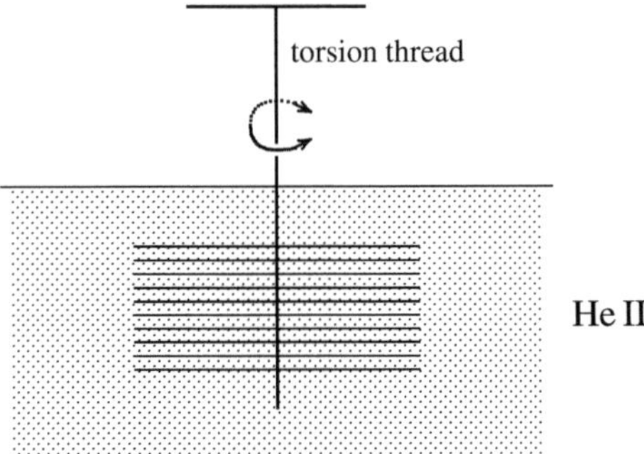

Figure 38.5 A rotatable pendulum consists of parallel circular metal disks; it is suspended by a torsion thread. The pendulum is immersed in helium liquid with a temperature below T_λ. During rotation, the normal density rotated together with these disks (due to its viscosity), but not the superfluid density. The effective mass of the pendulum can be used to measure the normal density and its temperature dependence.

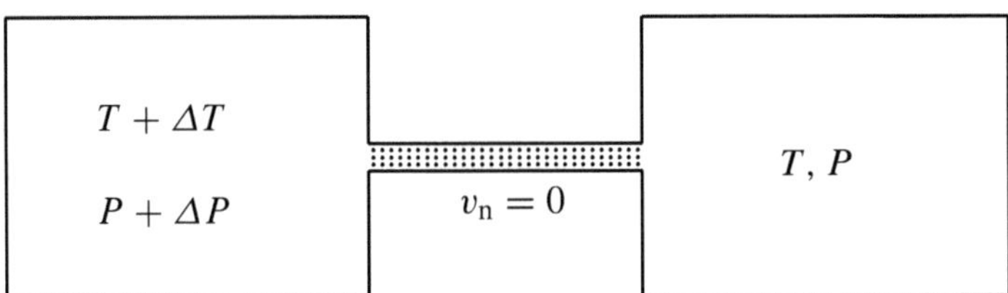

Figure 38.6 Fountain effect: Two containers with He II have a connection through which only the superfluid density can flow. Then a temperature difference implies a pressure difference, and vice versa. By measuring the ratio $\Delta P / \Delta T$, the entropy difference between the normal and superfluid phase can be determined.

Fountain effect

Two containers with He II are connected by a tube which is clogged with a fine powder (Figure 38.6). First let the temperature and pressure in both containers be the same. Now the pressure in the left container is increased. Thereby the helium liquid tries to flow into the right container. Due to the experimental conditions, this is only possible for the superfluid fraction. Particles with zero entropy flow to the right; this creates a temperature difference (mechanocaloric effect).

Conversely, a temperature difference creates a pressure difference (thermo-mechanical effect). In one of the first demonstrations of this effect (Allen and Jones 1938), helium was sprayed out of the heated container due to the increased pressure. Due to this experiment, the effect is known as the fountain effect.

In the arrangement shown in Figure 38.6, an equilibrium is established with respect to particle exchange, so that the chemical potential in both containers is the same, $\mu_1 = \mu_2$. By dT and dP we denote the temperature and pressure difference between the containers. Then $d\mu = -s\, dT + v\, dP = 0$ yields

$$\left(\frac{dP}{dT}\right)_{\mathrm{FP}} = \frac{S}{V} \qquad \text{(fountain pressure)} \tag{38.16}$$

The pressure difference caused by a temperature difference is called *fountain pressure* (FP).

The decisive reason for this effect is that the superfluid density does not carry entropy. The experimental confirmation of (38.16) therefore confirms (38.15).

Second sound

We first consider the ordinary sound (first sound) in a gas, a liquid or a solid. For given P and T, the density of the substance has a constant equilibrium value ϱ_0. Against deviations $\delta\varrho(\mathbf{r}, t) = \varrho - \varrho_0$ there are restoring forces. Such deviations from the equilibrium density therefore lead to oscillations. For small deviations, the oscillations are harmonic and can be described by the wave equation

$$\left(\frac{1}{c_1^2}\frac{\partial^2}{\partial t^2} - \delta\right)\delta\varrho(\mathbf{r}, t) = 0 \qquad \text{(1st sound)} \tag{38.17}$$

The solutions are of the form

$$\varrho(\mathbf{r}, t) = \varrho_0 + \delta\varrho = \varrho_0 + A\,\cos(\mathbf{k}\cdot\mathbf{r} - \omega t) \tag{38.18}$$

with $\omega = c_1 k$ and the speed of sound c_1. Here t denotes the time (and not the relative temperature).

Since He II behaves as if it consists of two independent densities, in addition to the fluctuations of the total density, variations of the density components (superfluid and normal) are possible. For $\varrho = \text{const.}$, i.e. without excitations of the 1st sound, $\delta\varrho_{\mathrm{s}}(\mathbf{r}, t) = -\delta\varrho_{\mathrm{n}}(\mathbf{r}, t)$ must hold. Also for the deviations $\delta\varrho_{\mathrm{s}}$ there are restoring

forces towards the (position and time independent) equilibrium value, which lead to
the wave equation

$$\left(\frac{1}{c_2^2} \frac{\partial^2}{\partial t^2} - \delta \right) \delta\varrho_s(r, t) = 0 \qquad \text{(2nd sound)} \qquad (38.19)$$

The solutions of this wave equation are again of the form (38.18). The speed of
sound c_2 is of the size 20 m/s (in the temperature range from 1 to 2 K) compared to
$c_1 \approx 220$ m/s for the 1st sound.

The density ϱ_n carries the entire entropy, and thus also the entire heat content of
the liquid. Therefore, an increased fraction ϱ_n/ϱ means a higher temperature. This
means that $\delta\varrho_n = A\cos(k \cdot r - \omega t)$ represents a *temperature wave*. Such a wave can
be excited by a heating element operated with an alternating current. The 2nd sound
is only weakly attenuated; this implies that He II is a very good heat conductor.

For the temperature waves, the division (38.12) with (38.15) is essential. The
existence of the 2nd sound therefore confirms these statements.

Rotating bucket with He II

We will now discuss an experiment demonstrating the vortex freedom of the super-
fluid density. A bucket filled with liquid rotates with the angular velocity $\boldsymbol{\omega} = \omega\, \boldsymbol{e}_z$
around its vertical axis of symmetry (in a gravitational field $\boldsymbol{g} = -g\, \boldsymbol{e}_z$). For an
ordinary viscous liquid and after some time, the liquid adjusts itself to the velocity
field

$$\boldsymbol{v}(r) = \boldsymbol{\omega} \times \boldsymbol{r}, \qquad \text{rot } \boldsymbol{v} = 2\,\boldsymbol{\omega} \qquad (38.20)$$

This is the velocity field of *rigid rotation*, i.e. the field that also results for a rotating
rigid body.

We now consider the same experiment with helium at $T = 0$, where the whole
liquid is superfluid. Then such a velocity field is not possible because of (38.15).
It would be conceivable that the helium liquid completely ignores the rotation of
the bucket due to $\eta_s = 0$. However, we can force the helium liquid to rotate by
first solidifying it by a pressure of 30 bar, then setting the bucket into rotation, and
return to normal pressure. The helium liquid then carries an angular momentum,
which must show up in the velocity field. We set $\boldsymbol{v} = v_s(\rho)\, \boldsymbol{e}_\varphi$ for the velocity field
in cylindrical coordinates ρ, φ and z where the z-axis is the axis of rotation. The
vortex freedom results in

$$\boldsymbol{v}_s = v_s(\rho)\, \boldsymbol{e}_\varphi \quad \xrightarrow[\text{vortex freedom}]{\text{rot } \boldsymbol{v}_s = 0} \quad v_s(\rho) = \frac{\text{const.}}{\rho} \qquad (38.21)$$

If the velocity field is not equal to zero (and this must be the case for the described
experimental setup), then it becomes singular at $\rho \to 0$. Now the term *velocity
field* (instead of the velocities of individual atoms) implies an averaging over a few
atomic distances; it does not apply on the Ångström scale. The velocity field (38.21)
therefore only applies in the range $\rho > 1$ Å. The configuration with $v_s = \text{const.}/\rho$

for $\rho > 1\,\text{Å}$ and const. $\neq 0$ represents a *vortex thread*; whereby the position of the vortex filament is the z-axis ($\rho = 0$).

In the framework of the ideal Bose gas, the superfluid velocity field is described by (38.8), i.e. by $v_\mathrm{s} = v_0 = (\hbar/m)\,\mathrm{grad}\,S(r)$. The comparison with (38.21) results in $S(r) = i\,n\,\varphi$ for the phase of the condensate wave function, and $v_\mathrm{s}(\rho) = n\,\hbar/(m\,\rho)$. Now the wave function (38.5) must be the same for φ and $\varphi + 2\pi$, i.e. n must be an integer. This means that the angular momentum of a vortex thread is quantized:

$$\oint d\boldsymbol{r} \cdot \boldsymbol{v}_\mathrm{s} = 2\pi\rho\, v_\mathrm{s}(\rho) = \frac{2\pi\,\hbar}{m}\, n \qquad \text{with } n = 0, \pm 1, \pm 2, \ldots \qquad (38.22)$$

The existence of vortex filaments and their quantization were postulated by Onsager in 1949, and the experimental confirmation was done in 1961. In the above described experimental setup, there will usually be many vortex filaments with angular momentum $\hbar$. The vortex filaments then carry the same total angular momentum as a rigidly rotating liquid. As a consequence, the surface of the liquid in the bucket is also curved like an ordinary liquid (we assumed a gravitational field $g = -g\,\boldsymbol{e}_z$). The vortex filaments end at the vessel walls and on the surface of the liquid. They are visible at the surface forming a regular pattern. This experiment is an impressive proof of the vortex-free nature of the superfluid flow.

Comparison between IBG and ^{4}He

The mere existence of the phase transition in the IBG and its absence in the ideal Fermi gas provides an *explanation* for the occurrence of the λ transition in ^{4}He and for its absence in ^{3}He.

In view of the complete neglect of the interactions in the IBG, the agreement of T_cr with T_λ can be considered as good. An effective mass $m \to m^* = 3m/2$ in (38.3) makes both temperatures almost equal. Such an effective mass can be understood, for example, in the model of a vortex-free flow around a sphere (Exercise 8.6 in [2]).

Due to the lack of interaction, it is not surprising that many properties of the λ transition cannot be described quantitatively by the IBG. This applies in particular to the energy and thus for the specific heat. The specific heats show at best a qualitative similarity (Figure 38.4).

For the condensate of the IBG, we have summarized some properties (38.4)–(38.11) which are in agreement with the experimental results (38.12)–(38.15) for He II. This comparison shows:

1. The macroscopic wave function of the IBG is the basis of a qualitative explanation of superfluidity.

2. The IBG leads directly to the two-fluid model which can explain various properties of He II. In particular, the division into two parts (condensed/non-

condensed or superfluid/normal) without any spatial separation becomes understandable. The temperature dependence of the condensate fraction shows is qualitatively similar to that of the superfluid fraction (Figure 38.3).

3. The IBG explains that the superfluid flow is vortex-free. It also explains the quantization of the vortex angular momenta.

4. The IBG explains that the entropy of the superfluid density disappears.

The IBG thus reproduces essential properties of liquid helium. In detail, however, the IBG leads to quantitatively incorrect statements:

1. The IBG does not explain the logarithmic singularity of the specific heat (Figure 38.4).

2. The temperature dependence of ϱ_0 and ϱ_s are quantitatively different. At the transition point we have $\varrho_s \propto |t|^{2/3}$ (liquid helium) as compared to $\varrho_0 \propto |t|$ (IBG), see Figure 38.3. Also for $T \to 0$ the temperature dependence are quite different: The normal density behaves like $\varrho_n \propto T^3$ in ^{4}He, whereas $\varrho_{n.c.} \propto T^{3/2}$ holds non-condensed density in the IBG).

The connection between the IBG and ^{4}He was first investigated by F. London (Phys. Rev. 54 (1938) 947); a more detailed introduction can be found in London's book *Superfluids*, Vol. II (Dover Publications, Inc., New York 1954). The explanation of the hydrodynamic properties of ^{4}He by the IBG is handled with in detail by S. J. Putterman in *Superfluid Hydrodynamics* (North Holland Publishing Company, London 1974). A detailed investigation of the IBG was carried out by R. M. Ziff, G. E. Uhlenbeck and M. Kac in Phys. Rep. 32 (1977) 169.

Because of the similarities between the IBG and ^{4}He it is inviting to add modifications to the IBG with the aim of reproducing the real properties of the λ transition and of He II. For early attempts at such modifications we refer to London's book *Superfluids*, and for a more recent approach to *A model for the λ-Transition of Helium*, https://arxiv.org/abs/cond-mat/0203353.

Quasiparticle model

The relevant excitations of the IBG are given by the single-particle energies $\varepsilon = \hbar^2 k^2 / 2m$. For $T \ll T_\lambda$, the lowest and therefore statistically relevant excitations of liquid helium are of a different form. These elementary excitations can be determined by inelastic neutron scattering; their dispersion relation is shown in Figure 38.7. The linear branch (small k) belongs to phonons, the excitations in the region of the minimum are called rotons according to Landau. Phonons are the quanta of density waves (1st sound, (38.17) and Chapter 33). The excitations of higher momenta correspond to single-particle excitations that are modified by the interaction. The historical term "roton" has no physical meaning.

The measured dispersion relation (Figure 38.7) applies to $T \lesssim 1\,\text{K}$. At 1 K, the width Γ of the roton excitations is comparable to their energy ε. The width is caused by interactions between the excitations and leads to a finite lifetime $\tau \sim \hbar/\Gamma$. With increasing Γ (especially for $\Gamma > \varepsilon$), the energy and momentum of the quasiparticles are less and less well-defined. The excitations of the system are then no longer simple elementary excitations; the definition as "quasiparticles" increasingly loses its meaning.

The *quasiparticle model* is an ideal Bose gas with the energies ε_p shown in Figure 38.7, and without a fixed number of particles. The statistical treatment of the quasiparticle model is analogous to the phonon gas discussed in Chapter 33. Because of the increasing width of the quasiparticles, the model should be limited to the range $T \lesssim 1\,\text{K}$; in practice, it still provides useful results up to about 2 K. At the λ transition itself, the quasiparticle model breaks down completely; therefore, it is not a basis for the description of the phase transition itself. In the validity range of the model, the excited quasiparticles constitute the normal density ϱ_n.

For low temperatures, the lowest excitations, i.e. the phonons, play the decisive role. As shown in Chapter 33, the specific heat is then proportional to T^3. This also applies to the number of phonons and therefore to ϱ_n:

$$c_V = c_\text{phon} \propto T^3 , \qquad \varrho_\text{n} = \varrho_\text{phon} \propto T^3 \qquad (T \ll 1\,\text{K}) \qquad (38.23)$$

Near 1 K, the roton excitations start to provide a substantial contribution. Their excitation probability is exponentially small (proportional to $\exp(-\Delta/k_\text{B}T)$, where $\Delta \approx 9\,k_\text{B}\,\text{K}$); however, their phase space (number of possible modes) is much larger than that of the phonons. Due to the exponential factor, mainly the excitations in the roton minimum play a role. It is therefore possible to pretend that the dispersion relation consists of the two parts

$$\varepsilon_\text{phon} = c_1 k \quad \text{and} \quad \varepsilon_\text{rot} = \Delta + \frac{\hbar^2}{2m^*} (k - k_0)^2 \qquad (38.24)$$

The comparison with the measured dispersion relation results in $c_1 \approx 220\,\text{m/s}$, $\Delta \approx 9k_\text{B}\text{K}$, $m^* \approx 0.16\,m$ and $k_0 \approx 2\,\text{Å}$. After the division (38.24), the respective parts of specific heat and normal density can be calculated separately:

$$c_V = c_\text{phon} + c_\text{rot} , \qquad \varrho_\text{n} = \varrho_\text{phon} + \varrho_\text{rot} \qquad (38.25)$$

This leads to a quantitatively good description.

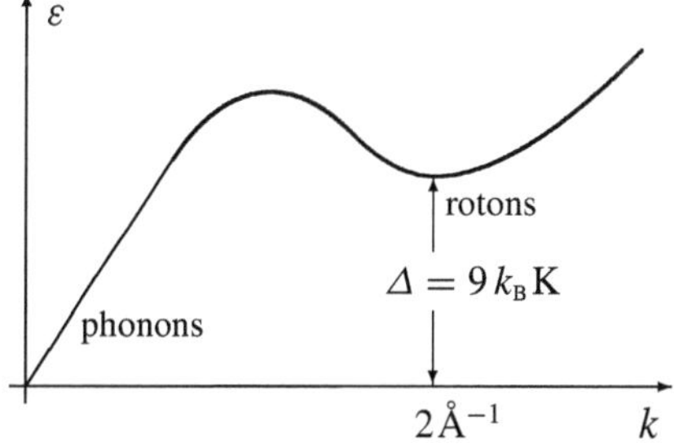

Figure 38.7   Quasiparticle spectrum of liquid ^{4}He at low temperatures.

39 Landau theory

In the Chapters 39 and 40 we discuss second order phase transitions in general. As examples we have presented the Weiss model of ferromagnetism (Chapter 36), the van der Waals gas at the critical point (Chapter 37), the Bose-Einstein condensation and the λ transition of helium (Chapter 38). The processes in the immediate vicinity of a phase transition (the critical point in Figure 37.5) are called "critical phenomena". Another example is the transition between a normal conducting and a superconducting state.

In this chapter, we generalize the free energy $\mathcal{F}$ of the Weiss model to the Landau and the Ginzburg-Landau approach. This approach represents a general, phenomenological model the critical phenomena. In the Landau theory, we investigate the specific heat, the susceptibility and the fluctuations.

At the phase transition, the thermodynamic quantities change in a striking way. The starting point of Landau theory is the choice of a suitable macroscopic quantity that changes in such a characteristic way. This could be the magnetization for the ferromagnetic transition or the volume for the condensation of a gas. This quantity is called *order parameter* ψ. The choice of ψ is obvious in in many cases (as in the case of magnetization); however, it is not unique. For example, for the transition liquid-gas transition, the volume or the density could be chosen.

For 2nd order phase transitions, the mean value is $\overline{\psi}(T)$ of the order parameter is continuous at the transition point T_{cr}, but its derivative displays a jump. The order parameter is chosen such that its mean value disappears at T_{cr}, i.e.

$$\overline{\psi(T_{\mathrm{cr}})} = 0 \tag{39.1}$$

Examples of such order parameters are

$$\psi = \begin{cases} M/M_0 & \text{paramagnetic-ferromagnetic} \\ \sqrt{\varrho_0/\varrho} & \text{Bose-Einstein condensation} \\ \sqrt{\varrho_s/\varrho} & \text{λ-transition} \\ (v - v_{\mathrm{cr}})/v_{\mathrm{cr}} & \text{gaseous–liquid} \end{cases} \tag{39.2}$$

In the latter case, $v = V/N$ is the volume per particle, and v_{cr} is its value at the critical point. Alternatively the order parameter $\psi = (n - n_{\mathrm{cr}})/n_{\mathrm{cr}}$ with the particle density $n = N/V$ can be used. In the second and third cases the root occurs because the macroscopic wave function is the fundamental quantity. We also know from

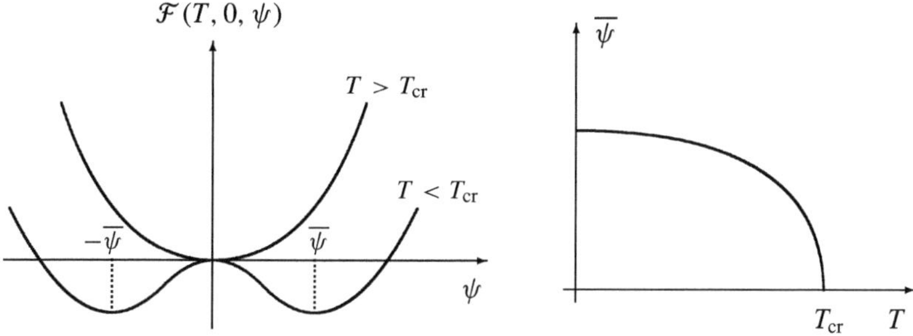

Figure 39.1   The left part shows the free energy (39.4) of the Landau theory as a function of the order parameter ψ; the external field h is zero here. The equilibrium value $\overline{\psi}$ results the condition from $\mathcal{F}$ = minimal. This equilibrium value is shown in the right as a function of temperature.

Equations (38.4) and (36.28) that $(\varrho_0/\varrho)^{1/2}$ and M/M_0 at T_{cr} have the same power behavior.

The order parameter ψ describes the change in structure at the phase transition by

$$\overline{\psi} = \begin{cases} 0 & T \geq T_{cr} \\ \neq 0 & T < T_{cr} \end{cases} \tag{39.3}$$

For the first three examples in (39.2), we have seen this in the previous chapters. The fourth example is discussed in more detail in the Exercises 40.1 and 40.2.

At $T = T_{cr}$, the mean order parameter $\overline{\psi}$ is continuous and has the value zero. It should therefore be possible to expand the free energy in the vicinity of T_{cr} into powers of ψ. The form of this expansion for $|T - T_{cr}| \ll T_{cr}$ is taken from (36.25):

$$\boxed{\mathcal{F}(T, h, \psi) = F_0(T) + V\left(a\,(T - T_{cr})\,\psi^2 + u\,\psi^4 - h\,\psi\right)} \tag{39.4}$$

Examples for the external field h which can influence the order parameter, are

$$h = \begin{cases} B & \text{magnetic field} \\ P - P_v & \text{pressure} \end{cases} \tag{39.5}$$

At the liquid-gas transition, the difference $P - P_v$ to the vapor pressure plays the role of the (small) external field (see Exercises 40.1 and 40.2). For the ideal Bose gas or liquid helium, no such field h is known; the corresponding term has a formal function only. The volume V shall be constant and is therefore not listed in the arguments of $\mathcal{F}$. For a given T and P, an analogous approach would start from the free enthalpy (instead of the free energy).

At the transition point, the mean value of the order parameter is zero, and it is continuous at this point. Therefore, ψ is small for $T \approx T_{cr}$ and the free energy can be expanded into powers of ψ. The coefficients of ψ^n generally depend on the temperature, but they can be expanded into powers of the small quantity $T - T_{cr}$. If the expansion of the coefficient of ψ^2 starts with the first power of $T - T_{cr}$ (as assumed), and if only even powers of ψ occur, this results in the behavior (39.3); this will be shown below. The linear term $h\psi$ implies the proportionality $\overline{\psi} \propto h$ (for example $M \propto B$) for $T > T_{cr}$.

Overall, (39.4) represents a plausible expansion of the free energy into powers of ψ and $T - T_{cr}$. The approach (39.4) was introduced by Landau in 1937. The description of the critical phenomena in this model is called *Landau theory*.

Figure 39.1 shows how the condition $\mathcal{F}$ = minimal leads to the equilibrium value $\overline{\psi}$ of the order parameter. The sign of the term $(T - T_{cr})\,\psi^2$ causes $\overline{\psi}$ to disappear for $T > T_{cr}$ (and $h = 0$), but to adopt a finite value for $T < T_{cr}$. By inserting $\overline{\psi} = \overline{\psi}(T, h)$ into $\mathcal{F}$, the equilibrium value F of the free energy is obtained:

$$F(T, h) = \mathcal{F}\left(T, h, \overline{\psi}(T, h)\right) \tag{39.6}$$

This is a crucial step: While all derivatives of $\mathcal{F}$ are continuous, the second derivative of F has a jump at T_{cr}, and the third derivative is singular.

In Landau's theory, we now calculate some quantities, in particular the specific heat and the susceptibility. The results are partly known from Chapter 36. The ferromagnetic case can be used throughout as an example where the calculated quantities have an obvious meaning.

First, we determine the minimum of $\mathcal{F}(T, 0, \psi)$ with respect to ψ:

$$\frac{1}{V}\frac{\partial \mathcal{F}(T, 0, \psi)}{\partial \psi} = 2a\,(T - T_{cr})\,\psi + 4u\,\psi^3 = 0 \qquad (h = 0) \tag{39.7}$$

From this we obtain the equilibrium value for $h = 0$ (denoted by an index 0):

$$\overline{\psi_0}^{\,2} = \begin{cases} 0 & (T \geq T_{cr}) \\[2mm] \dfrac{a}{2u}\,(T_{cr} - T) & (T < T_{cr}) \end{cases} \tag{39.8}$$

For ferromagnetism, this is known from (36.28). For the ideal Bose gas it is consistent with (38.4) (for $|T - T_c| \ll T_{cr}$). At the position of the minimum, $\mathcal{F}$ determines equilibrium value of the free energy:

$$F(T, 0) = \mathcal{F}(T, 0, \overline{\psi_0}) = F_0(T) + \begin{cases} 0 & (T \geq T_{cr}) \\[2mm] -\dfrac{V a^2}{4u}\,(T_{cr} - T)^2 & (T < T_{cr}) \end{cases} \tag{39.9}$$

The quantity $\overline{\psi_0}^{\,2}$ has a kink at T_{cr}; its second derivative is therefore singular. Inserting $\overline{\psi_0}^{\,2}$ turns the analytic function into $\mathcal{F}(T, h, \psi)$ into the non-analytical function $F(T, h)$. The specific heat at constant field results from

$$c_h(T, h) = \frac{C_h}{V} = \frac{T}{V}\frac{\partial S(T, h)}{\partial T} = -\frac{T}{V}\frac{\partial^2 F(T, h)}{\partial T^2} \tag{39.10}$$

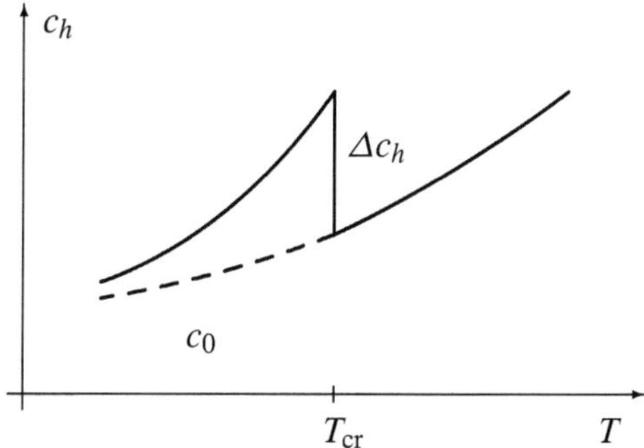

Figure 39.2 In the Landau theory, the specific heat c_h a jump of the size $\Delta c_h = a^2/2u$ at T_{cr}.

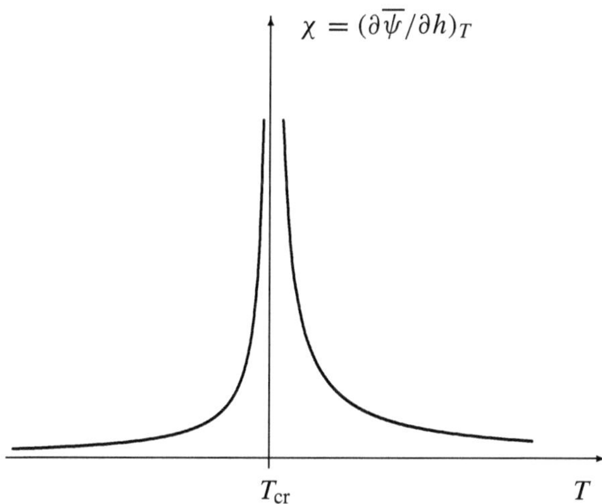

Figure 39.3 The susceptibility χ diverges in Landau theory at the critical point proportional to $1/|T - T_{\mathrm{cr}}|$. The prefactor for $T > T_{\mathrm{cr}}$ is twice as large than that for $T < T_{\mathrm{cr}}$. For the ferromagnetic transition, χ is the magnetic susceptibility χ_{m}. At the critical point of the phase transition gaseous-liquid, χ is the isothermal compressibility κ_T (up to a factor).

For (39.9) we obtain

$$c_h(T,0) = c_0(T) + \begin{cases} 0 & (T > T_{\mathrm{cr}}) \\[2ex] \dfrac{a^2}{2u}\, T & (T < T_{\mathrm{cr}}) \end{cases} \tag{39.11}$$

Here c_0 denotes the continuous component due to F_0. The contribution $c_h - c_0$ is associated with the phase transition. It increases linearly with the temperature and then jumps to zero at T_{cr}; the specific heat is therefore discontinuous. This behavior is sketched in Figure 39.2.

We now consider deviations $\delta\psi$ from the equilibrium value $\overline{\psi_0}$ caused by a weak external field h:

$$\psi = \overline{\psi_0} + \delta\psi \tag{39.12}$$

We insert this into (39.4) and determine the equilibrium value from

$$\frac{1}{V}\frac{\partial \mathscr{F}(T,h,\psi)}{\partial\psi} = 2a\,(T - T_{\mathrm{cr}})\left(\overline{\psi_0} + \delta\psi\right) + 4u\left(\overline{\psi_0} + \delta\psi\right)^3 - h = 0 \tag{39.13}$$

We restrict ourselves to the first order in $\delta\psi$:

$$\left[2a\,(T - T_{\mathrm{cr}})\,\overline{\psi_0} + 4u\,\overline{\psi_0}^{\,3}\right] + \left[2a\,(T - T_{\mathrm{cr}}) + 12\,u\,\overline{\psi_0}^{\,2}\right]\delta\psi - h = 0 \tag{39.14}$$

The first bracket is zero according to (39.7); in the second bracket we insert $\overline{\psi_0}^{\,2}$ from (39.8). The solution is the equilibrium value $\overline{\delta\psi}$. The quotient of $\overline{\delta\psi}$ and h is the susceptibility:

$$\chi = \left(\frac{\partial\overline{\psi}}{\partial h}\right) = \frac{\overline{\delta\psi}}{h} = \begin{cases} \dfrac{1}{2a\,|T - T_{\mathrm{cr}}|} & (T > T_{\mathrm{cr}}) \\[3ex] \dfrac{1}{4a\,|T - T_{\mathrm{cr}}|} & (T < T_{\mathrm{cr}}) \end{cases} \tag{39.15}$$

This susceptibility is sketched in Figure 39.3. In the ferromagnetic case, $\overline{\psi}$ is the magnetization $\overline{M} = M$ (bar usually omitted), h the magnetic field B and χ the magnetic susceptibility $\chi_{\mathrm{m}} = \partial M/\partial B$. For the critical point of the transition liquid–gaseous we use $\overline{\psi} = \overline{n - n_{\mathrm{cr}}/n_{\mathrm{cr}}} = (n - n_{\mathrm{cr}})/n_{\mathrm{cr}}$ (bar again omitted) and $h = P - P_{\mathrm{cr}}$. This yields the susceptibility

$$\chi = \frac{1}{n_{\mathrm{cr}}}\left(\frac{\partial\,(n - n_{\mathrm{c}})}{\partial\,(P - P_{\mathrm{cr}})}\right)_T = \frac{1}{n_{\mathrm{c}}}\left(\frac{\partial\,(N/V)}{\partial P}\right)_T = -\frac{n}{V n_{\mathrm{cr}}}\left(\frac{\partial V}{\partial P}\right)_T = \frac{n}{n_{\mathrm{cr}}}\,\kappa_T \tag{39.16}$$

Here, the susceptibility is the isothermal compressibility κ_T. The divergence of the compressibility at the critical point means that weak disturbances can cause large density fluctuations. These density fluctuations can also be thermally excited.

We now consider the case $T = T_{cr}$ and $h \neq 0$. For this we get $\overline{\psi_0} = 0$, and from (39.13) follows

$$\frac{1}{V}\frac{\partial \mathcal{F}(T_{cr}, h, \psi)}{\partial \psi} = 4u\,(\delta\psi)^3 - h = 0 \qquad (39.17)$$

The equilibrium value of the order parameter is then

$$\overline{\psi} = \overline{\psi_0 + \delta\psi} = \overline{\delta\psi} = \overline{\delta\psi} = \left(\frac{h}{4u}\right)^{1/3} \qquad (T = T_{cr}) \qquad (39.18)$$

This is for example the magnetization that at the transition point (where $M_s = 0$) is caused by an external magnetic field. In (36.17) we had obtained this result in the form $B = (k_B T/3\mu_B)(M/M_0)^3$.

Fluctuations

We now allow for a spatial dependence of the order parameter i.e. $\psi = \psi(r)$. The Landau approach (39.4) is then generalized to the Ginzburg-Landau approach:

$$\mathcal{F}(T, h, \psi) = F_0(T) + \int d^3r \left(A\,(\nabla\psi)^2 + a\,(T - T_{cr})\,\psi^2 + u\,\psi^4 - h\,\psi\right) \qquad (39.19)$$

Deviations from the equilibrium value are called *fluctuations*. In this section, we specifically examine position dependent fluctuations $\delta\psi(r)$ of the order parameter.

The free energy (39.4) was made plausible as an expansion into powers of quantities that are small in the vicinity of the critical point, namely the quantities ψ and $T - T_{cr}$. For a position dependent ψ, the factor V must first be replaced by the integral over the volume of the system. In addition, derivatives of ψ have to be taken into account. In this respect, the terms written on in (39.19) are the simplest possible terms. A linear term $\nabla\psi$ would give a direction dependent contribution; this is not allowed in the isotropic system. The coefficient A must be chosen positive so that the condition of minimal $\mathcal{F}$ yields a constant ψ.

The free energy (39.4) is a function of the field ψ; the necessary condition for the minimum was $\partial\mathcal{F}/\partial\psi = 0$, (39.7). In contrast, the free energy in (39.19) depends on the function $\psi(r)$; thus $\mathcal{F}$ is a *functional* of ψ. The minimum of $\mathcal{F}$ must be sought with respect to all functions $\psi(r)$; this task is known from the variational calculus (Part III in [1]). The necessary conditions for the existence of the minimum of such a functional are the respective Euler-Lagrange equations:

$$\delta\mathcal{F} = \delta\int d^3r\; f(\nabla\psi, \psi, r) = 0 \quad\longleftrightarrow\quad \sum_{i=1}^{3}\frac{\partial}{\partial x_i}\frac{\partial f}{\partial(\partial\psi/\partial x_i)} = \frac{\partial f}{\partial\psi} \qquad (39.20)$$

Here f is the free energy density

$$f(\nabla\psi, \psi, r) = A\,(\nabla\psi)^2 + a\,(T - T_{cr})\,\psi^2 + u\,\psi^4 - h(r)\,\psi \qquad (39.21)$$

The connection (39.20) is known from mechanics: The left-hand side corresponds to Hamilton's principle $\delta \int dt\, \mathcal{L} = 0$ with the Lagrange function $\mathcal{L}$, the right-hand side corresponds to the Lagrange equations. The Euler-Lagrange equations from (39.20) result in

$$2A\,\Delta\psi = 2a\,(T - T_{\mathrm{cr}})\,\psi + 4u\,\psi^3 - h \tag{39.22}$$

Here $\Delta = \sum (\partial/\partial x_i)^2$ is the Laplace operator. For a weak external field of the form

$$h(\boldsymbol{r}) = h_k\,\exp(\mathrm{i}\boldsymbol{k}\cdot\boldsymbol{r}) \tag{39.23}$$

we are looking for a solution of the form

$$\psi_k = \overline{\psi_0} + \delta\psi_k\,\exp(\mathrm{i}\boldsymbol{k}\cdot\boldsymbol{r}) \tag{39.24}$$

In first order in $\delta\psi$ we obtain from (39.22)

$$2a\,(T - T_{\mathrm{cr}})\,\overline{\psi_0} + 4u\,\overline{\psi_0}^{\,3} + \tag{39.25}$$

$$\left(2A\,k^2 + 2a\,(T - T_{\mathrm{cr}}) + 12u\,\overline{\psi_0}^{\,2}\right)\delta\psi_k\,\exp(\mathrm{i}\boldsymbol{k}\cdot\boldsymbol{r}) - h_k\,\exp(\mathrm{i}\boldsymbol{k}\cdot\boldsymbol{r}) = 0$$

Because of (39.7), the first line disappears. From the second line we read off the susceptibility

$$\chi_k = \left(\frac{\partial\overline{\psi_k}}{\partial h_k}\right)_T = \frac{\overline{\delta\psi_k}}{h_k} = \frac{1}{2A\,k^2 + 2a\,(T - T_{\mathrm{cr}}) + 12u\,\overline{\psi_0}^{\,2}} \tag{39.26}$$

We insert the equilibrium value $\overline{\psi_0}^{\,2}$ from (39.8):

$$\chi_k = \frac{\chi}{1 + k^2\,\xi^2} = \begin{cases} \dfrac{1}{2A\,k^2 + 2a\,|T - T_{\mathrm{cr}}|} & (T > T_{\mathrm{cr}}) \\[3mm] \dfrac{1}{2A\,k^2 + 4a\,|T - T_{\mathrm{cr}}|} & (T < T_{\mathrm{cr}}) \end{cases} \tag{39.27}$$

The result can be written in the form $\chi/(1 + k^2\,\xi^2)$ where $\chi = \chi_{k=0}$ is the susceptibility (39.15) of the homogeneous system. Thereby we have introduced the *correlation length* ξ,

$$\xi = \begin{cases} \sqrt{\dfrac{A}{a\,|T - T_{\mathrm{cr}}|}} & (T > T_{\mathrm{cr}}) \\[4mm] \sqrt{\dfrac{A}{2a\,|T - T_{\mathrm{cr}}|}} & (T < T_{\mathrm{cr}}) \end{cases} \tag{39.28}$$

determines the k-dependence of the susceptibility. For the interpretation of ξ, we consider an external field that is localized at a point:

$$h(\boldsymbol{r}) = h_0\,\delta(\boldsymbol{r}) = \frac{h_0}{(2\pi)^3}\int d^3k\,\exp(\mathrm{i}\boldsymbol{k}\cdot\boldsymbol{r}) = \int d^3k\,h_k\,\exp(\mathrm{i}\boldsymbol{k}\cdot\boldsymbol{r}) \tag{39.29}$$

The amplitude in k-space is constant, $h_k = h_0/(2\pi)^3$. This field causes the fluctuation

$$
\begin{aligned}
\overline{\delta\psi(r)} &= \int d^3k\; \overline{\delta\psi_k}\, \exp(i\boldsymbol{k}\cdot\boldsymbol{r}) = \int d^3k\; \chi_k\, h_k\, \exp(i\boldsymbol{k}\cdot\boldsymbol{r}) \\
&= \frac{h_0\,\chi(T)}{(2\pi)^3} \int d^3k\; \frac{\exp(i\boldsymbol{k}\cdot\boldsymbol{r})}{1+k^2\xi^2} \;\propto\; \exp\left(-\frac{r}{\xi}\right)
\end{aligned}
\tag{39.30}
$$

This fluctuation of the order parameter falls off on the length ξ, although the external perturbation (39.29) does not specify a length. Therefore, the length ξ is a characteristic length of the system. It determines in particular the length scale of thermally excited fluctuations.

Far away from the phase transition, the length ξ is of the size of the mean particle distance, i.e. about a few ångström. When approaching the critical point, ξ diverges according to (39.28). Then there are also thermally excited fluctuations with the range of the wavelength of visible light ($\lambda = 4000\ldots7500\,\text{Å}$). These fluctuations lead to a strong light scattering, which can be observed as the *critical opalescence*. When approaching the critical point, a milky turbidity can be seen in the substance.

We discuss the fluctuation for the ferromagnetic phase transition. The susceptibility describes how strong the response of the system (magnetization) to an external disturbance (magnetic field). A spatially periodic magnetic field as in (39.23) could be caused by an electromagnetic wave.

Above the phase transition there is no long-range order; on average the magnetization is zero. A $\delta\psi \neq 0$ over a range of size ξ means that neighboring spins align in such a range. That means that for $T > T_{\text{cr}}$ there is a local order on the length scale ξ. Globally, however, $\overline{\delta\psi} = 0$ (in each case for $h = 0$). For $T \to T_{\text{cr}}^+$ then correlation length ξ diverges. As the transition point is approached, there are larger and larger regions in which the spins align. However, the individual regions have different alignments, so the global magnetization is still zero. At T_{cr}, the correlation length ξ becomes infinite. Then it comes to a long-range order.

Below T_{cr}, the long-range order manifests itself by the spontaneous magnetization. Thermal fluctuations tend to disturb this order. The characteristic linear extension of these perturbations diverges for $T \to T_{\text{cr}}^-$; close to the transition point, the fluctuations become long-ranged. For $T \to T_{\text{cr}}^-$ the fluctuations become so strong that the spontaneous magnetization beaks down.

Thermal fluctuations

Near to T_{cr} the susceptibility χ_k is so large that even a very weak field can cause a finite fluctuation $\delta\psi_k$. Then such fluctuations are also excited thermally, i.e. without an external field. Spatially, these thermal fluctuations extend over regions of the size ξ. We calculate the strength of the thermal fluctuations

$$
\Delta\psi_{\text{therm}}^2 = \left(\Delta\psi_{\text{therm}}\right)^2 = \overline{\left(\psi(r) - \overline{\psi_0}\right)^2}
\tag{39.31}
$$

We first consider spatially constant fluctuations around the equilibrium value $\overline{\psi_0}$ for $h = 0$ (denoted by an index 0). For this we expand (39.4) around the value $\psi = \overline{\psi_0}$:

$$\mathcal{F}(T, 0, \psi) = \mathcal{F}_0 + \left.\frac{\partial \mathcal{F}}{\partial \psi}\right|_0 (\psi - \overline{\psi_0}) + \frac{1}{2}\left.\frac{\partial^2 \mathcal{F}}{\partial \psi^2}\right|_0 (\psi - \overline{\psi_0})^2 + \ldots \quad (39.32)$$

Since the equilibrium value occurs at the minimum of $\mathcal{F}$, the first derivative disappears. With

$$\left.\frac{\partial^2 \mathcal{F}}{\partial \psi^2}\right|_{\overline{\psi_0}} = V\left[2a(T - T_{\mathrm{cr}}) + 12\,u\,\psi^2\right]_{\overline{\psi_0}} = V\left\{\begin{array}{c} 2a\,(T - T_{\mathrm{cr}}) \\ -4a\,(T - T_{\mathrm{cr}}) \end{array}\right\} \overset{(39.15)}{=} \frac{V}{\chi}$$

$$(39.33)$$

and neglecting the higher terms, (39.32) becomes

$$\mathcal{F}(T, 0, \psi) = \mathcal{F}(T, 0, \overline{\psi_0}) + \frac{V}{2\chi}(\psi - \overline{\psi_0})^2 = \mathcal{F}(T, 0, \overline{\psi_0}) + \Delta\mathcal{F} \quad (39.34)$$

According to (10.9), the fluctuations of any macroscopic parameter lead to fluctuations of the entropy $\Delta S \sim k_{\mathrm{B}}$. This implies $\Delta F \sim k_{\mathrm{B}} T$, i.e.

$$\Delta F = \frac{V}{2\chi}\,\overline{(\psi - \overline{\psi_0})^2} \sim k_{\mathrm{B}} T \quad (39.35)$$

For (39.32)–(39.35) we started from (39.4), i.e. from a spatially constant ψ. From (39.35) it follows that *homogeneous* deviations $\psi - \overline{\psi_0}$ approach zero for $V \to \infty$; they therefore do not occur in the macroscopic system.

Spatially limited fluctuations, on the other hand, can be excited thermally. For a position dependent order parameter, we must first replace (39.4) by the more general approach (39.19). Then (39.35) becomes

$$\Delta F = \frac{1}{2\chi}\int d^3r\,\overline{(\psi - \overline{\psi_0})^2} \sim k_{\mathrm{B}} T \quad (39.36)$$

According to (39.30), the position dependent fluctuations extend over the correlation length ξ. For such a deviation

$$\overline{(\psi - \overline{\psi_0})^2} = \left\{\begin{array}{ll} \Delta\psi_{\mathrm{therm}}^2 & \text{in a range of size } \xi^3 \\ 0 & \text{otherwise} \end{array}\right. \quad (39.37)$$

we get from (39.36)

$$\Delta F \sim \frac{\Delta\psi_{\mathrm{therm}}^2\,\xi^3}{\chi} \sim k_{\mathrm{B}} T \quad (39.38)$$

For a more general discussion, we consider d spatial dimension instead of three:

$$\Delta\psi_{\mathrm{therm}}^2 \sim \frac{\chi\,k_{\mathrm{B}} T}{\xi^d} \qquad \text{(thermal fluctuation)} \quad (39.39)$$

Here we insert χ from (39.15) and ξ from (39.28):

$$\Delta\psi_{\text{therm}}^2 \sim \frac{\chi\, k_{\mathrm{B}}T}{\xi^d} \sim \frac{k_{\mathrm{B}}T}{A^{d/2}}\left|a\left(T - T_{\text{cr}}\right)\right|^{d/2 - 1} \tag{39.40}$$

This equation determines the strength of the thermal fluctuations as a function of the temperature. These fluctuations are deviations from the equilibrium value with a spatial extension of the size ξ.

Validity of the Ginzburg-Landau theory

In the Ginzburg-Landau theory, the fluctuations can be qualitatively discussed and understood. Close to T_{cr}, however, the thermal fluctuations become so strong that they question the validity of this model. The Ginzburg-Landau expansion (39.19) implicitly assumes that the thermal fluctuations $\Delta\psi_{\text{therm}}$ are small compared to the mean value, i.e.

$$\Delta\psi_{\text{therm}}^2 \ll \overline{\psi_0}^{\,2} = \frac{a}{2u}\left|T - T_{\text{cr}}\right| \tag{39.41}$$

Only if this holds, the expansion (39.19) into powers of the derivatives of ψ can be terminated after the term $(\nabla\psi)^2$; otherwise, the higher derivatives would be just as large or larger.

We insert (39.40) into (39.41):

$$\frac{u}{A^{d/2}}\, k_{\mathrm{B}}T \left|a\left(T - T_{\text{cr}}\right)\right|^{d/2 - 2} \ll 1 \tag{39.42}$$

This condition for the validity of the Ginzburg-Landau theory is violated for $d = 3$ when approaching T_{cr}. However, for $d > 4$ the condition will be fulfilled close to the transition point; for $d = 4$ it might be fulfilled.

For real systems ($d = 3$), we can expect that the Ginzburg-Landau theory fails in the vicinity of T_{cr}. In fact, here it leads to quantitatively incorrect results. Qualitatively, the Ginzburg-Landau theory describes the critical phenomena largely correctly; it is therefore also the starting point for further theories.

For the case $d = 4$, the Ginzburg-Landau theory is largely valid (as more detailed investigations show). It can also be used as a starting point for approximating real systems (with $d = 3$). This is one reason for the artificial generalization to d dimensions; moreover, in the next chapter we will introduce scaling laws which depend on the dimension d.

Summary

Notwithstanding the problems discussed in the last section, the Ginzburg-Landau theory provides a sensible first model for the critical phenomena. Quantitatively, however, it provides incorrect results; this applies in particular to the critical exponents (Chapter 40).

The scope of this book allows only for a brief introduction to critical phenomena and the Ginzburg-Landau theory. For the sake of simplicity, we have assumed exactly one order parameter, (39.2). In fact, there can be $n \geq 1$ order parameters. For example, the magnetization of the ferromagnet is described by a vector M, i.e. by three independent quantities ($n = 3$). The macroscopic wave function of the ideal Bose gas or He II is complex, i.e. it consists of two independent functions ($n = 2$).

A model as the Ginzburg-Landau theory emphasizes the commonalities of quite different phase transitions. Because of this commonalities, essential properties of the phase transition (like kind of the singularity, susceptibility χ or the correlation length ξ) should only depend on n, d and the symmetry of the system. Such a *universality* is actually found in real systems. In contrast, the parameters A, a and u of the Ginzburg-Landau approach depend on the respective system.

40 Critical exponents

For $T \to T_{\mathrm{cr}}$ thermodynamic quantities generally exhibit a power behavior that is characterized by critical exponents. The Ginzburg-Landau theory provides certain values for these exponents which, however, usually deviate from the experimental values. In this chapter the critical exponents and general relations between them are derived. These relations are called scaling laws.

Definition

At the phase transition, the relative temperature

$$t = \frac{T - T_{\mathrm{cr}}}{T_{\mathrm{cr}}} \tag{40.1}$$

goes to zero. For $|t| \to 0$, thermodynamic quantities often exhibit a power behavior that is defined by an exponent, the *critical exponent*. In the Ginzburg-Landau theory, we have found a power behavior for the following quantities:

$$
\begin{array}{lll}
\text{Specific heat} & c & \propto \ |t|^{-\alpha} \\[4pt]
\text{Order parameter for } t < 0 & \overline{\psi_0} & \propto \ t|^{\beta} \\[4pt]
\text{Susceptibility} & \chi & \propto \ |t|^{-\gamma} \\[4pt]
\psi\text{-}h \text{ relation at } t = 0 & h & \propto \ \overline{\psi}^{\,\delta} \\[4pt]
\text{Correlation length} & \xi & \propto \ |t|^{-\nu}
\end{array}
\tag{40.2}
$$

In the case of ferromagnetism, ψ may be identified with the magnetization M and h with the magnetic field B.

The specific heat, the susceptibility and the correlation length are defined below *and* above the transition point. For distinguishing both sides, the exponents for $t < 0$ are marked by a dash:

$$
c \propto \begin{cases} |t|^{-\alpha} \\ |t|^{-\alpha'} \end{cases}
\qquad
\chi \propto \begin{cases} |t|^{-\gamma} \\ |t|^{-\gamma'} \end{cases}
\qquad
\xi \propto \begin{cases} |t|^{-\nu} & (t > 0) \\ |t|^{-\nu'} & (t < 0) \end{cases}
\tag{40.3}
$$

The critical exponents on both sides of the transition are often equal.

We consider the critical behavior of the order parameter for the phase transition liquid-gaseous. As order parameter we choose $\psi = (n - n_{\mathrm{cr}})/n_{\mathrm{cr}}$ with the particle

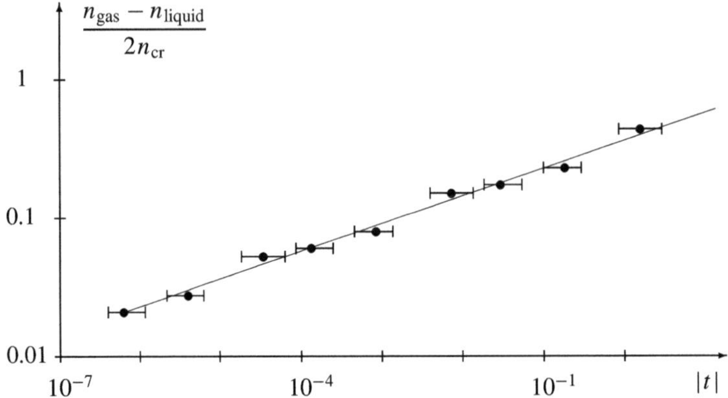

Figure 40.1 We consider a point on the vapor pressure curve just below the critical point, for example $t = -10^{-4} < 0$. When crossing of the vapor pressure curve, the density makes a jump from n_{gas} to n_{liquid}. In the logarithmic diagram the size of this jump is plotted versus the relative temperature $|t|$ for CO_2. The experimental data can be fitted by a straight line. The slope of the straight line defines the critical exponent $\beta \approx 1/3$.

density $n = N/V$. For $h = P - P_{\text{v}} = 0$ (i.e. on the vapor pressure curve) we obtain

$$\frac{n - n_{\text{cr}}}{n_{\text{cr}}} = \begin{cases} \dfrac{n_{\text{gas}} - n_{\text{cr}}}{n_{\text{cr}}} = C_1 \, |t|^{\beta} \\[2ex] \dfrac{n_{\text{liquid}} - n_{\text{cr}}}{n_{\text{cr}}} = C_2 \, |t|^{\beta'} \end{cases} \tag{40.4}$$

Both statements refer to $t < 0$; in this respect, the designation of the exponents with β and β' deviates from the convention (40.3).

In the experiment, we find a behavior as in (40.4) with $\beta = \beta' \approx 1/3$. Because $\beta = \beta'$ the difference $n_{\text{liquid}} - n_{\text{gas}}$ can be plotted (Figure 40.1) in order to determine β. In the Exercises 40.1 and 40.2, the critical behavior of the van der Waals gas will be investigated further.

Absolute values

We compare the values of the critical exponents of the Ginzburg-Landau theory (GL) with the most frequently occurring experimental values (Exp):

$$\text{GL:} \quad \alpha = \alpha' = 0, \quad \beta = 1/2, \quad \gamma = \gamma' = 1, \quad \delta = 3, \quad \nu = \nu' = 1/2$$

$$\text{Exp:} \quad \alpha \approx \alpha' \approx 0, \quad \beta \approx 1/3, \quad \gamma \approx \gamma' \approx 4/3, \quad \delta \approx 4.5, \quad \nu \approx \nu' \approx 1/3$$

$$\tag{40.5}$$

The Ginzburg-Landau values for α and α' are from (39.11), for β from (39.8), for γ and γ' form (39.15), for δ from (39.18) and for ν and ν' from (39.28). The specified experimental values are found in particular for magnetic systems and for the liquid-gas transition at the critical point.

We briefly discuss the meaning of $\alpha = 0$. For $\alpha > 0$ the specific heat c diverges. For $\alpha < 0$, the specific heat $c \propto |t|^{|\alpha|}$ goes to zero for $|t| \to 0$, and the specific heat is continuous. For $\alpha = 0$, we get $c \propto |t|^0 = $ const.; for prefactors are in general different above and below T_{cr}, this means a jump in the specific heat as in (39.11). However, size of the jump remains undefined; insofar a continuous behavior is not excluded. Because of

$$\lim_{\alpha \to 0} \frac{|t|^\alpha - 1}{\alpha} = \ln |t| \tag{40.6}$$

$\alpha = 0$ can also mean a logarithmic (i.e. a rather weak) singularity. This applies for example to the λ transition of helium.

The Ginzburg-Landau theory leads mostly to incorrect values for the critical exponents. We briefly mention the modern procedure that leads to realistic values. A phase transition may imply singularities in the thermodynamic quantities at the transition point. For such a singularity we note: First, it can only occur in the infinite system ($V \to \infty$, $N \to \infty$, see last part of Chapter 35). Secondly, an expansion like the Ginzburg-Landau approach is bound to fail because it assumes an analytical expression for the small quantities ψ and t. However, such an analytical expansion (Ginzburg-Landau) may well be assumed for the free energy in a *finite* volume range; because in a finite volume no singularities can occur. This finite volume shall initially extend over a number of microscopic units (lattice constants, particle distances). One then investigates how the free energy changes when this volume is increased by a certain factor. Such an enlargement is equivalent to a renormalization of the unit length and leads to a corresponding renormalization of the parameters of the Ginzburg-Landau approach. One then calculates how the parameters behave when the volume approaches infinity. The resulting *renormalization group theory* (also known as Landau-Wilson theory) was developed by K. G. Wilson; an overview can be found in Rev. Mod. Phys. 55 (1983) 583. The renormalization group theory leads to non-trivial (deviating from the Ginzburg-Landau values in (40.5)) critical exponents. As further literature reference is Shang-Keng Ma, *Modern Theory of Critical Phenomena*, (Benjamin/Cummings Publishing Company 1976).

Scaling laws

The basic idea of the theories that go beyond the Ginzburg-Landau model is the *scale invariance*: For $T \to T_c$ the natural length scales of the system (such as the lattice spacing) lose their meaning. Then the singular components (which survive for $T \to T_{\mathrm{cr}}$) of the thermodynamic potentials are *homogeneous functions*. We use this homogeneity property of the thermodynamic potentials as a hypothesis. From this *scaling hypothesis* we derive the power behavior (40.2) and various relations (scaling laws) between the critical exponents.

Homogeneous functions

We first explain the concept of a *homogeneous function*. As an example, we first consider a cubic volume $V(x)$ depending on the edge length x. Under the transformation $x \to \mu x$, the function $V(x)$ *scales* according to $V(\mu x) = \mu^3 V(x)$. More generally, a function of a variable can scale like $F(\mu x) = \lambda(\mu) F(x)$, where $\lambda(\mu)$ is initially indeterminate. For two subsequent scalings, scalings $x \to \mu_1 x$ and $x \to \mu_2 x$, the homogeneity property yields $\lambda(\mu_1 \mu_2) = \lambda(\mu_1) \lambda(\mu_2)$. This applies to any μ_1 and μ_2. Therefore λ can only depend on a power of μ, i.e. $\lambda(\mu) = \mu^{1/a}$ or $\mu = \lambda^a$ where a is a real exponent. This property defines a *homogeneous function*

$$F(\lambda^a x) = \lambda F(x) \tag{40.7}$$

If we set $\lambda = x^{-1/a}$, the power behavior $F(x) = F(1) x^{1/a}$ follows. Correspondingly, we obtain for a homogeneous function of two variables:

$$F(\lambda^a x, \lambda^b y) = \lambda F(x, y) \tag{40.8}$$

There are two exponents, a and b. With $\lambda = x^{-1/a}$ we write the homogeneity property in a slightly different form:

$$F(x, y) = x^{1/a} F(1, y/x^{b/a}) \equiv x^{1/a} G(y/x^{b/a}) \tag{40.9}$$

The newly introduced function G depends only on the combination $y/x^{b/a}$ of the variables. Homogeneous functions of more than two variables can be defined analogously.

Free energy as a homogeneous function

We now assume that the free energy $\mathcal{F}$ is a homogeneous function of the variables t and ψ:

$$
\begin{aligned}
\mathcal{F}(t, h, \psi) &= F_0(T) + V\left[f(-t, \psi) - h\,\psi\right] \\
&= F_0(T) + V\left[(-t)^{1/a} G(\psi/(-t)^{b/a}) - h\,\psi\right] \tag{40.10}
\end{aligned}
$$

We start with the case $h = 0$. The necessary condition for the minimum of $\mathcal{F}$ is

$$\frac{\partial \mathcal{F}}{\partial \psi} = 0 \quad \Longrightarrow \quad G'(\psi/(-t)^{b/a}) = 0 \tag{40.11}$$

This condition determines the equilibrium value

$$\overline{\psi_0} = \text{const.} \cdot (-t)^{b/a} = \begin{cases} \text{const.} \cdot (-t)^\beta & (t < 0) \\ 0 & (t > 0) \end{cases} \tag{40.12}$$

This yields a power behavior with the critical exponent $\beta = b/a$. In our examples (such as the Weiss model), the constant disappears for $t > 0$.

For $h \neq 0$, the condition $\partial F/\partial \psi = 0$ leads to

$$h = (-t)^{1/a-\beta}\, G'\big(\bar{\psi}/(-t)^{\beta}\big) \xrightarrow{(-t)\to 0} \text{const.} \cdot (-t)^{1/a-\beta-\delta\beta}\, \bar{\psi}^{\delta} \tag{40.13}$$

In the limiting case $(-t) \to 0$, this must result in $h \propto \bar{\psi}^{\delta}$ must result. From this follows $1/a = \beta(1 + \delta)$ and (40.10) becomes

$$\mathcal{F}(t, h, \psi) = F_0(T) + V\left((-t)^{\beta\delta+\beta}\, G\big(\psi/(-t)^{\beta}\big) - h\,\psi \right) \tag{40.14}$$

We write on once again the condition $\partial F/\partial \psi = 0$:

$$h\big(t, \bar{\psi}\big) = (-t)^{\beta\delta}\, G'\big(\bar{\psi}/(-t)^{\beta}\big) \tag{40.15}$$

Here we consider h as a function of t and $\bar{\psi}$; physically, one would rather consider $\bar{\psi}$ as a function of t and h. The equations (40.14) and (40.15) are the starting point for deriving the scaling laws.

The Landau theory and the Weiss model are special cases of (40.14) and (40.15). For example, (40.15) becomes in the Weiss model

$$B(t, M) \overset{(36.17)}{=} t\,M + M^3 = (-t)^{3/2}\left(-\frac{M}{(-t)^{1/2}} + \frac{M^3}{(-t)^{3/2}} \right) \tag{40.16}$$

Here $\beta = 1/2$, $\delta = 3$ and $G'(x) = -x + x^3$; the occurring constants were suppressed.

Specific heat

We calculate the specific heat c_h for $h = 0$. We use the thermodynamic potential of the free energy $F(T, h)$ as a starting point:

$$F(T, 0) = \mathcal{F}(t, 0, \overline{\psi_0}) = F_0 + V\,(-t)^{\beta\delta+\beta}\, G\big(\overline{\psi_0}/(-t)^{\beta}\big) \tag{40.17}$$

By inserting the equilibrium values $\overline{\psi_0} = 0$ for $t > 0$ and $\overline{\psi_0} \propto (-t)^{\beta}$ for $t < 0$, the function G yields the constants

$$G_+ = G(0), \qquad G_- = G\big(\overline{\psi_0}/(-t)^{\beta}\big) = G(\text{const.}) \tag{40.18}$$

Then (40.17) becomes

$$F(T, 0) = F_0(T) + V\,|t|^{\beta\delta+\beta} \cdot \begin{cases} G_+ & (t > 0) \\ G_- & (t < 0) \end{cases} \tag{40.19}$$

Any sign that may occur can be included into the constants. In the Landau theory we get $G_+ = 0$, (39.9).

The heat capacity follows from the free energy

$$C_h(t, h) = T \frac{\partial S(t, h)}{\partial T} = \frac{T}{T_{\mathrm{cr}}} \frac{\partial S}{\partial t} = -\frac{T}{T_{\mathrm{cr}}^2} \frac{\partial^2 F(t, h)}{\partial t^2} \tag{40.20}$$

With (40.19) and $\partial^2/\partial t^2 = \partial^2/\partial |t|^2$ we then obtain the specific heat

$$c_h(t, 0) - c_0(T) = -\frac{V}{N} \frac{T}{T_{\mathrm{cr}}^2} G_\pm \frac{\partial^2 |t|^{\beta\delta+\beta}}{\partial |t|^2} \overset{|t| \to 0}{\propto} G_\pm |t|^{\beta\delta+\beta-2} \propto |t|^{-\alpha} \tag{40.21}$$

Here c_0 is the contribution from F_0 and $c_h - c_0$ is the critical part. The coefficients G_+ and G_- apply for $t > 0$ and for $t < 0$, respectively. In the last expression, the definition of α was used. The exponent $\beta\delta + \beta - 2$ applies for both sides of the phase transition. From this follows

$$\alpha = \alpha' \tag{40.22}$$

The critical exponents of the specific heat are therefore the same below and above the phase transition. From (40.21) follows also the scaling law

$$2 - \alpha = \beta\delta + \beta \tag{40.23}$$

Susceptibility

From (40.15) we calculate the susceptibility

$$\frac{1}{\chi} = \left(\frac{\partial h}{\partial \overline{\psi}} \right)_t = (-t)^{\beta\delta} \left(\frac{\partial G'(\overline{\psi}/(-t)^\beta)}{\partial \overline{\psi}} \right)_t = (-t)^{\beta\delta-\beta} G''(\overline{\psi}/(-t)^\beta) \tag{40.24}$$

In the second derivative G'' of the function G we insert the equilibrium solutions for $h \to 0$, namely $\overline{\psi} = 0$ for $t > 0$ and $\overline{\psi} = \overline{\psi}_0$ for $t < 0$. This results in constants G''_+ and G''_-, as in (40.18), which no longer depend on t. Then (40.24) becomes

$$\frac{1}{\chi} = G''_\pm \cdot |t|^{\beta\delta-\beta} \propto |t|^\gamma \tag{40.25}$$

In the last expression, the definition (40.2) of γ was used. From this follows

$$\gamma = \gamma' \tag{40.26}$$

and the scaling law

$$\gamma = \beta(\delta - 1) \tag{40.27}$$

The elimination of δ from (40.23) and (40.27) yields the scaling law

$$\alpha + 2\beta + \gamma = 2 \tag{40.28}$$

Scale invariance

We obtain another scaling law from the assumption of *scale invariance*. The scaling invariance states that for $|t| \to 0$ there is only one relevant length scale, namely the correlation length ξ. On this scale, the fluctuations $\Delta\psi_{\text{therm}}$ must then be of the same size as $\overline{\psi_0}$, i.e.

$$\Delta\psi_{\text{therm}} \sim \overline{\psi_0} \qquad (\text{for } |t| \to 0, \text{ scale invariance}) \tag{40.29}$$

The size of the thermal fluctuations on the scale of ξ is given in (39.39). This yields

$$\Delta\psi_{\text{therm}}^2 \sim \frac{\chi\, k_{\mathrm{B}} T}{\xi^d} \propto |t|^{-\gamma+d\nu} \propto \overline{\psi_0}^{\,2} \propto |t|^{2\beta} \tag{40.30}$$

Here we have used (40.29) and the definitions (40.2) of γ, ν and β. From (40.30) we read off the scaling law $d\nu = \gamma + 2\beta$. With (40.28) this results in

$$2 - \alpha = d\nu \tag{40.31}$$

Because of (40.22), $\nu = \nu'$ also applies.

Summary

Our considerations led to

$$\boxed{\alpha = \alpha', \qquad \gamma = \gamma', \qquad \nu = \nu'} \tag{40.32}$$

and to the following three *scaling laws*:

$$\boxed{\alpha + 2\beta + \gamma = 2, \qquad \delta = 1 + \frac{\gamma}{\beta}, \qquad 2 - \alpha = d\nu} \tag{40.33}$$

This means that two of the five exponents under consideration (α, β, γ, δ and ν) are independent.

The scaling laws are usually well fulfilled, as can be seen from the experimental results given in (40.5). The values of the Ginzburg-Landau theory fulfill all scaling laws except (40.31). The relation (40.31) would be valid for $d = 4$.

Exercises

40.1   Critical exponents of the van der Waals gas

Expand the equation of state $P = k_{\mathrm{B}} T/(v - b) - a/v^2$ (with $v = V/N$) of the van der Waals gas around the critical point, i.e. into powers of the variables

$$t = \frac{T - T_{\mathrm{cr}}}{T_{\mathrm{cr}}}, \qquad v = \frac{v - v_{\mathrm{cr}}}{v_{\mathrm{cr}}}, \qquad p = \frac{P - P_{\mathrm{cr}}}{P_{\mathrm{cr}}} \tag{40.34}$$

Neglect terms of the order tv^2 and tv^3. Show

$$p = 4t - 6tv - \frac{3}{2} v^3 + \mathcal{O}(v^4) \tag{40.35}$$

Determine the vapor pressure $p_{\mathrm{V}}(t)$ using Maxwell's construction. Calculate v_{gas} and v_{liquid} and the critical exponents β and β' for $t < 0$. Determine the isothermal compressibility κ_T from

$$\frac{1}{\kappa_T} = -v \left(\frac{\partial P}{\partial v} \right)_T \approx -P_{\mathrm{cr}} \left(\frac{\partial p}{\partial v} \right)_t$$

Specify the critical exponents γ and γ'. Finally, examine the p-v relation at $t = 0$, and use it to calculate the critical exponent δ.

40.2   Landau energy for the van der Waals gas

Consider the free enthalpy as a function of the variables T, P and V,

$$\mathcal{G}(T, P, V) = F(T, V) + P V \tag{40.36}$$

For given T and P, the equilibrium is given by the minimum of $\mathcal{G}(T, P, V)$ as a function of V. The equilibrium value of the free enthalpy is then $G(T, P) = \mathcal{G}(T, P, V_{\min})$. Show that $\mathcal{G}(T, P, V) = $ minimal yields to the thermal equation of state equation of state $P = P(T, V)$.

Now look specifically at the van der Waals gas. Start from the free energy

$$F(T, V) = F_0(T) - N \left[k_{\mathrm{B}} T \, \ln \left(\frac{v - b}{v_{\mathrm{cr}} - b} \right) + a \left(\frac{1}{v} - \frac{1}{v_{\mathrm{cr}}} \right) \right]$$

and set up (40.36). Use the variables t, v and p from (40.34) and expand $\mathcal{G}$ up to the order $\mathcal{O}(v^4)$ around the critical point. Neglect the term of order tv^3. Check that this expansion is consistent with that of the thermal equation of state (40.35).

The part that depends on order parameter v is of the standard form (39.4) of the Landau theory,

$$\mathcal{G}(T, P, V) = \ldots + C \left[hv + 3tv^2 + \frac{3}{8} v^4 \right] \qquad (C = \text{const.}) \tag{40.37}$$

Specify the external field h. Discuss the equilibrium value of the order parameter v as a function of t for $h = 0$.

VII Non-equilibrium processes

41 Approach to equilibrium

In Part VII, we undertake a foray over non-equilibrium processes. These include in particular transport processes that are described by the respective transport equations (such as the heat conduction or diffusion equation).

Chapters 41 and 42 present the fundamental equations for treating transport phenomena; these chapters can also be skipped. In Chapter 43, various transport equations are derived in an elementary kinetic model.

In this chapter, we introduce the master equation as a simple quantum mechanical balance equation. From the master equation follows the increase of the entropy in a closed system (H theorem), i.e. the approach to equilibrium. We discuss the derivation of the master equation from the von Neumann equation.

Master equation

We consider a closed quantum mechanical system with the microstates r (Chapter 5). The microstates may be given by the eigenstates of a Hamiltonian operator H_0 for the eigenvalue E_r. The Hamiltonian operator $H = H_0 + V$ of the considered system shall differ from H_0 by a small perturbation V. For the closed system the perturbation V is time-independent. Such a perturbation leads to the transition probabilities (Chapter 41 in [3]):

$$W_{rr'} = \frac{\text{probability for } r \to r'}{\text{time}} = \frac{2\pi}{\hbar} \left| \langle r | V | r' \rangle \right|^2 \delta(E_r - E_{r'}) \tag{41.1}$$

These probabilities are symmetrical:

$$W_{rr'} = W_{r'r} \tag{41.2}$$

The macrostate of a system is given by a statistical ensemble i.e. by the specification of the probabilities $\{P_1, P_2, P_3, \ldots\}$. Here, P_r is the probability that a system of the ensemble (or a physical system at a certain point in time) is in the state r.

For the temporal change of the probabilities $P_r(t)$, the balance equation follows from (41.1)

$$\frac{dP_r(t)}{dt} = -\sum_{r'} W_{rr'} P_r + \sum_{r'} W_{r'r} P_{r'} = \sum_{r'} W_{rr'} (P_{r'} - P_r) \qquad (41.3)$$

This equation is called *master equation*. On the right-hand side there are the probabilities per time that a system of the ensemble will reach or leave the state r.

H-theorem

The approach to equilibrium was stated in Chapter 5 as a fact of experience. Formally, this approach to equilibrium follows from the master equation. To show this we consider the quantity

$$H(t) = \sum_r P_r \ln P_r \qquad (41.4)$$

and calculate its temporal change using (41.3):

$$\frac{dH}{dt} = \sum_r \left(\dot{P}_r \ln P_r + \dot{P}_r \right) = \frac{1}{2} \left(\sum_r \dot{P}_r \ln(e\, P_r) + \sum_{r'} \dot{P}_{r'} \ln(e\, P_{r'}) \right)$$

$$= -\frac{1}{2} \sum_{r,r'} W_{rr'} (P_r - P_{r'}) \left(\ln(e\, P_r) - \ln(e\, P_{r'}) \right)$$

$$= \frac{1}{2} \sum_{r,r'} W_{rr'} P_r \underbrace{\left(1 - \frac{P_{r'}}{P_r} \right) \ln\left(\frac{P_{r'}}{P_r} \right)}_{\leq 0} \qquad (41.5)$$

From $W_{rr'} P_r \geq 0$ and $(1 - x) \ln x \leq 0$ follows

$$\frac{dH(t)}{dt} \leq 0 \qquad (H\text{-theorem}) \qquad (41.6)$$

This result, known as the H theorem, distinguishes a *time direction*. The master equation is therefore not time-reversal invariant.

In (23.36) we have introduced the entropy $S = -k_\mathrm{B} \sum_r P_r \ln P_r$ of an arbitrary macrostate $\{P_r\}$. With

$$S = -k_\mathrm{B} H = -k_\mathrm{B} \sum_r P_r \ln P_r \qquad (41.7)$$

(41.6) becomes the second law of TD for closed systems. The equilibrium state is defined as the macrostate in which the macroscopic quantities (such as entropy) do no longer change, i.e.,

$$\frac{dH(t)}{dt} = 0 \qquad (\text{equilibrium}) \qquad (41.8)$$

In non-equilibrium, H always decreases according to (41.6); therefore the equilibrium value of H represents a minimum, or

$$S = -k_\mathrm{B}\, H = \text{maximum} \qquad \text{(equilibrium)} \qquad (41.9)$$

This statement has been discussed in detail in Chapter $9-12$. From (41.8) and (41.5) follows for the equilibrium

$$P_r = P_{r'} \quad \text{for all } r,\, r' \text{ with } E_r = E_{r'} \qquad (41.10)$$

The condition $P_r = P_{r'}$ only follows for the states with $E_r = E_{r'}$, because only for such states the transfer matrix elements $W_{rr'}$ in non-zero. In a closed system, only states of the same energy $E_r = E$ can be reached; for all other states, we obtain $P_r = 0$. This gives us the final result:

$$P_r = \begin{cases} \text{const.} & \text{for } E_r = E \\ 0 & \text{otherwise} \end{cases} \qquad \text{(equilibrium)} \qquad (41.11)$$

In equilibrium, all accessible microstates r are equally probable. Accessible means, that a state can be reached via the transitions (41.1). This is the fundamental postulate introduced in Chapter 5.

The path from a non-equilibrium state to equilibrium is associated with $dS > 0$, (41.6). Therefore, the master equation describes irreversible processes.

Von Neumann equation

The master equation is a plausible balance equation for the probabilities P_r. In this respect, the above discussion makes the statement $\Delta S \geq 0$ plausible.

However, the real problem lies in the derivation of the master equation from the basic laws of physics. These fundamental laws (or laws of nature) such as Newton's axioms, Maxwell's equations or the Schrödinger equation are time-reversal invariant. The master equation is not time-reversal invariant because it implies $dS/dt \geq 0$; this defines a time direction. A derivation of the master equation from the Schrödinger equation (or from the von Neumann equation) must therefore contain steps that violate the time reversal symmetry. This section (it is somewhat more formal and can be skipped) outlines the possible path of such a derivation.

We consider an ensemble of N systems that are in the quantum mechanical states $|i\rangle = |1\rangle, |2\rangle, \ldots, |N\rangle$. These states are normalized, but otherwise arbitrary. Several systems of the considered ensemble can be in the same state. The *density operator* (also called statistical operator)

$$\hat{\rho}(t) = \frac{1}{N} \sum_{i=1}^{N} |i\rangle\langle i| \qquad (41.12)$$

defines the statistical ensemble, because it determines the mean values of all operators:

$$\overline{F} = \frac{1}{N} \sum_{i=1}^{N} \langle i \,|\, \hat{F} \,|\, i \rangle = \mathrm{tr}\left(\hat{F} \,\hat{\rho} \right) \tag{41.13}$$

Here tr denotes the trace of the operator or matrix. The simplest example for studying the density operator is a system with two states, such as a particle with a spin $1/2$. For this, consider Exercise 37.4 in [3].

The states $|i\rangle$ are generally time-dependent and satisfy the Schrödinger equation $i\hbar\,(\partial/\partial t)|i\rangle = H|i\rangle$. From this follows the time derivative of the density operator,

$$i\hbar\,\frac{\partial \hat{\rho}}{\partial t} = \left[H,\, \hat{\rho}(t) \right] \qquad \text{(von Neumann equation)} \tag{41.14}$$

The solution $\hat{\rho}(t)$ of this *von Neumann equation* determines the probabilities

$$P_r(t) = \langle r \,|\, \hat{\rho}(t) \,|\, r \rangle \tag{41.15}$$

As in (41.1), the $\{|r\rangle\}$ shall form a complete, orthonormalized set of states. The von Neumann equation defines the temporal development of the $P_r(t)$ and therefore that of the macrostate $\{P_1(t),\ P_2(t),...\}$. It describes in particular the temporal development of a non-equilibrium state.

The von Neumann equation should not be confused with the equation which is obtained for the temporal change of operators in the Heisenberg picture (Chapter 35 in [3]). In contrast to the Heisenberg picture, we consider time-dependent states here. The von Neumann equation is the Schrödinger equation for the density operator.

It is easy to show that

$$\hat{\rho}\,(t+dt) = \exp(-\mathrm{i}\,H\,dt/\hbar)\,\hat{\rho}(t)\,\exp(+\mathrm{i}\,H\,dt/\hbar) \tag{41.16}$$

is equivalent to the von Neumann equation. We use this to calculate

$$\begin{aligned}
P_r(t+dt) &= \langle r \,|\, \hat{\rho}(t+dt) \,|\, r \rangle \tag{41.17}\\[2mm]
&= \sum_{r',r''} \langle r \,|\, \exp(-\mathrm{i}\,H\,dt/\hbar) \,|\, r' \rangle\, \rho_{r'r''}\, \langle r'' \,|\, \exp(+\mathrm{i}\,H\,dt/\hbar) \,|\, r \rangle
\end{aligned}$$

Here we have used (41.16) and inserted a complete sets of states ($1 = \sum_{r'} |r'\rangle\langle r'|$ and $1 = \sum_{r''} |r''\rangle\langle r''|$) on the left and right of $\hat{\rho}$. The matrix element $\langle r' | \hat{\rho}(t) | r'' \rangle$ has been denoted by $\rho_{r'r''}$.

For a macroscopic system, (41.17) contains sums over a large number of intermediate states $|r'\rangle$ and $|r''\rangle$. We assume[1] now that the phases of the matrix elements

---

[1]For an investigation of the question, under which conditions this assumption is justified, we refer to G. V. Chester, Rep. Progr. Phys. 26 (1963) 411.

$\langle r | \exp(-\mathrm{i}\,H\,dt/\hbar)\,|r'\rangle$ are statistically distributed and that therefore all contributions with $r' \neq r''$ average away. This gives us

$$P_r(t + dt) \approx \sum_{r'} \big| \langle r |\exp(-\mathrm{i}\,H\,dt/\hbar)\,r'\rangle \big|^2 P_{r'}(t) \qquad (41.18)$$

We denote the absolute value of the matrix element by

$$W_{rr'} = \frac{|\langle r |\exp(-\mathrm{i}\,H\,dt/\hbar)\,|r'\rangle|^2}{dt} \qquad (41.19)$$

The states $|r\rangle$ are eigenstates of a (model) Hamiltonian operator H_0, i.e. $H_0|r\rangle = E_r|r\rangle$. The actual Hamilton operator $H = H_0 + V$ contains an additional perturbation operator V, which leads to transitions between the states $|r\rangle$. As an example, the Hamilton operator might describe an ideal gas, whereas the perturbation V describes the elastic collisions in the gas. In the context of the time-dependent perturbation theory of quantum mechanics, it is shown (Chapter 41 in [3]) that the $W_{rr'}$ in (41.19) are equal to the transition probabilities per time given in (41.1).

The following applies to $W_{rr'}$

$$\sum_{r'} W_{rr'} = \sum_{r'} \frac{\langle r |\exp(-\mathrm{i}\,H\,dt/\hbar)\,r'\rangle\langle r' |\exp(+\mathrm{i}\,H\,dt/\hbar)\,|r\rangle}{dt} = \frac{1}{dt} \qquad (41.20)$$

Here, $1 = \sum_{r'} |r'\rangle\langle r'|$ is used. We calculate the time derivative of P_r:

$$\frac{dP_r}{dt} = \frac{P_r(t + dt) - P_r(t)}{dt} = \sum_{r'} W_{rr'} P_{r'}(t) - \sum_{r'} W_{rr'} P_r(t) \qquad (41.21)$$

In the first term we have used (41.18) with (41.19), in the second (41.20). The result is the master equation (41.3). The decisive step in this derivation of the master equation is the phase averaging in (41.17), which leads to (41.18).

In the classical limit, the probabilities P_r are replaced by the probability density $\rho(q_1,...,q_f,p_1,...,p_f,t)$ in phase space. For this ρ, Newton's axioms yield the Liouville equation (see for example Appendix A.13 in [6]). The Liouville equation is the classical pendent of the von Neumann equation.

42 Boltzmann equation

For a classical, dilute gas, the master equation becomes the Boltzmann equation. Using the example of the electrical conductivity, we demonstrate how transport phenomena can be treated with the Boltzmann equation.

The microstate of a classical gas consisting of N particles (atoms or molecules) is given by

$$r = \left(r_1, v_1, r_2, v_2, \ldots, r_N, v_N \right) \tag{42.1}$$

Internal degrees of freedom of the gas particles (excitations of the electrons, rotations or vibrations of the molecules) are not taken into account.

For a statistical treatment of N particles of the same kind it is sufficient to calculate the probability distribution $f(r, v, t)$ for a selected particle. The function f can be normalized to the number of particles N. Then we get

$$f(r, v, t)\, d^3r\, d^3v = \left\{ \begin{array}{l} \text{number of particles in the phase} \\ \text{space volume } d^3r\, d^3v \text{ at } r, v. \end{array} \right. \tag{42.2}$$

The probability density $f(r, v, t)$ determines the macroscopic state of the classical diluted gas; it takes the place of the probabilities $P_r(t)$ considered in the last chapter.

In the following we restrict ourselves to a diluted gas; then it is sufficient to take two-particle collisions into account. Under this restriction, the balance equation for $f(r, v, t)$ is the *Boltzmann equation* (also called Boltzmann transport equation):

$$\boxed{ \left(v \cdot \frac{\partial}{\partial r} + \frac{F}{m} \cdot \frac{\partial}{\partial v} + \frac{\partial}{\partial t} \right) f(r, v, t) = \int d^3v_1 \int d\Omega \; V \, \frac{d\sigma(\Omega)}{d\Omega} \cdot \\ \left(f(r, v', t)\, f(r, v_1', t) - f(r, v, t)\, f(r, v_1, t) \right) }$$

$$\tag{42.3}$$

The left-hand side takes into account the change due to the motion of the particles, and due to an external force field $F(r, t)$. The right side takes into account the change due to collisions between two particles (Figure 42.1). Here, Ω is the scattering angle in the center of mass system, $V = |v - v_1|$ is the relative velocity, and $d\sigma/d\Omega$ is the differential cross-section.

The two sides of the balance equation (42.3) correspond to that of the master equation (41.3). The master equation is the balance equation for the probabilities

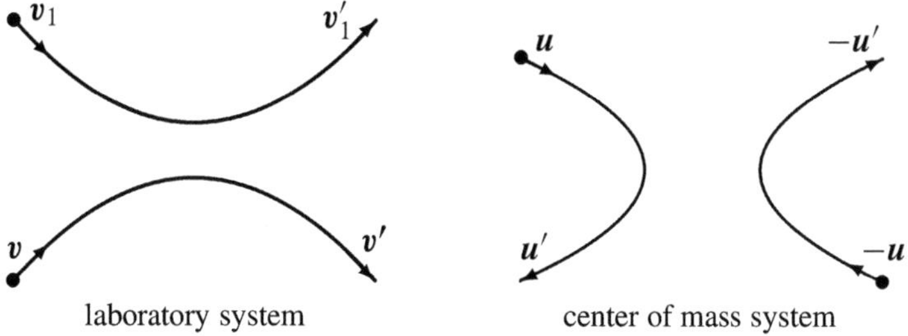

Figure 42.1   Collisions between two particles in the laboratory system and in the center of mass system. In the center of mass system, the process can be specified by the velocity u and the scattering angle $\Omega = (\theta, \phi)$.

P_r of the quantum mechanical microstates of the whole system. The Boltzmann equation is the balance equation for the probability density $f(r, v, t)$ of the classical states (r, v) of a single particle.

In the following we present a qualitative justification of the Boltzmann equation. First of all, we recall that we consider a diluted classical gas (without vibrations or rotations). Classical means that the microstates are given by (42.1), and the conditions "diluted" allows us to consider two-particle collisions only. Without the internal degrees of the gas particles, the scattering is elastic (without energy loss).

The left-hand side of (42.3) describes the change in f *without* collisions. To see this, we consider the particles in (42.2). At a slightly later time $t + dt$, these particles are at the position $r + v\,dt$ and have the velocity $v + F\,dt/m$. As their number does not change, they make the same contribution to f at the phase space point:

$$f(r, v, t)\, d^3r\, d^3v \;=\; f(r', v', t')\, d^3r'\, d^3v'$$

$$\;=\; f(r + v\,dt, \; v + F\,dt/m, \; t + dt)\, d^3r\, d^3v \qquad (42.4)$$

Using $r' = r + v\,dt$ and $v' = v + (F/m)\,dt$ one can determine the Jacobi determinant $|J|$ in $d^3r'\, d^3v' = |J|\, d^3r\, d^3v$. Neglecting the terms $\mathcal{O}(dt^2)$ one obtains $|J| = 1$.

We now expand the right-hand side in (42.4) into the infinitesimal quantities. Then we obtain from (42.4) the *collisionless Boltzmann equation*

$$\left(v \cdot \frac{\partial}{\partial r} + \frac{F}{m} \cdot \frac{\partial}{\partial v} + \frac{\partial}{\partial t} \right) f(r, v, t) = 0 \qquad \text{(without collisions)} \qquad (42.5)$$

Here we used the notation $\partial/\partial a = e_x\, \partial/\partial a_x + e_y\, \partial/\partial a_y + e_z\, \partial/\partial a_z$.

The right-hand side of (42.3) is called the collision term. The collision term takes into account that collisions cause particles to enter the phase space volume $d^3r\, d^3v$ at r, v, or to leave it. In Figure 42.1 on the left, a particle with velocity v scatters off another particle; after that, the particle is generally no longer in the considered volume d^3v at v. We first calculate this loss term that takes these scattering processes into account.

The number of scattering processes per time is equal to the cross section times the current density of the colliding particles and times the number of target particles. In the phase space cell $d^3r\, d^3v$ there are $f(\boldsymbol{r}, \boldsymbol{v}, t)\, d^3r\, d^3v$ (target) particles, (42.2). The density of the colliding particles is $f(\boldsymbol{r}, \boldsymbol{v}_1, t)\, d^3v_1$, where we consider a finite velocity interval. The number of collisions depends on the relative velocity $V = |\boldsymbol{v} - \boldsymbol{v}_1|$, so that the current density is equal to $V f(\boldsymbol{r}, \boldsymbol{v}_1, t)\, d^3v_1$. For cross-section, we consider a finite solid angle range, $d\sigma = (d\sigma/d\Omega)\, d\Omega$. We now form the product of the number of target particles, the current density and the cross section $d\sigma$. As on the left-hand side of the Boltzmann equation, we omit the phase space volume $d^3r\, d^3v$. Finally, we must sum over all possible scattering processes, i.e. integrate over $\boldsymbol{v}_1$ and Ω. Here is $\Omega = (\theta, \phi)$ is the angle between $\boldsymbol{V}' = \boldsymbol{v}'_1 - \boldsymbol{v}'$ and $\boldsymbol{V} = \boldsymbol{v}_1 - \boldsymbol{v}$. This is also the scattering angle in the center of mass system; for given velocities $\boldsymbol{v}$ and $\boldsymbol{v}_1$, it determines the elastic scattering process.

The argumentation just described leads to the loss term on the right-hand side of (42.3). If we reverse the arrows in Figure 42.1 we obtain a scattering into the considered range d^3v at $\boldsymbol{v}$. The resulting gain term is analogous to the loss term; the size of the relative velocity $|\boldsymbol{v} - \boldsymbol{v}_1| = |\boldsymbol{v}' - \boldsymbol{v}'_1|$ is again V. The arguments $\boldsymbol{v}'$ and $\boldsymbol{v}'_1$ in the gain term are connected with $\boldsymbol{v}$, $\boldsymbol{v}_1$ and Ω (due to the momentum and energy conservation).

Maxwell distribution

We show that the Boltzmann equation for the force free case ($\boldsymbol{F} = 0$) leads to Maxwell's velocity distribution. We restrict ourselves to a position independent distribution $f = f(\boldsymbol{v}, t)$. This holds for the equilibrium distribution, because of the homogeneity of the space (in the force-free case).

In the classical system under consideration, all macroscopic quantities can be calculated from $f(\boldsymbol{v}, t)$. The equilibrium is the macrostate in which macroscopic quantities no longer change. We denote the corresponding equilibrium distribution by $f_0(\boldsymbol{v})$.

We derive the equilibrium distribution $f_0(\boldsymbol{v})$ from the Boltzmann equation: Because of $\boldsymbol{F} = 0$ and $\partial f(\boldsymbol{v}, t)/\partial \boldsymbol{r} = 0$, the left side of (42.3) reduces to $\partial f/\partial t$. For the equilibrium distribution $\partial f_0(\boldsymbol{v})/\partial t = 0$ holds. Therefore, the right-hand side of (42.3) must disappear. We consider velocities $\boldsymbol{v}'$ and $\boldsymbol{v}'_1$, which may result from an elastic scattering of particles with $\boldsymbol{v}$ and $\boldsymbol{v}_1$. For these velocities, $d\sigma/d\Omega \neq 0$ holds. Therefore, the bracket expression in the integral in (42.3) must be zero:

$$f_0(\boldsymbol{v})\, f_0(\boldsymbol{v}_1) - f_0(\boldsymbol{v}')\, f_0(\boldsymbol{v}'_1) = 0 \tag{42.6}$$

We take the logarithm:

$$\ln f_0(\boldsymbol{v}) + \ln f_0(\boldsymbol{v}_1) = \ln f_0(\boldsymbol{v}') + \ln f_0(\boldsymbol{v}'_1) \tag{42.7}$$

The left side is a function of the velocities before the collision, the right side is the same function after the collision (or vice versa). Equation (42.7) means that this function is a *conserved quantity*. Conserved quantities for the elastic collision are the (kinetic) energy, the momentum and the angular momentum. Only the energy and the momentum can be expressed by velocities alone. For these conserved quantities the following applies

$$m\,(\boldsymbol{v}+\boldsymbol{v}_1)=m\,(\boldsymbol{v}'+\boldsymbol{v}_1'),\qquad \frac{m}{2}\left(v^2+v_1^{\,2}\right)=\frac{m}{2}\left(v'^2+v_1'^2\right) \tag{42.8}$$

From (42.7) we see, that the wanted distribution f_0 must be additive (with respect to the two particles), and it must be a conserved quantity (because of the equality of both sides). From this conditions and from (42.8) follows the general form

$$\ln f_0(\boldsymbol{v})=a+\boldsymbol{b}\cdot\boldsymbol{v}+c\,v^2 \tag{42.9}$$

This yields the mean value of the velocity, $\bar{\boldsymbol{v}}\propto\int d^3v\,\boldsymbol{v}\,f_0\propto\boldsymbol{b}$. We now go to the inertial frame in which this mean value disappears, i.e. $\boldsymbol{b}=0$; formally, this can be achieved by a Galilean transformation. For $\boldsymbol{b}=0$ Equation (42.9) yields the Maxwell's velocity distribution:

$$f_0(\boldsymbol{v})=f_0(v)=A\,\exp(-\beta m v^2/2) \tag{42.10}$$

We replaced the constant c in (42.9) by $-\beta m/2$ by another, initially unknown constant β. From (42.10) follows for the mean energy per particle $m\,\overline{v^2}/2=3/(2\beta)$. This determines the physical meaning of β, in accordance with the previously introduced relation $\beta=1/k_{\mathrm{B}}T$.

From the master equation we have obtained a microcanonical equilibrium (41.11), and from the Boltzmann equation a canonical equilibrium (42.10). This is due to the fact that the P_r refers to the distribution of the microstates of the closed system, while $f(\boldsymbol{v})$ represents the velocity distribution of the individual particles (in the heat bath of all other particles).

Relaxation time approximation

We introduce a simple approximation for the collision term of the Boltzmann equation. We consider small deviations from the equilibrium distribution f_0 from (42.10):

$$f(\boldsymbol{r},\boldsymbol{v},t)=f_0(v)+\delta f(\boldsymbol{r},\boldsymbol{v},t) \tag{42.11}$$

We approximate the right-hand side of the Boltzmann equation by

$$\int d^3v_1\int d\Omega\,V\,\frac{d\sigma}{d\Omega}\Big(f(\boldsymbol{r},\boldsymbol{v}',t)\,f(\boldsymbol{r},\boldsymbol{v}_1',t)-f(\boldsymbol{r},\boldsymbol{v},t)\,f(\boldsymbol{r},\boldsymbol{v}_1,t)\Big)\approx-\frac{\delta f(\boldsymbol{r},\boldsymbol{v},t)}{\tau} \tag{42.12}$$

For $f=f_0$ the collision term disappears; for small deviations it will therefore be proportional to δf. For the coefficient $1/\tau$ of δf we carry out a rough estimate: We approximate $\int d\Omega\,d\sigma/d\Omega(\dots)$ by the cross section σ, $\int d^3v\,f_0(v)\dots$ by the mean particle density n, and V by the mean velocity $\overline{v}$. This estimation neglects numerical factors and the terms that are quadratic in δf. However, it correctly takes into account the dimension of the factors. For the coefficient $1/\tau$ we thus obtain

$$\frac{1}{\tau} \approx \overline{v}\, n\, \sigma \tag{42.13}$$

Elementary kinetic considerations in the next chapter show that τ is roughly the average collision time between the gas particles. For the non-equilibrium distribution (42.1), the approach to equilibrium is achieved by a (few) collisions of each particle. Therefore, the time τ is called *relaxation time*. It characterizes the time needed for the system to relax into the equilibrium.

Electrical conductivity

The Boltzmann equation is the starting point for the calculation of various transport coefficients, such as viscosity, electrical conductivity or thermal conductivity. Using the example of electrical of the electrical conductivity, we demonstrate how such a calculation works.

We consider a classical gas consisting of charged particles (charge q) in a homogeneous electric field. For this, the force on the left side of the Boltzmann equation reads

$$\boldsymbol{F} = q\, E_z\, \boldsymbol{e}_z = \text{const.} \tag{42.14}$$

The force shall be constant in time and space, and we are looking for a time and position independent correction to the equilibrium distribution, i.e.

$$f(\boldsymbol{r}, \boldsymbol{v}, t) = f_0(\boldsymbol{v}) + \delta f(\boldsymbol{v}) \tag{42.15}$$

Since the right-hand side of the Boltzmann equation is of the order δf, this also applies to the left-hand side. It is therefore sufficient to use f_0 on the left-hand side. Since f_0 does not depend on the position and time, the left-hand side yields $(\boldsymbol{F}/m) \cdot \partial f_0/\partial \boldsymbol{v}$. On the right-hand side, we use the relaxation time approximation (42.12):

$$\frac{q\, E_z}{m} \frac{\partial f_0(v)}{\partial v_z} \approx -\frac{\delta f(v)}{\tau} \tag{42.16}$$

For the considered classical gas, $f_0(v)$ is the Maxwell distribution (42.10) for which

$$\frac{\partial f_0}{\partial v_z} = -\beta\, f_0\, m\, v_z \tag{42.17}$$

holds. From the last two equations follows

$$\delta f(v) \approx \tau\, q\, E_z\, \beta\, f_0(v)\, v_z \tag{42.18}$$

For $f = f_0 + \delta f$ we calculate the mean current density:

$$
\begin{aligned}
j_z &= q \int d^3v \; v_z \left(f_0(v) + \delta f(v) \right) \approx \tau q^2 E_z \beta \int d^3v \; v_z^2 \, f_0 \\
&= \frac{\tau q^2 E_z}{k_{\rm B} T} \, n \, \overline{v_z^2} = \frac{n q^2 \tau}{m} \, E_z
\end{aligned}
\tag{42.19}
$$

In the last step we used the relation $m \, \overline{v_z^2}/2 = k_{\rm B} T /2$ of the equipartition theorem. From the result, we can read off the *electrical conductivity* $\sigma_{\rm el}$:

$$
\sigma_{\rm el} = \frac{j_z}{E_z} = \frac{n q^2}{m} \, \tau
\tag{42.20}
$$

In the next chapter, we derive this and other transport coefficients directly from elementary kinetic considerations. Here we have sketched the principle way for a more precise calculation based on the Boltzmann equation.

Exercises

42.1   Continuity equation for particle density

A system of $N \gg 1$ particles has the classical Hamiltonian function

$$
H = \sum_{i=1}^{N} \frac{p_i^2}{2m} + V(r_1,...,r_N, t)
$$

The probability density $\varrho(r_1,...,r_N, p_1,..., p_N, t)$ in the $6N$-dimensional phase space shall be time independent, $d\varrho/dt = 0$. This condition can be written out as the *Liouville equation*

$$
\frac{\partial \varrho}{\partial t} + \sum_{i=1}^{N} \left(\frac{\partial \varrho}{\partial r_i} \cdot \dot{r}_i + \frac{\partial \varrho}{\partial p_i} \cdot \dot{p}_i \right) = 0
\tag{42.21}
$$

Here $\partial/\partial r_i$ and $\partial/\partial p_i$ are the gradients with respect to the coordinates r_i and the momenta p_i.

From this Liouville equation, derive a continuity equation for the particle density $n(r, t)$,

$$
n(r, t) = \int d^{3N}r \; d^{3N}p \; \varrho(r_1,...,r_N, p_1,..., p_N, t) \sum_{i=1}^{N} \delta(r - r_i)
\tag{42.22}
$$

Here, $d^{3N}r \; d^{3N}p$ is the volume element in the $6N$-dimensional phase space.

43 Kinetic gas model

In an elementary kinetic gas model we derive the transport equations for electric current, diffusion, viscosity and heat conduction. In this model, the gas particles are assumed to have locally the velocity distribution of an ideal gas. However, in contrast to the ideal gas, the collisions between the particles are taken into account explicitly.

The transport equations contain coefficients that determine the strength or the speed of the process. We calculate these *transport coefficients* in the kinetic gas model; thereby we often neglect factors of the size $\mathcal{O}(1)$. In general, these transport coefficients will in depend on the temperature and the pressure.

Average collision time

We consider a classical gas consisting of atoms or molecules; however, most considerations can be transferred to other systems (for example electrons). The occurring quantities are evaluated numerically for air under normal conditions.

In scattering experiments, an incoming current density j (incident particles per area and time) falls on a target. The number of scattered particles per time is then measured, dN_{scatt}/dt. The quotient of these quantities is the cross section σ:

$$\sigma = \frac{dN_{\text{scatt}}/dt}{j} \tag{43.1}$$

The cross section σ is the area at which the incoming current density j is effectively scattered. We now apply these concepts to the atoms of a gas. We imagine the atoms of the gas as spheres with the diameter d. The cross-section for classical scattering of hard spheres is equal to the geometric cross section

$$\sigma = \pi d^2 \tag{43.2}$$

An atom traveling the path l collides with all particles in the volume σl (Figure 43.1). For a given density $n = N/V$ we obtain therefore

$$N_{\text{scatt}} = n\,\sigma\,l = \text{number of collisions} \tag{43.3}$$

During the time t, a picked out particle traverses the path $l = \overline{v}\,t$. As the particles move relative to each other, the effective path is $l = \overline{v}_{\text{rel}}\,t$, where $\overline{v}_{\text{rel}}$ is the mean relative velocity of two particles.

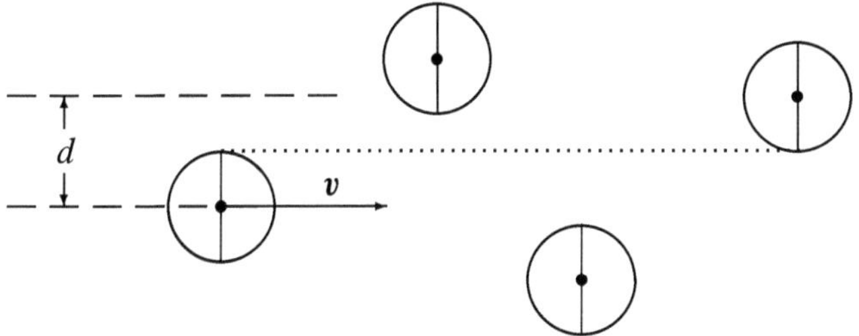

Figure 43.1 Without collisions, the particles move along straight lines. We consider specifically a straight path for the particle shown on the left. All other particles whose distance is less than d (particle diameter) from this path will be hit. On the path l, a particle will hit all other particles whose center of mass lies in a cylinder with the volume σl. These are $N_{\text{scatt}} = n\,\sigma\,l$ particles. For $N_{\text{scatt}} = 1$, this yields the mean free path $\lambda \approx 1/(n\,\sigma)$.

The *average collision time* τ is determined by the condition that there is exactly one collision on average during this time. We insert $N_{\text{scatt}} = 1$ and $l = \overline{v}_{\text{rel}}\,\tau$ into (43.3) in (43.3):

$$1 = N_{\text{scatt}} = n\,\sigma\,\overline{v}_{\text{rel}}\,\tau = \sqrt{2}\,n\,\sigma\,\overline{v}\,\tau \tag{43.4}$$

The last step $\overline{v}_{\text{rel}} = \sqrt{2}\,\overline{v}$ results from the solution of Exercise 24.5, where a Maxwell velocity distribution was used. This gives us the *average (or mean) collision time*,

$$\boxed{\tau = \frac{1}{\sqrt{2}\,n\,\sigma\,\overline{v}} \qquad \text{average collision time}} \tag{43.5}$$

During the time τ a particle travels on average the distance $\lambda = \overline{v}\,\tau$, called *mean free path*,

$$\boxed{\lambda = \frac{1}{\sqrt{2}\,n\,\sigma} \qquad \text{mean free path}} \tag{43.6}$$

We estimate τ and λ for air at normal pressure ($P \approx 1\,\text{bar}$) and room temperature ($T \approx 300\,\text{K}$). For these normal conditions the particle density is

$$n = \frac{N}{V} = \frac{6\cdot 10^{23}}{22\,\text{l}} = 2.7\cdot 10^{25}\,\frac{1}{\text{m}^3} \tag{43.7}$$

All air molecules are counted here, i.e. essentially the N_2 and O_2 molecules. Together they give the pressure P and the density $n \approx P/k_B T$. On average, each particle has the volume $v = V/N = 1/n$ at its disposal. This results in the mean distance $\langle r \rangle$ of two molecules:

$$\langle r \rangle \approx \frac{1}{n^{1/3}} = 30\,\text{Å} \tag{43.8}$$

From (24.24) we obtain the mean velocity

$$\overline{v} = \sqrt{\frac{8}{\pi}\frac{k_{\mathrm{B}}T}{m}} \approx 440\,\frac{\mathrm{m}}{\mathrm{s}} \tag{43.9}$$

Here, $k_{\mathrm{B}}T \approx \mathrm{eV}/40$ and $m \approx 30\,\mathrm{GeV}/c^2$ was used for N_2 and O_2. The air molecules are not spherical as assumed in Figure 43.1. Averaging over all directions, we assume a geometrical cross section $\sigma = \pi d^2$ where d is the effective molecule diameter. For this diameter we use the plausible value $d \approx 3\,\text{Å}$. This gives us

$$\lambda = \frac{1}{\sqrt{2}\,\pi\,d^2 n} \approx 10^{-7}\,\mathrm{m} = 1000\,\text{Å}\,, \qquad \tau = \frac{\lambda}{\overline{v}} \approx 2\cdot 10^{-10}\,\mathrm{s} \tag{43.10}$$

These values are realistic. The comparison between $d \approx 3\,\text{Å}$, $\langle r\rangle \approx 30\,\text{Å}$ and $\lambda \approx 1000\,\text{Å}$ illustrates the size relations on the microscopic scale. The volume available per particle is larger than the intrinsic volume by a factor of $10^3 \approx \langle r\rangle^3/d^3$; the air is a dilute gas. Consequently, the mean free path is much larger than the mean distance between the particles.

The transition from a non-equilibrium state to equilibrium takes place locally (in a range of several λ's) by a few collisions; in this respect, the collision time also determines the approach to a local equilibrium. The approach to the global equilibrium is described by the transport equations to be discussed.

Electrical conductivity

We consider a classical gas with an average collision time τ. The particles shall carry the charge q. We investigate the transport of electric charges caused by an external electric field $\boldsymbol{E}$. These considerations can be transferred to the electron gas of a metal.

The equation of motion for the ith particle (mass m, charge q) in the time *between two collisions* is

$$m\,\dot{\boldsymbol{v}}_i = q\,\boldsymbol{E} \tag{43.11}$$

From this follows

$$\boldsymbol{v}_i = \frac{q\,\boldsymbol{E}}{m}\,t_i + \boldsymbol{v}_{i,0} \tag{43.12}$$

Here $\boldsymbol{v}_{i,0}$ is the velocity of the particle immediately after the last collision, and t_i is the time that has passed since this collision.

We now consider a group of $N_0 \gg 1$ particles (approximately all particles in a volume element) immediately after a collision. The collision probability per time is the same for each particle, namely $1/\tau$. The number of particles that have not yet collided again after the time span t is $N(t) = N_0 \exp(-t/\tau)$ (this follows from $dN(t)/dt = -N(t)/\tau$). The probability for the occurrence of a t_i-value in (43.12) is proportional to $N(t_i)$; from this follows the probability density $w(t) \propto \exp(-t/\tau)$

for the distribution of the time spans $t_1, t_2, ..., t_{N_0}$. The normalized ($\int_0^\infty w(t)\,dt = 1$) probability density is $w(t) = \exp(-t/\tau)/\tau$. We use this to calculate the mean velocity of the particle ensemble due to the external field:

$$v_{\text{drift}} = \frac{qE}{m}\,\overline{t_i} = \frac{qE}{m}\frac{1}{N_0}\sum_1^{N_0} t_i = \frac{qE}{m}\int_0^\infty dt\,w(t)\,t = \frac{qE}{m}\,\tau \qquad (43.13)$$

We assume that this *drift velocity* is small compared to the mean velocity, i.e.

$$\left|v_{\text{drift}}\right| = \frac{qE}{m}\,\tau \ll \left|v(0)\right| \qquad (43.14)$$

As an example, we consider a singly ionized O_2 molecule ($m = 32\,\text{GeV}/c^2$ and $q = e = 1.6 \cdot 10^{-19}$ coulomb) in air ($\tau = 2 \cdot 10^{-10}\,\text{s}$) for a field strength $E = 10\,000$ volt/meter. For this, $v_{\text{drift}} = qE\tau/m \approx 6\,\text{m/s}$ is small compared to $|v(0)| \sim \overline{v} \approx 440\,\text{m/s}$.

Since the additional velocity v_{drift} is small, we can assume that the velocities $v_{i,0}$ immediately after a collision are isotropically distributed, i.e. $\overline{v_{i,0}} \approx 0$. With this we obtain the mean velocities

$$\overline{v_i} = \frac{qE}{m}\,\overline{t_i} + \overline{v_{i,0}} \approx v_{\text{drift}} \qquad (43.15)$$

The electric current density is equal to the product of the charge density $n\,q$ and the average velocity:

$$j = \sigma_{\text{el}}\,E = n\,q\,\overline{v_i} \approx q\,n\,v_{\text{drift}} \qquad (43.16)$$

The proportionality coefficient σ_{el} between the current density j and the field E is called *electrical conductivity*. In the presented kinetic gas model we obtain

$$\boxed{\;\sigma_{\text{el}} \approx \frac{n\,q^2\,\tau}{m} \qquad \begin{array}{l}\text{electrical} \\ \text{conductivity}\end{array}\;} \qquad (43.17)$$

The statement $j = \sigma_{\text{el}}\,E$ is known as *Ohm's law*; it implies that σ_{el} is independent of E. In a circuit, this means that the current is proportional to the applied voltage. In real systems, σ_{el} is only approximately independent of the applied field.

In the Gaussian unit system, charge has the dimension $[q] = \text{esu} = \text{cm}^{3/2}\,\text{g}^{1/2}/\text{s}$ (esu = electrostatic unit), $[m] = \text{g}$, $[n] = \text{cm}^{-3}$ and $[\tau] = \text{s}$; from this follows $[\sigma] = \text{s}^{-1}$. In the SI system on the other hand, $[\sigma] = \text{A}/(\text{Vm})$; this follows from $[j] = \text{ampere}/\text{m}^2 = \text{A}/\text{m}^2$ and $[E] = \text{volt}/\text{m} = \text{V}/\text{m}$.

We had obtained the result (43.17) already in (42.20) by a rough approximation of the Boltzmann equation. The time τ introduced in the relaxation time approximation (42.12) has now got a simple physical meaning (average collision time).

A classical gas will generally be only partially ionized. The density of the ions must then be used for n, while τ results from the collisions of all gas particles. The actual value of σ_{el} depends thus on the degree of ionization.

An important case is the electron gas ($q = -e$, $m = m_e$) in a metal. The electrons can approximately be treated as an ideal Fermi gas. According to Chapter 32, only the electrons of the softened Fermi edge contribute to the heat capacity. In contrast to this, *all* electrons contribute to the electrical conductivity. Due to the applied field all electrons receive the additional drift velocity v_{drift}. This shifts the entire Fermi distribution; this is not hindered by the Pauli principle. Likewise as in the classical gas, the applied field tends to accelerate the electrons. As in the classical gas, a finite drift velocity results due to collisions.

We apply (43.17) to the mobile electrons of a of a metal. Here, the collision time is not determined by electron-electron collisions, but rather by collisions with the crystal lattice, i.e. by electron-phonon interaction. The time then depends on the temperature-dependent phonon density (Chapter 33). Equation (43.17) can be used to determine the mean collision time τ from the easily measurable conductivity. The following applies specifically to copper

$$\sigma_{el} \approx 6 \cdot 10^{17}\,\mathrm{s}^{-1}, \qquad \tau \approx 2 \cdot 10^{-14}\,\mathrm{s} \qquad (\text{copper, } T = 0^\circ\mathrm{C}) \qquad (43.18)$$

The collision time $\tau = m\,\sigma_{el}/(n\,e^2)$ was calculated from the experimental value of the conductivity σ_{el}. It was assumed that there is one free (at least mobile) electron per atom; for the atomic density in the copper crystal we used $n \approx 1/(12\,\text{Å}^3)$.

For electrons, the drift velocity (43.14) is of the same size as for the above considered ions. This is due to the fact that for electrons both, the collision time and the mass, are smaller by a factor of about 10^4. The velocities of the electrons are of the size of the Fermi velocity (Chapter 32), i.e. $|v(0)| \sim c/100$. These velocities exceed the possible values of v_{drift} by many orders of magnitude.

Friction

Due to the collisions, a constant force leads to a finite drift velocity $v_{drift} = \tau\,F/m$, (43.13), and not to a constant acceleration. Such behavior can be described by a friction term in Newton's equation of motion. Through

$$m\,\dot{v} = F - \gamma\,v \qquad (43.19)$$

we define a *friction coefficient* γ (also called friction constant). For $t \to \infty$, the solution of this equation of motion becomes $v(\infty) = F/\gamma$. Identifying $v(\infty)$ with the drift velocity $v_{drift} = \tau\,F/m$ yields $\gamma = m/\tau$ or

$$\boxed{\gamma = \frac{m\,\bar{v}}{\lambda} \qquad \text{friction coefficient}} \qquad (43.20)$$

The measurement of the ratio F/v_{drift} determines the friction coefficient γ; its dimension is $[\gamma] = \mathrm{kg/s}$. Just like all other transport coefficients, the friction coefficient will in general depend on the temperature and the pressure.

Diffusion

The collisions between the gas particles lead to the equalization (homogenization) of an initially inhomogeneous density. This process is called *diffusion*.

Let the density of the gas depend only on the z-coordinate only, i.e. $n = n(z)$. In Figure 43.2 we identify X with the density n. We consider the particles that fly through the area $z = z_0$ (neglecting factors of size 1). The particles at z_0 the velocity directions are isotropic resulting in a zero net flow through the area $z = z_0$. The particles at $z_0 \pm \lambda$ also have all possible velocity directions. If their velocity points in the $\mp z$ direction, they result in a contribution to the current through the area $z = z_0$. About $1/6$ of the particles will have this direction. The size of the velocity is $\overline{v}$; it is also assumed that $\overline{v}$ does not depend on z. The particles that are further than λ away from z_0 scatter before they reach the area $z = z_0$; they do not contribute to the current at the considered time. At z_0 we obtain thus the particle current density

$$j_z = \frac{\text{number of particles}}{\text{time} \cdot \text{area}} \approx \frac{1}{6}\,\overline{v}\left(n(z_0 - \lambda) - n(z_0 + \lambda)\right) = \frac{\overline{v}}{6}\frac{\partial n}{\partial z}\left(-2\lambda\right) \quad (43.21)$$

This shows that the strength of the diffusion current is proportional to the negative density gradient. The transport equation

$$j_z = n\,v_{\text{diff}} = -D\,\frac{\partial n}{\partial z} \quad (43.22)$$

defines the *diffusion coefficient* also called *diffusion constant*; in addition, we have introduced the diffusion velocity v_{diff}. From the last two equations we obtain

$$\boxed{\,D \approx \frac{\overline{v}\,\lambda}{3} \qquad \text{diffusion coefficient}\,} \quad (43.23)$$

For air we use (43.9) and (43.10):

$$D \approx 1.5 \cdot 10^{-5}\,\frac{\text{m}^2}{\text{s}} \qquad \text{(air)} \quad (43.24)$$

Experimentally one finds $D \approx 2 \cdot 10^{-5}\,\text{m}^2/\text{s}$ for O_2 in air under normal conditions.

Diffusion is the process of density equalization in a medium at rest. In the real world, a density equalization is more likely to occur through convection (like winds in the atmosphere).

Using (43.22) with (43.23), the density equalization of a trace gas (for example a fragrance in air) can be treated. Then $n(z)$ is the density of the trace gas. The mean free path λ is determined by the collisions of the trace gas particles with all gas particles; thereby the collisions of the trace gas particles with each other play a minor role. For the distribution of the trace gas (e.g. fragrance in a room), convection and not diffusion is usually the dominant process.

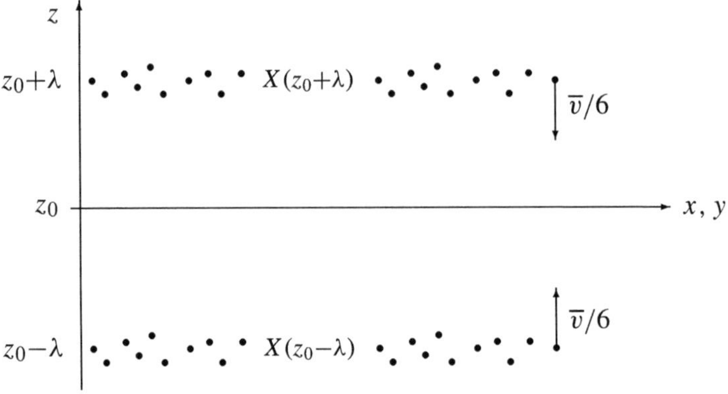

Figure 43.2   A quantity X (density, temperature, mean velocity) is inhomogeneous in the z-direction. Then the statistical velocities of the gas particles lead to a net flow that tends to equalize (homogenize) this quantity. The calculation of this flow is based on the observation that the particles maintain their direction only over the mean free path λ. For a rough statistical estimation, it can be assumed that 1/6 of the particles travel with the velocity $\overline{v}$ in one of the six different directions ($\pm x$-, $\pm y$ and $\pm z$-direction).

Diffusion equation

We derive the diffusion equation. Since the number of particles is conserved, the diffusion flow must satisfy the continuity equation $\dot{n} = -\operatorname{div} \boldsymbol{j}$, thus

$$\frac{\partial n(z, t)}{\partial t} = -\frac{\partial (n \, v_{\mathrm{diff}})}{\partial z} \tag{43.25}$$

We insert (43.22) into this and obtain the *diffusion equation*

$$\frac{\partial n(z, t)}{\partial t} = D \, \frac{\partial^2 n(z, t)}{\partial z^2} \tag{43.26}$$

The generalization to three dimensions reads

$$\boxed{\frac{\partial n(\boldsymbol{r}, t)}{\partial t} = D \, \Delta n(\boldsymbol{r}, t) \qquad \text{diffusion equation}} \tag{43.27}$$

Here Δ is a Laplace operator. It is easy to check that

$$n(z, t) = \frac{N}{\sqrt{4\pi D t}} \, \exp\left(-\frac{z^2}{4Dt}\right) \tag{43.28}$$

solves the diffusion equation (43.26). This solution is normalized, $\int dz \, n(z, t) = N$. The solution of (43.27) shall be determined in Exercise 43.1.

These considerations can also be applied to the molecules or small particles dissolved in a liquid or to a trace gas in air. The mean free path λ results in this case from the scattering of the particles under consideration with those of the medium. The mean velocity is given by the temperature of the medium. The density gradient refers solely to the dissolved particles or the trace gas. For these applications, the solution (43.28) shows how an initially localized distribution broadens. The measure for this broadening is

$$(\Delta z)^2 = \overline{(z - \bar{z})^2} = \overline{z^2} = \frac{1}{N} \int_{-\infty}^{\infty} dz \, z^2 \, n(z,t) = 2Dt \qquad (43.29)$$

We use (43.24) to estimate how far a picked out molecule in air moves away from its initial position:

$$\Delta z = \sqrt{2Dt} \approx \begin{cases} 5\,\text{mm} & (t = 1\,\text{second}) \\ 1\,\text{m} & (t = 1\,\text{day}) \end{cases} \qquad (43.30)$$

This applies to an air molecule in air, but roughly also to the molecule of a trace gas or a fragrance molecule in air. In view of the relatively high velocities ($\bar{v} \approx 440\,\text{m/s}$) of the air molecules, the distances achieved are rather modest. This is due to the frequent collisions that lead to a zigzag motion (random walk). The spread of fragrances in a room due to diffusion will require a relatively long time. In practice, the distribution takes place much faster by air convection.

Brownian motion

A special system for which diffusion and friction can be studied is the *Brownian motion*. Small (Brownian) particles are floating on the surface of a liquid and perform visible zigzag movements. This two-dimensional random walk is the result of collisions of the Brownian particle with the thermally moving molecules of the liquid.

For Brownian motion, γ and D can be determined experimentally as follows. One charges the Brownian particles electrically and applies an electric field $E = E\,e_z$. This leads to a drift motion with $v_{\text{drift}} = qE/\gamma$, i.e. to the mean displacement

$$\bar{z} = \frac{qE}{\gamma} t \qquad (43.31)$$

In addition to this enforced displacement, there is an undirected diffusion motion with

$$(\Delta z)^2 = \overline{(z - \bar{z})^2} = 2Dt \qquad (43.32)$$

This was calculated in (43.29) for $\bar{z} = 0$; the result is also valid for $\bar{z} \neq 0$. The fluctuation Δz can now be measured with and without a field.

The quantities $\bar{z}$ and $\overline{\Delta z^2}$ can be determined experimentally. The ith particle shall start at $z_i(0) = 0$. One now follows the motion of this particle and determines its position $z_i(t)$ after a time period t (with $t \gg \tau$). The observation of N particles ($i = 1,..., N$, with $N \gg 1$) then results in the mean values

$$\bar{z} = \frac{1}{N} \sum_{i=1}^{N} z_i(t), \qquad (\Delta z)^2 = \overline{(z - \bar{z})^2} = \frac{1}{N} \sum_{i=1}^{N} \left(z_i(t) - \bar{z}(t) \right)^2 \qquad (43.33)$$

Plotted over t, these variables result in straight lines, (43.31) and (43.32). The slopes of the respective straight lines determine γ and D.

Einstein relation

Both, friction (more generally *dissipation*) as well as diffusion (more general *fluctuation*) have their origin in the collisions between the particles. Therefore, a relation between γ and D exists. This relation is called the Einstein relation or, in a more general context, fluctuation dissipation theorem.

From (43.20) and (43.23) we obtain $\gamma D \approx m \bar{v}^2/3 = \mathcal{O}(1) k_{\mathrm{B}} T$. By a special consideration, we determine the factor $\mathcal{O}(1)$ to 1 and obtain the *Einstein relation*

$$\boxed{\gamma D = k_{\mathrm{B}} T \qquad \text{Einstein relation}} \qquad (43.34)$$

To determine the numerical factor, we start from the barometric formula

$$n(z) = n(0) \, \exp\left(-\frac{mgz}{k_{\mathrm{B}} T} \right) \qquad (43.35)$$

This can be understood as an equilibrium resulting from the following processes:

1. The gravity force $F = -mg$ leads according to (43.19) to the mean drift (sinking) velocity v_{drift} of the particles:

$$v_{\mathrm{drift}} = -\frac{mg}{\gamma} \qquad (43.36)$$

2. On the other hand, according to (43.22), an inhomogeneous density causes a diffusion flow with the velocity

$$v_{\mathrm{diff}} = -\frac{D}{n} \frac{\partial n}{\partial z} = D \frac{mg}{k_{\mathrm{B}} T} \qquad (43.37)$$

We consider only the z-component of both velocities. In equilibrium, the net current must be zero, i.e.

$$v_{\mathrm{drift}} + v_{\mathrm{diff}} = -\frac{mg}{\gamma} + D \frac{mg}{k_{\mathrm{B}} T} = 0 \qquad (43.38)$$

From this follows (43.34). In this derivation, only the definitions $\gamma = F/v_{\mathrm{drift}}$ and $D = -j_z/(\partial n/\partial z)$ were used, but not the approximation made in the kinetic gas model.

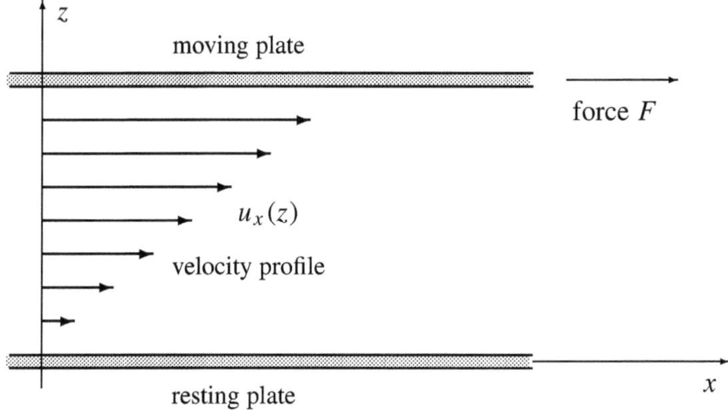

Figure 43.3 Two parallel plates are moving against each other with constant speed. Between the plates there is a gas or a liquid. On the surface of the plates, the medium is carried along. Under suitable experimental conditions (low velocities), it comes to the depicted laminar flow. In order to maintain this flow, a force F must be exerted on the moving plate. The force per plate area A is proportional to the viscosity of the medium and to the velocity gradient.

Viscosity

The viscosity η of a liquid or a gas can be determined by the schematic experimental setup shown in Figure 43.3. In order to maintain the velocity gradient $\partial u_x/\partial z$, a force F per area A is required:

$$\frac{F}{A} = -\eta\, \frac{\partial u_x}{\partial z} \tag{43.39}$$

The proportionality constant η is called *viscosity* or viscosity coefficient. The velocity u_x is the flow velocity of the gas (or liquid). The flow velocity is equal to the mean value of the velocities of the individual gas particles, $\boldsymbol{u} = \overline{\boldsymbol{v}}$ where $|\boldsymbol{u}| \ll \overline{v}$ shall hold.

Without external forces, such a velocity profile decays; left to its own devices, the system moves into equilibrium with $\boldsymbol{u} = \overline{\boldsymbol{v}} = 0$. Such a decay is comparable to the exponential decay of the solutions of (43.19) due to the friction term. The viscosity is therefore also referred to as internal friction.

In practice, the viscosity is determined via the force $\boldsymbol{F} = 6\pi\eta\, r\boldsymbol{u}$ that must be exerted on a sphere moving through a viscous medium (sphere radius r, sphere velocity $\boldsymbol{u}$, laminar flow). This formula is called Stokes' law (also Stokes' law of friction). Another important application is the fluid flow current $I = $ volume/time through a cylindrical tube with the radius r. For this one obtains $I = \pi r^4(\Delta P/\Delta l)/(8\eta)$, where $\Delta P/\Delta l$ is the pressure gradient along the tube (Hagen-Poiseuille equation).

We derive (43.39). To do this, we look at Figure 43.2 with $X = u_x$; the further considerations are analogous to diffusion. We consider the particles that cross the area $z = z_0$. On average, the particles cross a distance λ without collision. The particles coming from above carry the momentum $m\,u_x\,(z_0 + \lambda)$, the particles coming from below $m\,u_x\,(z_0 - \lambda)$. (These momenta refer to the mean extra velocity due to the macroscopic flow; individual velocities are much larger and isotropically distributed.) This results in the momentum current density

$$J_p = \frac{\text{momentum}}{\text{time} \cdot \text{area}} \approx \frac{1}{6}\, n\, m\, \bar{v}\left(u_x(z_0 - \lambda) - u_x(z_0 + \lambda)\right) \tag{43.40}$$

In order to maintain the velocity profile, the external force per area must be equal to this momentum transfer per time and area:

$$\frac{F}{A} = J_p \approx \frac{1}{6}\, n\, m\, \bar{v}\, \frac{\partial u_x}{\partial z}\, (-2\lambda) \tag{43.41}$$

This shows that F/A is proportional to the velocity gradient, as required by the viscosity definition (43.39). The last equation yields our model estimate for the viscosity coefficient:

$$\boxed{\ \eta \approx \frac{n\,\lambda\,\bar{v}\,m}{3} \qquad \text{viscosity}\ } \tag{43.42}$$

We insert $m = 5 \cdot 10^{-26}$ kg for N_2- or O_2 molecules and (43.7)–(43.10):

$$\eta \approx 2 \cdot 10^{-5}\, \frac{\text{N s}}{\text{m}^2} \qquad \text{(air)} \tag{43.43}$$

Under normal conditions, the experimental value is $1.7 \cdot 10^{-5}\,\text{Ns/m}^2$. For water, one measures $10^{-3}\,\text{Ns/m}^2$, for glycerol about $1\,\text{Ns/m}^2$.

With (43.6) and (43.9), Equation (43.42) becomes

$$\eta \approx \frac{m}{3\sqrt{2}\,\sigma}\sqrt{\frac{8}{\pi}\frac{k_{\mathrm{B}}T}{m}} = \text{const.} \cdot \sqrt{T} \tag{43.44}$$

This describes the temperature dependence of the viscosity. Because of $\lambda \propto 1/n$, (43.6), the viscosity of a gas is approximately pressure independent.

Heat conduction

A density gradient leads to a particle flow, and a velocity gradient leads to a momentum flow. Analogous to this, a temperature gradient results in a heat flow. The mechanism for these transport processes is sketched in Figure 43.2. In the first case we have to set $X = n$ (density) and in the second $X = u_x$ or $X = m\,u_x$ (velocity profile). In the third we must set $X = T$ (temperature) or $X = c\,T$, where c is the heat capacity per particle.

The particles flying downwards in Figure 43.2 carry the thermal energy $c\,T\,(z_0 + \lambda)$ with them, the particles flying upwards $c\,T\,(z_0 - \lambda)$. We assume that c, n and $\bar{v}$ do not depend on z. This results in the heat current density

$$J_Q = \frac{\text{heat}}{\text{time} \cdot \text{area}} \approx \frac{1}{6}\, n\, \bar{v}\, c\left(T(z_0 - \lambda) - T(z_0 + \lambda)\right) = -\frac{1}{3}\, n\, \bar{v}\, \lambda\, c\, \frac{\partial T}{\partial z} \tag{43.45}$$

The result shows that the heat current density J_Q is proportional to the temperature gradient:

$$J_Q = -\kappa \, \frac{\partial T}{\partial z} \qquad (43.46)$$

This proportionality holds in the limiting case of a small temperature gradient, and under the exclusion of other processes such as convection or radiation. Equation (43.46) defines the *thermal conductivity* (also thermal conductivity coefficient) κ. Our estimation (43.45) yields

$$\boxed{\quad \kappa \approx \frac{n \, \lambda \, \bar{v} \, c}{3} \qquad \text{thermal conductivity} \quad} \qquad (43.47)$$

The comparison with (43.42) shows

$$\kappa = \frac{c}{m} \, \eta = \text{const.} \cdot \eta \qquad (43.48)$$

For the proportionality it was assumed that the specific heat c per particle does not depend on the temperature. Considering a gas with $\lambda \propto 1/n$, (43.6), η as well as κ are approximately independent of the pressure. At very low pressure (i.e. very low density n), however, λ of (43.6) can be larger than the distance L of the vessel walls. Then L takes over the role of the mean free path and has to be used in (43.47). Because of $\lambda \approx L$ and $n \approx P/k_B T$, the thermal conductivity in this case is proportional to the pressure. A vacuum in a double-walled thermos flask therefore improves the insulating property.

Using (43.7)–(43.10) and $c = 7k_B/2$ for an ideal diatomic gas, we estimate κ from (43.47) for air:

$$\kappa \approx \frac{7}{6} \, n \, \lambda \, \bar{v} \, k_B \approx 2 \cdot 10^{-2} \, \frac{W}{Km} \qquad \text{(still air)} \qquad (43.49)$$

Experimentally, one finds $2.4 \cdot 10^{-2}\,\mathrm{W/(Km)}$ under normal conditions. In most cases, convection constitutes the main contribution to heat transport in air; therefore the restriction *at rest* (or still) is important. Similarly, low values as in (43.49) can be achieved with glass wool or polystyrene ($\kappa \approx 4 \cdot 10^{-2}\,\mathrm{W/(Km)}$). These materials prevent a convection.

The result (43.47) can also be applied to solids in which the phonons are responsible for heat transport. In the classical limiting case (Dulong-Petit law, see (33.31)), $nc = 3k_B N/V$ holds where N is the number of atom. A rule of thumb is that the thermal conductivity of solid bodies is proportional to their mass density mN/V. Therefore, for example, concrete house walls insulate worse than hollow blocks, and lightweight materials are usually good insulators (wood, polystyrene, glass wool). For a metal, one had also to consider the contribution of the mobile electrons.

The thickness d of a thermal insulation determines the temperature gradient $\partial T/\partial z \approx -|T_1 - T_2|/d$. The heat flow through such a layer is then

$$J_Q = k\,|T_1 - T_2|, \qquad k = \frac{\kappa}{d}, \qquad [k] = \frac{W}{m^2\,K} \qquad (43.50)$$

This coefficient k is called heat transfer coefficient or k-value. It is used to characterize the thermal insulation of windows or walls. For double-glazed windows, the values are between $k \approx 3\,W/(m^2\,K)$ for older panes and around $k \approx 1\,W/(m^2\,K)$ for newer panes. In winter ($\Delta T = 20\,K$), one may expect a heat flow of about 20 watt through a $1\,m^2$ large modern window. For the actual energy balance of a window, one must take into account further effects, like the convection (such as the exchange due to leaky windows) and thermal radiation (see greenhouse effect in Chapter 34).

Heat conduction equation

The three-dimensional generalization of (43.46) yields

$$J_Q = -\kappa\ \mathrm{grad}\,T(r, t) \qquad (43.51)$$

Here we allow for a time dependence of the temperature field. If we disregard external heat sources, the continuity equation applies to the heat content of the medium under consideration, i.e.

$$\frac{\partial(nc\,T)}{\partial t} + \mathrm{div}\,J_Q = 0 \qquad (43.52)$$

We assume position and time independent material parameters (κ, n, c). Then, from the last two equations, we obtain the *heat conduction equation*,

$$\boxed{\ \frac{\partial T(r, t)}{\partial t} = \frac{\kappa}{nc}\,\Delta T(r, t) \qquad \text{heat conduction equation}\ } \qquad (43.53)$$

This equation has the same structure as the diffusion equation (43,27). Inhomogeneous heat distributions tend to balance each other out similarly as inhomogeneous density distributions do.

The material parameter κ/nc is called *thermal diffusivity*. From (43.47) with (43.9) – (43.10) we obtain

$$\frac{\kappa}{nc} \approx \frac{\lambda\,\overline{v}}{3} \approx 1.5 \cdot 10^{-5}\ \frac{m^2}{s} \qquad \text{(still air)} \qquad (43.54)$$

In the kinetic gas model, this quantity coincides with the diffusion constant D. The measured value is $1.8 \cdot 10^{-5}\ m^2/s$.

The heat conduction equation describes the spatial-temporal temperature equalization. From it, we may estimate the relaxation time τ for the temperature equalization. For a range with the linear extension R we get

$$\tau_{\mathrm{relax}} \approx \frac{nc}{\kappa}\,R^2 \qquad (43.55)$$

Periodic temperature changes (frequency ω) then fall off on the length $R \approx (\kappa/nc\,\omega)^{1/2}$ (Exercise 43.2).

Exercises

43.1   Solution of the diffusion equation

The diffusion equation

$$\frac{\partial n(\boldsymbol{r}, t)}{\partial t} = D\,\Delta n(\boldsymbol{r}, t)$$

is considered. Carry out a Fourier transformation for the spatial coordinates. Integrate the remaining time dependent differential equation. Specialize the solution to the initial condition $n(\boldsymbol{r}, 0) = N\,\delta(\boldsymbol{r})$, and determine the solution $n(\boldsymbol{r}t)$.

43.2   Temperature fluctuation in the ground

At the Earth's surface, the seasonal variation of the temperature is

$$T(z = 0, t) = T_0 + A\,\sin(\omega_0 t)$$

We assume $A = 10\,^\circ\mathrm{C}$ and $2\pi/\omega_0 = 1$ year. The temperature distribution $T(z, t)$ in the ground shall be investigated where z denotes the depth. Solve the heat conduction equation by a suitable separation approach. At what depth must a wine cellar be placed so that the temperature fluctuation is less than $1\,^\circ\mathrm{C}$? At what depth is it warmest in winter?

Geothermal heat can be disregarded, as only a depths of a few meters are relevant here. The thermal conductivity coefficient for the ground is $\kappa/nc \approx 7 \cdot 10^{-7}\,\mathrm{m^2/s}$.

Index

Abbreviations

IBG	ideal Bose gas (model)
irrev	irreversible
p.i.	partial integration
q.s.	quasi-static
rev	reversible
TD	thermodynamics

Units

Å	ångström, $1\,\text{Å} = 10^{-10}\,\text{m}$
bar	$1\,\text{bar} = 10^5\,\text{N/m}^2$
°C	degree celsius
cal	calorie, $1\,\text{cal} \approx 4.2\,\text{J}$
eV	electron volt
fm	fermi, $1\,\text{fm} = 10^{-15}\,\text{m}$
GeV	giga electron volt $1\,\text{GeV} = 10^9\,\text{eV}$
K	kelvin, $T_t = 273.16\,\text{K}$
Pa	pascal, $1\,\text{Pa} = 1\,\text{N/m}^2$

Symbols

$= \text{const.}$	equal to a constant
$\equiv$	identical equal or defined by
$\stackrel{(5.20)}{=}$	yields with the help of equation (5.20)
$\widehat{=}$	corresponds to
$\propto$	proportional to
$\approx$	approximately equal
$\sim$	of similar size
$= \mathcal{O}(\ldots)$	of the order of magnitude

A

absolute zero, 66, 107
acoustic phonons, 278
adiabat (adiabatic curve)
$\quad$ P-V-relation, 123–124
adiabatic demagnetization, 210
adiabatic process ($\mathchar'26\mkern-12mu dQ = 0$), 48, 74, 141
air, transport coefficients, 365–366
Andronikashvili's experiment, 327

antisymmetry, 220, 235–236
average collision time, 364–366
average free path, 364–366
average occupation number, 239–242
Avogadro constant, 109

B

barometric formula, 134, 192–372
binomial distribution, 13
black body radiation, 284
Bohr's magneton, 205
boiling point delay, 316
boiling point elevation, 168–169
boiling temperature, 161
Boltzmann, 61, 65
$\quad$ constant, 64, 108–109
$\quad$ equation, 358–363
$\quad$ factor, 175
Bose gas (ideal), 244, 249–259
$\quad$ in oscillator, 256–257
$\quad$ in magnetic trap, 257–259
Bose statistics, Bose-Einstein statistics, 237, 241–242
Bose-Einstein condensation, 249–259
boson, Bose particle, 235, 241, 249–259, 275, 282
Boyle-Mariotte law, 123
Brownian motion, 11, 196, 371–372

C

cal, calorie, 103
caloric equation of state, 99
canonical ensemble, 173–177
canonical partition function, 175
Carnot process, 150–151, 154
cavity radiation, 280, 289
Celsius scale, 105, 107
central limit theorem, 22–27
chemical equilibrium, 169
chemical potential, 155–159, 162, 294–296
chemical reaction, 171
classical microstate, 30–32
classical partition function, 189